ISAAC ASIMOV won acclaim as the world's leading
writer of both science fiction and nonfiction, having
written more than 475 books over a period spanning 42
years. His recent nonfiction titles include *The Exploding
Suns* and *Frontiers*. Dr. Asimov died in April 1992, at age
72, in New York City.

ATOM
Journey Across the
Subatomic Cosmos

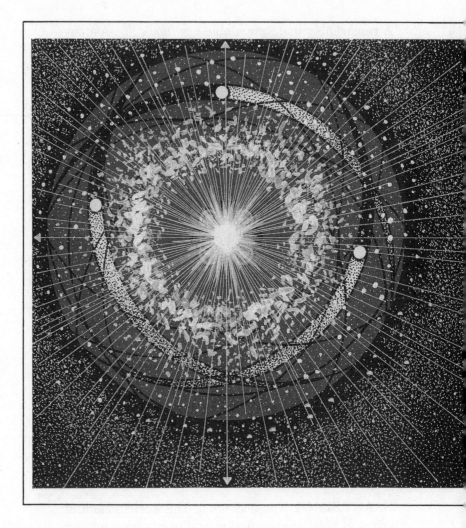

Illustrated by D. F. Bach

ISAAC ASIMOV

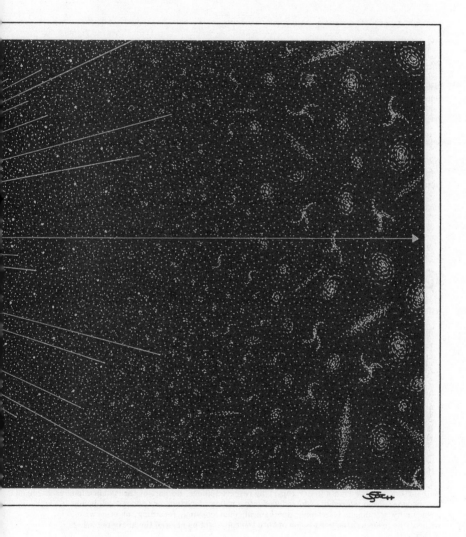

TT TRUMAN TALLEY BOOKS / **PLUME** / NEW YORK

TRUMAN TALLEY BOOKS/PLUME
Published by the Penguin Group
Penguin Books USA Inc., 375 Hudson Street, New York, New York 10014, U.S.A.
Penguin Books Ltd, 27 Wrights Lane, London W8 5TZ, England
Penguin Books Australia Ltd, Ringwood, Victoria, Australia
Penguin Books Canada Ltd, 10 Alcorn Avenue, Toronto, Ontario, Canada M4V 3B2
Penguin Books (N.Z.) Ltd, 182-190 Wairau Road, Auckland 10, New Zealand

Penguin Books Ltd, Registered Offices: Harmondsworth, Middlesex, England

Published by Truman Talley Books/Plume, an imprint of Dutton Signet,
a division of Penguin Books USA Inc. Previously published in a Truman Talley Books/
Dutton edition.

First Plume Printing, August, 1992

25 26 27 28 29 30

Copyright © Nightfall, Inc., 1991
Illustrations copyright © D. F. Bach, 1991
All rights reserved

 REGISTERED TRADEMARK—MARCA REGISTRADA

LIBRARY OF CONGRESS CATALOGING-IN-PUBLICATION DATA
Asimov, Isaac, 1920-
    Atom : Journey across the subatomic cosmos / Isaac Asimov :
illustrated by D.F. Bach.
        p.   cm.
    "Truman Talley Books".
    Originally published: New York : Dutton, 1991.
    ISBN 0-452-26834-6
    1. Atoms—Popular works. 2. Atoms—Miscellanea. 3. Nuclear
physics—Popular works. 4. Nuclear physics—Miscellanea.
5. Electromagnetism—Popular works. 6. Electromagnetism—
Miscellanea. I. Title.
[QC173.A778 1992]
539.7—dc20   92-3301
                                                  CIP

Printed in the United States of America

BOOKS ARE AVAILABLE AT QUANTITY DISCOUNTS WHEN USED TO PROMOTE PRODUCTS OR
SERVICES. FOR INFORMATION PLEASE WRITE TO PREMIUM MARKETING DIVISION, PENGUIN
BOOKS USA INC., 375 HUDSON STREET, NEW YORK, NEW YORK 10014.

To Truman "Mac" Talley
Who represents book publishing at its best.

CONTENTS

Contents

Contents

1
MATTER

Dividing Matter

Suppose you had a large heap of small, smooth pebbles—thousands of them. If you had nothing better to do, you might decide to divide it into two smaller heaps, approximately equal in size. You could discard one of these heaps, keep the other, and divide it in two again. Of these two still smaller heaps, you could discard one and keep the other for further division, and repeat the process over and over.

You might wonder how long you could keep that up. Forever? You know better than that. No matter how large the heap was to begin with, you would eventually be left with a tiny "heap" made up of just two pebbles. (This would happen surprisingly quickly. Even if you started with a million pebbles, you would be down to two pebbles after about twenty divisions.) If you divided a heap of two pebbles once again, you would end up with one heap consisting of a single pebble, and the game would be over. You can't divide one pebble.

But wait! You can. You could place the pebble on an anvil and pound it with a hammer. You would break it up into fragments, and you could divide this into smaller and smaller heaps until you were down to a single fragment. You could then pound the fragment into dust and then divide the heap of dust until you ended up with a single, hardly visible dust particle. You could break that up, and keep on going.

This is not really a practical game because it's very hard to handle a grain of dust and try to break it up further. But you can *imagine*. Imagine that you can break up the dust into still finer particles, which you can break up yet further, getting it ever finer. Now ask yourself: is there any end to this?

It might not seem like a very important question, or even a particularly sensible question, in that you can't really try the experiment in any practical way. You find yourself dealing, very quickly, with objects that are too small to see, so that you don't even know whether or not you're breaking the heap down any further. Nevertheless, certain ancient Greek philosophers asked themselves this question and started a chain of thought that is still occupying thinkers to this day, twenty-five centuries later.

The Greek philosopher Leucippus (490– ? B.C.) is the first person we know of by name supposed to have considered this problem of dividing matter, and to have come to the conclusion that the process could *not* continue forever. He insisted that, sooner or later, one had to reach a fragment of matter so small that it could not be broken down into anything smaller.

A younger man, Democritus (460–370 B.C.), was one

of Leucippus's pupils. He accepted the notion of fragments of matter so small as to be unbreakable. He called such fragments *atomos,* which in Greek means "unbreakable," and such a fragment has come to be called an atom in English. To Democritus, all matter consisted of a collection of atoms, and if there was space between the atoms, that space contained nothing (the "void").

Democritus is supposed to have written sixty books expounding his theories, including his notions of what we now call atomism. In those days, though, when there was no printing and all books had to be hand copied, there were hardly ever very many copies; and, partly because his views were unpopular, the books were not copied many times. Over the centuries many books vanished. None of Democritus's books has survived.

Most philosophers of the time felt that it didn't make sense to suppose that some tiny individual particle was indivisible. They thought it made more sense to suppose that everything could be broken up into smaller and smaller bits of matter, endlessly.

In particular, the Greek philosophers Plato (ca. 427–347 B.C.) and Aristotle (384–322 B.C.) didn't accept atoms. Because they were the most profound and mentally wide-ranging of the ancient philosophers, their views tended to carry the day. But the argument was not unanimous. The influential Greek philosopher Epicurus (341–270 B.C.) took up atomism as the central core of his teachings. Epicurus is supposed to have written 300 books (ancient books tended to be short, incidentally), but none of them has survived.

The most important of the Epicureans in this connection was a Roman, Titus Lucretius Carus (96–55 B.C.), usually known simply as Lucretius. In 56 B.C., he published a long poem in Latin entitled *De Rerum Natura* (Latin for *On the Nature of Things*). In it, he explained the Epicurean view of atomism in great detail.

The book was very popular in its time, but in later

centuries, after Christianity had grown popular, Lucretius was denounced for what was considered to be atheism. He was no longer copied, and what copies already existed were destroyed or lost. Even so, one copy (only one!) survived through the Middle Ages and was discovered in 1417. It was recopied and then, half a century later, when printing came into use, Lucretius's poem was one of the first items to be printed.

The poem spread throughout western Europe and was the chief source of knowledge of the ancient theories of atomism. The French philosopher Pierre Gassendi (1592–1655), having read Lucretius, adopted the atomistic view himself, and wrote it up persuasively, thus spreading the doctrine.

In all the two thousand years between Leucippus and Gassendi, however, atomism, pro and con, was simply a subject of endless discussion among scholars. There was no *evidence* either for or against atomism. Various scholars accepted atoms or rejected them, according to which point of view pleased them better, or seemed more sensible. There was no way of forcing one view on someone who held the other view firmly. It was a subjective decision, and there was no arguing with taste.

About this time, however, some scholars were beginning to perform experiments; to set questions to nature, so to speak, and to study the results. In this way, evidence could be produced that was scientifically "compelling"; that is, it was evidence that compelled others to accept a point of view that they were subjectively against (provided they were intellectually honest).

The first to perform experiments that seemed to have a connection with the question of atomism was the British scientist Robert Boyle (1627–1691), who was strongly influenced by Gassendi's writings, and who was consequently an atomist.

In 1662, Boyle made use of a glass tube shaped like the letter J. The short arm was closed and the long arm open. He poured mercury into the opening and it filled the bottom, trapping air in the closed short arm. Boyle then poured additional mercury into the tube, the weight of which compressed the air in the short arm, decreasing the volume of the air as a result. If he doubled the height of the mercury column in the long arm, the volume of air in the short arm was halved. When the mercury was removed and the pressure released, the volume of air increased. This inverse relationship between pressure and volume has been called Boyle's law ever since.

This behavior of air under pressure was easily explained if one made use of atoms. Suppose the air is made up of atoms that are widely separated, with nothing in between—as Democritus had suggested. (This would account for the fact that a volume of air weighs so much less than the same volume of water or marble, where the atoms might be in contact.) Placing the air under pressure would force the atoms close together, squeezing out some of the nothingness, so to speak, and would decrease the volume. Relieving the pressure would allow the atoms to spread outward.

For the first time, atomism began to gain an upper hand. Someone might think that it wasn't sensible, or perhaps that it wasn't esthetic, to suppose the existence of atoms, but one could not argue with Boyle's experiment. This was especially true in that anyone could run the experiment himself and come up with the same observations.

If we must accept Boyle's experiment, then atomism offers a simple and logical explanation of his findings. Explaining the results without atoms is much more difficult.

From that point on, then, more and more scientists were atomists, but the issue was not yet completely settled. (We'll get back to the subject later.)

Elements

The ancient Greek philosophers wondered what the world was made of. Clearly, it was made of innumerable types of things, but scientists have always felt the urge to simplify. There was the feeling, therefore, that the world was made of some basic material (or some very few basic materials), of which everything else was one variation or another.

Thales (ca. 640–546 B.C.) is the first Greek philosopher supposed to have suggested that water was the basic material out of which everything was formed. Another, Anaximenes (570–500 B.C.), thought it was air. Still another, Heraclitus (ca. 535–475 B.C.), thought it was fire, and so on.

There was no way of deciding among these suggestions for there was no real evidence one way or another. The Greek philosopher Empedocles (495–435 B.C.) settled the issue by compromise. He suggested that the world was made of several different basic substances: fire, air, water, and earth. To this Aristotle added aether (from a Greek word for "blazing") as a special substance out of which the luminous heavenly bodies were composed.

These basic substances are called elements in English, from a Latin word of unknown origin. (We still describe storms by speaking of "the raging of the elements" as water pours down, air blows about, and fire burns as lightning.)

To those people who accepted the notion of the various elements, and who were atomists, it made sense to suppose that each element was composed of a different type of atom, so that the world consisted of four different types of atoms altogether, with a fifth type for the heavenly aether.

Even with only four types of atoms, it was possible to account for the great variety of objects on Earth. One only had to imagine that the various substances were made up

6

of combinations of different numbers of different types of atoms in different arrangements. After all, with only twenty-six letters (or with just two symbols, a dot and a dash), it is possible to build up hundreds of thousands of different words in English alone.

However, the doctrine of the four (or five) elements began to fade even as atomism began to move ahead. In 1661, Boyle wrote a book, *The Skeptical Chemist,* in which he took up the position that it was useless to guess at what the basic substances of the world might be. One had to determine what they were by experiment. Any substance that could not be broken down by chemical manipulation into any simpler substance was an element. Any substance that *could* be broken down into simpler components was not an element.

This is indisputable in principle, but it is not entirely easy in practice. Some substances cannot be broken down into anything simpler and might seem to be elements, but then the time might come when advances in chemistry will make it possible to break them down. And again, when one substance is converted into another it isn't always easy to decide which of the two is simpler.

Nevertheless, beginning with Boyle and continuing for over three centuries, chemists have labored to find substances that can be identified as elements. Examples of familiar substances that have been recognized as elements in this way are gold, silver, copper, iron, tin, aluminum, chromium, lead, and mercury. Gases such as hydrogen, nitrogen, and oxygen are elements. Air, water, earth, and fire are *not* elements.

At the present time, 106 elements are known. Eighty-three of them occur naturally on Earth in reasonable quantities, and the remaining twenty-three occur either in traces or only after having been manufactured in a laboratory. This means there are 106 different types of atoms known.

Atomism Triumphant

Most substances as they occur on Earth are not elements, but can be broken down into the various elements that make them up. Those substances that are put together out of a combination of elements are known as compounds (from Latin words meaning "to put together").

Chemists grew increasingly interested in trying to determine how much of each element might exist in a particular compound. Beginning in 1794, the French chemist Joseph Louis Proust (1754–1826) worked on this problem, and made a crucial discovery. There is a compound we now call copper carbonate. Proust began with a pure sample of this substance and broke it down into the three elements that made it up: copper, carbon, and oxygen. He found, in 1799, that in every sample he worked with, no matter how it was prepared, there were present for every five parts of copper (by weight) four parts of oxygen and one part of carbon. If he added additional copper to the mixture in preparing copper carbonate, the additional copper was left over. If he began with a shortage of copper, only the proportionate amount of carbon and oxygen combined with it to form copper carbonate, and the rest of the carbon and oxygen was left over.

Proust showed that this was also true for a number of other compounds he worked with. The elements of which they were composed were always present in definite proportions. This was called the law of definite proportions.

The law of definite proportions offered strong support for atomism. Suppose, for instance, that copper carbonate is made up of little groups of atoms (called molecules, from Latin words meaning "a small mass"), each group consisting of one copper atom, one carbon atom, and three oxygen atoms. Suppose also that the three oxygen atoms, taken

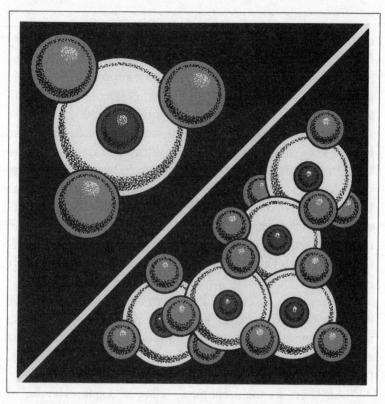

In 1799, Joseph Louis Proust broke down a pure sample of a substance into the three elements that comprised it: copper, carbon, and oxygen. He found that in every sample he worked with there were present for every five parts of copper (by weight) four parts of oxygen and one part of carbon. The atomic elements that compose a compound are always present in definite proportions.

together, are four times as heavy as the carbon atom, and that the copper atom is five times as heavy as the carbon atom. If every molecule of that compound is made up of that combination, then copper carbonate would always be made up of five parts copper, four parts oxygen, and one part carbon.

If it were possible to include in the molecule 1½ atoms of copper, or 3⅓ atoms of oxygen, or only ⅚ of an atom of carbon, the proportions of the three substances might vary from sample to sample of copper carbonate. However, the proportions *don't* vary. This not only supports the idea of atoms, but Democritus's suggestion that an atom is indivisible. It exists as an intact piece or as nothing.

The difference between the work of Democritus and Proust was this, however: Democritus had only a suggestion; Proust had *evidence*. (This is not to be taken as meaning that Proust was necessarily a greater or wiser man than Democritus. Proust had the benefit of twenty-one additional centuries of thought and work that he could draw upon. You might easily argue that it was much more remarkable that Democritus could hit on the truth so early in the game.)

Even with evidence, Proust did not necessarily have it all his own way. After all, it was possible that Proust's analyses were wrong, or that he was so eager to prove his own idea that he unconsciously twisted his observations. (Scientists are only human, and such things happen.)

Another French chemist, Claude Louis Berthollet (1748–1822), fought Proust every step of the way. He insisted that *his* analyses showed that compounds could be made up of elements in varying proportions. In 1804, however, the Swedish chemist Jöns Jakob Berzelius (1779–1848) began meticulous analyses that backed Proust's notion, and proved to the chemical world that the law of definite proportions was right.

At the same time, the English chemist John Dalton (1766–1844) was also working on the problem. He found that it was possible for compounds to be made up of elements in widely different proportions. Thus, in one gas, with molecules made up of carbon and oxygen, the proportions were three parts carbon to four parts oxygen. In another gas, with molecules made up of carbon and oxygen, the proportions were three parts carbon to eight parts oxygen. These, however, were two different gases with two different sets of properties, and for each one the law of definite proportions held.

Dalton suggested that in one gas the molecule was made up of an atom of carbon and an atom of oxygen, whereas in the other it was made up of an atom of carbon and *two* atoms of oxygen. It eventually turned out that he was correct, and the two gases came to be called carbon monoxide and carbon dioxide, respectively. (The prefix mon- is from the Greek word for "one," and di- is from the Greek word for "two.")

Dalton found this sort of thing was true in other cases, and in 1803 he announced this as the law of multiple proportions. He pointed out that this fit the notion of atoms, and it was he who called them atoms, deliberately going back to the old term as a tribute to Democritus.

Dalton said that to account for what was being found out about the proportion of elements contained in compounds, one had to decide that each element is made up of a number of atoms, all with the same fixed mass; that different elements have atoms of different masses; and that molecules are made up of a small, fixed number of different intact atoms.

In 1808, Dalton published a book entitled *New System of Chemical Philosophy*, in which he gathered all of the evidence he could find in favor of atomism and showed how it all fit together. With this book, Dalton established the modern atomic theory—modern, as opposed to that of the Greeks.

As it happens, the word *theory* is not properly understood by the general public, which tends to think of a theory as a "guess." Even dictionaries do not properly describe what the word means to scientists.

Properly speaking, a theory is a set of basic rules, supported by a great many confirmed observations by many scientists, that explains and makes sensible a large number of facts that, without the theory, would seem to be unconnected. It is as though the facts and observations are a number of dots representing cities, and lines representing country and state boundaries, distributed higgledy-piggledy on paper, making no sense. A theory is a *map* that puts each dot and line into the right place and makes a connected and sensible picture out of it all.

Theories are not necessarily correct in every detail, to begin with, and might never be entirely correct in every detail, but they are sufficiently correct (if they are good theories) to guide scientists in understanding the subject the theory deals with, in exploring further observations, and, eventually, in improving the theory.

Each of the basic rules Dalton set up for his atomic theory was not *quite* right. It turned out, eventually, that an element could have atoms of different mass, that two elements might have some atoms that were of the same mass, and that not all molecules were made up of small numbers of atoms. Dalton's rules were sufficiently close to right, however, to be very useful, and, as chemists learned more and more about atoms, they were able to correct the rules, as we shall see later on.

No scientific theory is instantly accepted by scientists. There are always those scientists who are suspicious of anything new—and this is perhaps a good thing. Theories should not slide into acceptance too easily; they should be questioned and tested vigorously. In this way, weak spots in the theory will be uncovered and, perhaps, strengthened.

As it happened, some of the most eminent chemists of Dalton's day were suspicious of the new theory, but it turned out to be so useful in helping to understand the observations of chemistry that chemist after chemist fell into line, and the entire scientific world eventually became atomists.

The Reality of Atoms

However well atomic theory worked, and however ingeniously it was improved, and however it managed to point the way to new discoveries, one disturbing fact remained: no one could see atoms or detect them in any way. All of the evidence in favor of atoms was indirect. You *inferred* that they existed from this fact, and *deduced* that they existed from that observation, but all of the inferences and deductions might be wrong. Atomic theory seemed to set up a scheme that worked, but it might have been just a simple model for something that was actually much more complicated. The working mode of the time was analogous to playing poker with chips. The chips can be used to bet with and to show how much money is being lost and won, and will be absolutely accurate in every way—but those chips are *not* money. They just symbolize the money.

Suppose, then, that the idea of atoms is merely a case of playing chemistry with chips. Atomism worked, but the atoms merely represented a truth that was much more complicated. There were some chemists, even a hundred years after Dalton, who were cautiously aware of this, and who warned against taking atoms too literally. Use them by all means, they would say, but don't think that they are necessarily really there in the shape of minute billiard balls. One scientist who thought this way was the Russian-German chemist Friedrich Wilhelm Ostwald (1853–1932).

The answer to this problem had long been on the way, however, and it started with an observation that seemed to have nothing to do with atoms, by a scientist who wasn't interested in atoms. (It's important to remember that all knowledge is of a piece and that any observation can have an unexpected and surprising connection to something that apparently has nothing to do with it.)

The vibration of a grain of pollen in water demonstrates the movement of the invisible molecules of water surrounding it.

In 1827, the Scottish botanist Robert Brown (1773–1858) was using a microscope to study pollen grains suspended in water. He noticed that each pollen grain was moving slightly and erratically, first in one direction then in another, as though it were shivering. He made sure that this wasn't the result of currents in the water or of motions set up by the fact that the water was evaporating. Brown concluded it had to be something else that caused the movement.

Brown tried other types of pollen, found that all of the grains moved in this fashion, and wondered if it was because the pollen grains had the spark of life in them. He tried pollen grains from herbariums, grains that were at least a century old. They moved in just the same way. He went on to try small objects in which there was no question of life existing—bits of glass, coal, or metals—and they *all* moved. This came to be called Brownian motion, and no one, at first, could explain it.

In the 1860s, however, the Scottish mathematician James Clerk Maxwell (1831–1879) tried to explain the behavior of gases on the basis that the atoms and molecules that made them up were in constant motion. Such constant motion of atoms had been suspected by early atomists, but Maxwell was the first to succeed in working the theory out mathematically. The way in which moving atoms and molecules bounced off each other, and off the walls of a container, as mathematically modeled by Maxwell, completely explained the behavior of gases. It explained Boyle's law, for instance.

Maxwell's work also produced a new understanding of temperature, for it turned out that temperature was the measure of the average speed of motion of the atoms and molecules making up not only gases, but liquids and solids. Even in solids, where atoms or molecules are frozen in place and can't move bodily from one point to another, those atoms or molecules vibrate about their average position, and the average speed of vibration represents the temperature.

In 1902, the Swedish chemist Theodor Svedberg (1884–1971) pointed out that one might explain Brownian motion by supposing that an object in water is bombarded from all sides by moving water molecules. Ordinarily, the bombardment from all sides is equal, so that the object remains at rest. To be sure, by sheer chance, a few more molecules might strike from one direction or another, but so many molecules strike all together that a small deviation from exact equality (two or three out of trillions) does not produce measurable movement.

If an object suspended in water is very small, however,

the number of molecules striking it from all sides is comparatively small, too, and if there is a small deviation, that might represent a fairly large effect, comparatively. The particle responds to the push of a few extra molecules from one particular direction by jerking slightly in the direction of the push. In the next moment, there are extra collisions in another direction, and the particle is pushed in that new direction. The particle moves randomly and erratically in response to the random motion of the surrounding molecules.

Svedberg was only speculating, but in 1905, the German-Swiss mathematician Albert Einstein (1879–1955) applied Maxwell's theory to the bombardment of small particles and showed quite conclusively that those particles would jiggle exactly as the pollen grains were observed to do. In other words, he presented mathematical equations that described Brownian motion.

In 1908, the French physicist Jean Baptiste Perrin (1870–1942) set about checking Einstein's equations against actual observations. He placed a fine powder of gum resin in water. If there were no bombardment by water molecules, then all of the particles of gum resin ought to have gone to the bottom of the container and remained there. If there were bombardment, some of the particles would be kicked upward against the pull of gravity. To be sure, those particles would settle again, but they would then be kicked up again, too. Some that were already up would be kicked up still further.

At any given time, the particles of gum resin would be spread upward. Most would be at the bottom, but some would be a little distance above, a few a greater distance above, still fewer a still greater distance above, and so on.

The mathematical equation worked out by Einstein showed what numbers of particles there should be at every height, depending upon the size of the particles and the size of the water molecules striking them. Perrin counted the number of particles at various heights and found that they followed Einstein's equation exactly. From this he calculated what size the water molecules must be, and what size the atoms that made them up must be.

Perrin published his results in 1913. The atoms, he had calculated, were roughly a hundred-millionth of a centimeter across. Put it another way: 100 million atoms placed side by side would stretch across a centimeter (250 million atoms placed side by side would stretch across an inch).

This was the nearest thing yet to an actual observation of atoms. If they could not quite be seen, the effects of their collisions could be seen and their actual size could finally be worked out. The most hard-nosed scientists had to give in. Even Ostwald admitted that atoms were real, that they weren't just make-believe models.

In 1936, the German physicist Erwin Wilhelm Mueller (1911–1977) got the idea of a device that would make it possible to magnify the point of a fine needle to such an extent that one could make pictures of it, with the atoms that compose it lined up as little luminous dots. By 1955, such atoms could actually be seen.

Yet people still speak of the atomic *theory*, because that is what it is—an intellectual map of large aspects of science that can be neatly explained by the existence of atoms. A theory, remember, is not a "guess," and no sane and qualified scientist can doubt that atoms exist. (This aspect of the proof that atoms exist is also true of other well-established scientific theories. The fact that they are theories does not make them uncertain, even when various fine details are still under dispute. This is particularly true of the theory of evolution, which is under constant attack from people who are either ignorant of science or, worse, who allow their superstitions to overcome what knowledge they might have.)

The Differences Among Atoms

It seems reasonable to suppose that if there are different types of atoms they must differ among themselves, somehow, in their properties. If this were not so, and if all atoms were identical in their properties, then why should some atoms, when heaped together, form gold, while others formed lead?

The ancient Greeks had their greatest intellectual success with the development of a rigorous form of geometry, so it was natural for some among them to think in terms of shapes when they thought of the atoms making up their "elements." To the Greeks, atoms of water might be viewed as spherical bodies that slipped over each other easily, which was why water poured. Atoms of earth would be cubic and stable so that earth *didn't* flow. Atoms of fire would be jagged and sharp, which was what made fire so painful, and so on.

The ancient Greeks also did not have it quite clear in their mind that one type of atom did not change into another. This was especially true if you considered that gold and lead were both varieties, in the main, of the element earth. Perhaps it was only necessary to pull apart the earth atoms in lead and put them into another arrangement that would make them gold; or one might modify the earth atoms in lead to change them slightly into a form that would make them gold.

For about two thousand years, various people, some of whom were earnest and science-minded, while a great many others were outright fakers and charlatans, kept trying to change base metals such as lead into the noble metal gold. This is called transmutation, from Latin words meaning "to change across." They always failed.

By the time the modern atomic theory was advanced,

To the ancient Greeks, atoms of water might be viewed as spherical bodies that slipped water over each other easily, which was why water poured.

it seemed clear that atoms were not only different from each other, but that one type of atom could not be changed into another. Each atom was fixed and permanent in its properties, so that an atom of lead could not be changed into an atom of gold. (The time was to come, as we shall see, when this was found to be not *quite* true, under very special conditions.)

But if different types of atoms are different from one another, of just what does the difference consist? Dalton reasoned as follows. If the water molecule is made up of eight parts oxygen to one part hydrogen, and if the molecule is made up of one atom of oxygen and one atom of hydrogen, then it must be that the individual oxygen atom weighs eight times as much as the individual hydrogen atom. (To be more precise, one should say that the individual oxygen atom has eight times the "mass" of the individual hydrogen atom. The weight of an object is the force with which the Earth attracts it, whereas the mass of an object is, roughly speaking, the amount of matter it contains. Mass is the more fundamental of the two concepts.)

Of course, Dalton had no way of knowing the mass of either a hydrogen or an oxygen atom, but whatever it was, the oxygen atom had a mass eight times that of a hydrogen atom. You could say that a hydrogen atom had a mass of 1, without saying 1 *what*. You could then say that an oxygen atom has a mass of 8. (Actually, we now say the hydrogen atom is 1 dalton, in honor of the scientist, but it is customary simply to leave it as 1.)

Dalton went to work with compounds containing other elements and worked out a system of numbers representing the relative masses of them all. He called them *atomic weights*, and the term is still used today, even though we should speak of atomic masses. (It frequently happens that scientists begin to use a particular term and then decide that another term would have been better, but find it is too late to change because people have grown far too accustomed to the poorer term. We'll come across other cases of the sort in this book.)

The trouble with Dalton's method of determining atomic weights was that he was forced to make assumptions that could too easily be wrong. He assumed that a water molecule consisted of one atom of hydrogen and one of oxygen, but he didn't have any evidence for that.

In that case, one must look for evidence. In 1800, the British chemist William Nicholson (1753–1815) passed an electric current through acidified water and obtained bubbles of both hydrogen and oxygen. Continued investigation of this phenomenon showed that the volume of hydrogen formed was just twice that of the oxygen, although the mass of oxygen liberated was eight times the mass of the double volume of hydrogen.

Why was twice the volume of hydrogen produced as compared to oxygen? Could it be that the water molecule was composed of *two* hydrogen atoms and one oxygen atom, instead of one of each? Could it be that the oxygen atom was eight times as heavy as both hydrogen atoms put together, or sixteen times as heavy as a single hydrogen atom? In other words, if hydrogen had an atomic weight of 1, was the atomic weight of oxygen 16, rather than 8?

Dalton refused to accept this notion. (It often happens that a great scientist, having taken a giant step forward, refuses to take other steps—as though the great first effort had exhausted him—and leaves it to others to continue to march forward.)

In this case, it was Berzelius who took the forward step, placing hydrogen at 1 and oxygen at 16. He continued with other elements and, in 1828, published a table of atomic weights that was much better than Dalton's had been. From the work of Berzelius, it seemed clear that every element had a different atomic weight, and that each atom of a particular element had the same atomic weight. (I must remind you again that these conclusions eventually proved to be not quite right, but they were near enough to right to be useful to chemists for nearly a century. Eventually, as more knowledge was gained, these views were modified in ways that slightly changed and immeasurably strengthened the atomic theory. This improvement of theories happens over and over and is the pride of science. To suppose that this should not happen and that theories should be absolutely correct to begin with is to suppose that a stairway stretching upward for five stories should consist of a single five-story-high step.)

Well, then, the volume of hydrogen produced when water is broken down by an electrical current is twice the volume of oxygen. How do we know from this that there are two hydrogen atoms to one oxygen atom in the molecule? It seemed sensible to Berzelius to suppose so, but he didn't know for sure. It, too, was an assumption, even though there was more evidence behind it than there was behind Dalton's assumption that there was one hydrogen atom and one oxygen in the water molecule.

In 1811, the Italian physicist Amedeo Avogadro (1776–1856) made a more general assumption. He suggested that in the case of *any* gas, a given volume always contains the same number of molecules. If one gas has twice the volume of another gas, the first gas has twice as many molecules as the other. This is called Avogadro's hypothesis. (A hypothesis is an assumption that is sometimes advanced just to see what the consequences would be. If the consequences go against known observations, then the hypothesis is wrong and it can be dismissed.)

Naturally, when a competent scientist advances a hypothesis that he thinks might be true, there is a good chance it will turn out to be true. One way of testing Avogadro's hypothesis, for instance, is to study a great many gases and to work out the number of each of the different types of atoms in the molecules of those gases on the basis that the hypothesis is true.

If one does that and ends by violating known observations, or ends by producing a contradiction—as when one line of argument based on the hypothesis shows that a particular molecule must have a certain atomic composition, and another line of argument shows it must have a different atomic composition—then Avogadro's hypothesis would have to be thrown out.

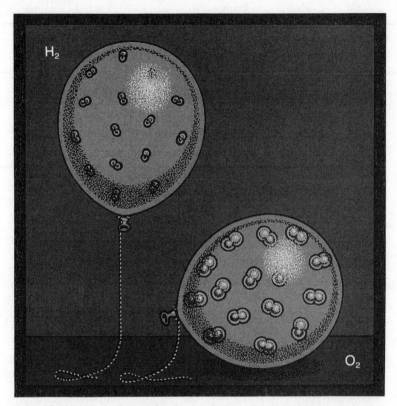

Avogadro's law: equal volumes of all gases under identical conditions of temperature and pressure contain equal numbers of molecules. For example, it might take .1 gram of hydrogen gas to fill a child's balloon. It would take approximately 1.6 grams of oxygen gas to inflate an identical balloon to an equal size, but both balloons would contain approximately the same number of molecules.

Actually, no one has ever found a case in which Avogadro's hypothesis is truly misleading, and the theory is no longer a hypothesis but is considered a fact, although there are conditions under which it must be modified. It is still *called* Avogadro's hypothesis, however, because chemists are so accustomed to calling it that.

One problem, however, was that when Avogadro's hypothesis was first advanced, very few chemists paid any attention to it. They either didn't hear of it, or dismissed it as either ridiculous or unimportant. Even Berzelius didn't make use of the hypothesis, so that his table of atomic weights was wrong in places.

In 1858, however, the Italian chemist Stanislao Cannizzaro (1826–1910) came across Avogadro's hypothesis and saw that that was what was needed to make sense out of figuring out how many atoms of each element there were in a compound, and getting the correct figures for atomic weight.

In 1860, there was a great international chemical congress, which chemists from all over Europe attended (it was the first of such international congresses). At that congress, Cannizzaro convincingly explained the hypothesis.

This at once improved the entire notion of atomic weight. About 1865, the Belgian chemist Jean-Servais Stas (1813–1891) put out a new table of atomic weights that was better than Berzelius's. Forty years later, the American chemist Theodore William Richards (1868–1928) made even more refined observations and got the very best values one could get before (as we shall see) the entire subject of atomic weight had to be modified because of new discoveries. By Richards's time, Nobel prizes were being handed out, and for his work on atomic weights he got the Nobel prize for chemistry in 1914.

As it happens, the element with the lowest atomic weight is hydrogen. If its atomic weight is set arbitrarily at 1, then the atomic weight of oxygen is a little bit less than 16. (That it is not exactly 16 is a point we will consider later on.) However, oxygen easily combines with a great many other elements, and it is much simpler to compare the atomic weight of some particular element to oxygen than to hydrogen. It is convenient, then, to set oxygen's atomic weight at some exact figure. It shouldn't be set at

1 because that would give seven elements atomic weights of less than 1, which would be inconvenient in making chemical calculations.

It became customary, then, to set the atomic weight of oxygen at exactly 16, which made the atomic weight of hydrogen just a little bit greater than 1. That meant that no element had an atomic weight of less than 1. Stas's list was made that way and that set the fashion. (However, the situation has been changed very slightly in recent years for reasons that will be explained later.)

If the elements are listed in order of increasing atomic weights, then it is possible to arrange them in a rather complicated table that demonstrates that certain properties of the elements repeat themselves periodically. If the table is arranged correctly, elements with similar properties fall into the same column. This is called the periodic table, and a workable version of it was first presented by the Russian chemist Dmitri Ivanovich Mendeleev (1834–1907) in 1869.

The periodic table was quite tentative at first because Mendeleev didn't know all of the elements. Many had not yet been discovered. In arranging the table so that similar elements were in the proper columns, Mendeleev was forced to leave gaps. He felt that these gaps represented undiscovered elements and, choosing three of those gaps, stated in 1871 that those undiscovered elements, once discovered, would have certain properties, which he described in detail. By 1885, all three elements were discovered and Mendeleev was proven precisely correct in each case. This offered very strong proof that the periodic table was a legitimate phenomenon, but no one could explain *why* it worked. (We will return to this later on.)

2

LIGHT

Particles and Waves

If we are prepared to admit that all matter is composed of atoms, then it is reasonable to ask if there is anything in the world that isn't matter and, therefore, isn't composed of atoms. The first possibility that might spring to mind is light.

It has always seemed obvious that light is immaterial. Solids and liquids can be touched; have mass, and therefore weight; and take up space. Gases cannot be felt in the same way that solids and liquids can, but a moving gas can be felt. We have all experienced high winds and we well know what a tornado can do. Then, too, air will take up room so that if an "empty" beaker (actually full of air) is plunged, open end down, into a tank of water, the water does not fill the beaker unless, somehow, the air is allowed to escape. In 1643, the Italian physicist Evangelista Torricelli (1608–1647) showed that air had weight and that this weight could support a column of mercury 76 centimeters (30 inches) high.

Light, however, has none of these properties. It cannot be felt, even though the heat it might produce can. It has never been found to have perceptible mass or weight, and it does not appear to take up space.

This doesn't mean that light was dismissed as unimportant because it was insubstantial. The first words of God, as given in the Bible are: "Let there be light." What's more, under the name of fire, it was the fourth of the ancient Earthly elements, on a par with the three material ones of air, water, and earth.

Sunlight was naturally considered to be light at its purest. It was white light, unchanging and eternal. If sunlight were made to pass through colored glass, it would pick up the color of the glass, but that would be an earthly impurity. Again, when objects burned on earth and gave off light, that light might be yellow, orange, or red. In some cases, if certain powders were cast into the fire, it might even burn green or blue. But again, these were earthly impurities that gave rise to color.

The one colored object that seemed to be divorced from anything earthly was the rainbow, which was sufficiently awe-inspiring to give rise to myths and legends. It was thought to be the bridge between heaven and earth, used by divine messengers. (The Greek messenger of the gods is given the name Iris, which is Greek for "rainbow.") It was also a divine guarantee that the world would never again be destroyed by flood, so that it appears at the end of rainstorms, indicating that God has remembered and stopped the rain.

In 1665, however, the English scientist Isaac Newton

(1642–1727) produced his own rainbow. In a darkened room, he allowed a beam of sunlight to enter through a hole in a shutter, and passed that beam through a three-dimensional triangular wedge of glass called a prism. The beam of light spread out and produced a band of colors on the white wall beyond, the colors being red, orange, yellow, green, blue, and violet, in that order—just the order in which they occur in the rainbow.

A rainbow, we now know, is caused by sunlight passing through the innumerable droplets of rain still in the air after a rainshower. These droplets have the same effect on light rays as a glass prism.

Apparently, then, sunlight is not "pure" light, after all. Its whiteness is merely the effect produced on the eye by a mixture of all of these colors. By having the light pass through a prism and then pass through another prism held in the reverse position, the separated colors will rejoin and form white light again.

In that these colors are thoroughly immaterial, Newton called the rainbow band a spectrum, from the Latin word for "ghost." Newton's spectrum created a problem, however. For the colors to be separated on passing through the prism, Newton believed, each one must have its ordinary straight-line path bent (refracted) as it passed into and out of the glass—each color bent to a different extent (red the least and violet the most), so that they were separated and seen each by itself when the beam hit the wall. What, then, could light be made of that would account for the separation of light into a spectrum?

Newton was an atomist and so it naturally occurred to him that light was made up of tiny particles, like the atoms of matter, except that the particles of light did not have mass. He had no clear notion, however, as to how the particles of colored light might differ among themselves, and why some should be refracted by a prism to a greater extent than others.

Furthermore, when two beams of light crossed each other, one remained unaffected by the other. If both consisted of particles, should not those particles collide and bounce off one another randomly so that the beam would grow fuzzy and spread outward after collision?

The Dutch physicist Christiaan Huygens (1629–1695) had an alternate suggestion. He thought light consisted of tiny waves. In 1678, he advanced arguments for showing that an entire series of waves might advance in what looked like a straight line, just as a beam of particles would, and that two beams, each made up of waves, would cross each other without either being, in the end, disturbed.

The trouble with the wave suggestion was that people thought of the types of waves produced in water, such as when a pebble is dropped into a still pond. As those water waves expand, they tend to move around an obstruction such as a piece of wood (diffraction) and join again on the other side. In that case, wouldn't light waves curve around an obstruction and cast no shadows, or at least fuzzy ones? Instead, as is well known, light casts sharp shadows if the light source is small and steady. Such sharp shadows are exactly what you would expect if light were a beam of minute particles, and this was considered a strong argument against waves.

It is interesting to note that the Italian physicist Francesco Maria Grimaldi (ca. 1618–1663) had noticed that a beam of light passing through two narrow openings, one behind the other, widened a little bit, indicating it had diffracted outward very slightly as it passed through the openings. His observation was published in 1665, two years after his death, but somehow it didn't attract attention. (In science, as in many other types of human endeavor, important discoveries or events sometimes get lost in the shuffle.)

Huygens, nevertheless, showed that light, if composed of waves, might well have waves of different lengths. Those portions of light with the longest waves would be least refracted. The shorter the waves, the greater the refraction. In this way, one could explain the spectrum, in that it might be that red had the longest waves and that orange, yellow, green, and blue were made up of successively shorter waves, while violet was made up of the shortest.

On the whole, as we look back on it, Huygens had the better of the argument, but Newton's reputation was growing rapidly (he was undoubtedly the greatest scientist who had ever lived) and it was hard to take up a position against him. (Scientists, in that they are as human as anyone else, are sometimes swayed by personalities as well as by logic.)

Throughout the 1700s, then, most scientists accepted the fact that light consisted of little particles. This might have helped the growth of atomism in connection with matter, and as atomism gained, that, in turn, strengthened the particle view of light.

In 1801, however, the English physicist Thomas Young (1773–1829) performed a crucial experiment. He let light fall upon a surface containing two closely adjacent slits. Each slit served as the source of a cone of light, and the two cones overlapped before falling on a screen.

If light were composed of particles, the overlapping region should receive particles from both slits and be brighter than the outlying regions that received particles from only one slit or the other. This was not so. What Young found was that the overlapping portions consisted of stripes—bright bands and dim bands alternating.

There seemed no way of explaining this phenomenon by the particle hypothesis. With waves, however, there were no problems. If the waves from one slit were in phase with those from the other slit, both keeping perfect step, then the ups and downs of one set of waves (or the ins and outs) would be reinforced by those of the other set, and the oscillation of the two combined would be stronger than of either separately. Brightness would increase.

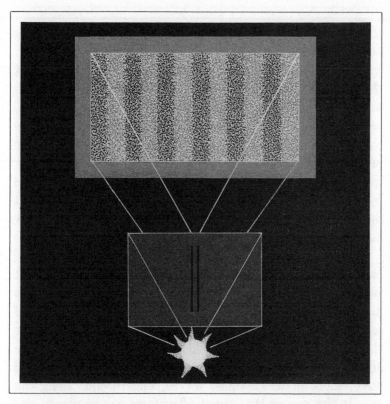

In 1801, Thomas Young let light fall on a surface containing two closely adjacent slits. The wedge of light from each slit fell on a screen and overlapped, resulting in a pattern of stripes—bright bands and dim bands alternating. There seemed no way of explaining this phenomenon by the particle hypothesis.

On the other hand, if the waves from one slit were out of phase with those from the other slit—if one set of waves went up while the other went down (or one went in while the other went out)—then the two waves would cancel each other, at least in part, and the two combined would be weaker than either separately. Brightness would decrease.

Young was able to show that, under the conditions of his experiment, the two sets of waves would be in phase in one region, out of phase in the next, in phase again in the next, and so on, alternately. The bright and dim bands that were observed would be exactly what would be expected of waves.

Because one set of waves interferes with, and cancels, the other set in specific places, these bands are called interference patterns. Such interference patterns are observed when one set of waves on a calm water surface overlaps another. They are also observed when two beams of sound (known to consist of waves) intersect each other. The wave nature of light thus appeared to be demonstrated by Young's experiment (although, as we might expect, that didn't mean that those who believed in the particle view surrendered easily—because they didn't).

It was even possible, from the width of the interference bands, to calculate the length of a single wave of light (wavelength). It turned out that light waves had lengths in the neighborhood of $\frac{1}{20,000}$ of a centimeter (or $\frac{1}{50,000}$ of an inch). The wavelength of red light was a little longer than that, while the wavelength of violet light was a little shorter. This means that a ray of light an inch long will have, more or less, 50,000 waves, end to end, along the ray. It also means that about fifty atoms can be placed end to end along a single wavelength of light.

That explained why light cast sharp shadows despite being made of waves. Waves bend around obstacles only when the obstacles are not much longer than the wave in question. A wave would not bend around anything substantially longer than itself. Sound waves are very long and can move around most ordinary obstacles.

Almost anything we can easily see, however, is much, much longer than a light wave, so there's virtually no turning for them, and the shadows they cast are sharp. There is a very *slight* turning effect, however, and where the objects are quite small, the shadow's edge is inclined to be slightly fuzzy. That explains the diffraction effect that Grimaldi had discovered 130 years before Young's time.

The issue wasn't settled, however. People knew of two

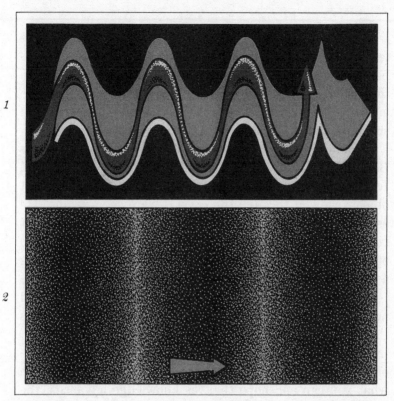

1. *Water waves spread outward, and the particles of water move up and down in a direction at right angles to the direction in which the wave progresses. This type of wave is called a* transverse wave.
2. *Sound waves also spread outward, but the particles of air move in and out in a direction parallel to the direction in which the wave progresses. This type is called a* longitudinal wave.

types of waves. There were water waves, in which the wave spread outward, but the particles of water moved up and down in a direction at right angles to the direction in which the wave progressed. This is called a transverse wave. There were also sound waves, in which the wave also spread outward, but the particles of air moved in and out, in a direction parallel to the direction in which the wave progressed. This is called a longitudinal wave.

Which of these two describes light waves? Huygens,

when he first elaborated the wave hypothesis, might have felt that light and sound, both being the cause of sense perceptions, should be similar in nature. Sound was known to be a longitudinal wave, so he suggested that light was a longitudinal wave, too. Young, when he demonstrated the wave nature of light, also thought so.

Earlier, in 1669, however, the Danish scholar Erasmus Bartholin (1625–1698) received a transparent crystal from Iceland, of a type now called Iceland spar. He noted that objects viewed through the crystal were seen double. He assumed that light passing through the crystal was refracted at two different angles, so that some emerged in one place and the rest in a slightly different place, producing a double image.

Bartholin could not explain why this should be, and neither could Newton or Huygens. The phenomenon was therefore pushed to one side as temporarily inexplicable. (Not everything can be explained at some particular stage of knowledge. The only sensible thing to do is to explain what you can and hope that, as knowledge advances, the time will come when the temporarily unexplainable can also be explained.)

In 1817, Young realized that double refraction could not be easily explained if light consisted of particles or of longitudinal waves. It could be explained quite easily, however, if light consisted of transverse waves.

The French physicist Augustin Jean Fresnel (1788–1827) adopted this point of view and worked out a careful theoretical study of light as transverse waves, one that explained all that was known about the behavior of light at that time. That settled it. For the next eighty years, physicists were quite satisfied that light was made up of tiny transverse waves and that that was the whole of the answer.

The Four Phenomena

It is a rare answer that is *completely* satisfactory, and this seems especially true in science, where every answer seems to uncover a more subtle question. If we grant that light exists in waves, as sound and a disturbed pond surface do, then there remains the problem that light waves travel easily through a vacuum, whereas sound waves and water waves do not.

Water waves exist because water molecules move up and down regularly. If water did not exist, then neither would water waves. Sound waves exist because air molecules (or the molecules of any medium through which sound travels) move in and out regularly. If air or any other medium does not exist, sound waves would not exist either.

In the case of light waves, however, what is it that is moving up and down? It can't be any type of ordinary matter, for light waves can pass through a vacuum where, apparently, there is no matter.

Newton had a similar problem when he worked out the law of universal gravitation in 1687. The Sun held the Earth in its gravitational grip across 150 million kilometers (93 million miles) of vacuum. How could the gravitational effect, whatever it was, travel across a vacuum?

Newton wondered if, perhaps, vacuum was not really *nothing* but consisted of a type of matter more subtle than ordinary matter and therefore not easily detectable. This vacuum matter came to be called ether, in homage to the "aether" Aristotle imagined as making up the heavenly bodies. The gravitational attraction pulled at the ether, and this pull was conducted from one bit of ether to the next until, finally, the Sun was pulling at the Earth.

Perhaps it was this ether (or another type) that waved up and down as light passed through. It had to fill all of space because we could see even the most distant stars. What's more, it had to be so fine and rarefied a type of matter that it did not in any way interfere with the passage of the Earth, or any other heavenly body, however light, as it progressed through space. Fresnel suggested that ether permeated the very body of the Earth and of all other heavenly bodies.

The particles of ether, however, when moved up, must experience a restoring force that moved them down, past the equilibrium point, then up again. The more rigid a medium, the more rapidly it vibrates up and down, and the more rapidly a wave progresses through it.

Light travels at a speed of 299,792 kilometers (186,290 miles) per second. This was first determined, very approximately, by the Danish astronomer Olaus Roemer (1644–1710) in 1676. To allow light to travel at such a speed, the ether must be more rigid than steel.

To have the vacuum made of something so fine that it allowed bodies to pass through it freely and without measurable interference, and at the same time so stiff as to be more rigid than steel, was rather puzzling, but scientists didn't seem to have any choice but to suppose that this was the case.

In addition to light and gravity, two other phenomena were known that could make themselves felt across a vacuum. They were electricity and magnetism. Both were first studied, according to tradition, by Thales. He studied a certain piece of iron ore, first found near the town of Magnesia on the eastern shore of the Aegean Sea. It had the property of attracting pieces of iron and he is supposed to have called it *ho magnetes lithos* ("the Magnesian rock"). Objects with the property of attracting iron have been called magnets ever since.

Thales also found that lumps of amber (a fossilized resin), if rubbed, attracted not iron particularly, but *any* light object. This difference in behavior meant the attraction was not that of magnetism. The Greek word for amber is *elektron* and, eventually, this phenomenon came to be called electricity as a consequence.

Sometime in the eleventh century, in China—but exactly where and by whom and under what circumstances is unknown—it was discovered that if a needle made of magnetic ore, or of steel that had been magnetized by being stroked by magnetic ore, was allowed to turn freely, it would align itself north and south. In addition, if the ends were marked in some way, it would be seen that the same end always turned north.

That end was called the magnetic north pole, and the other the magnetic south pole. In 1269, the French scholar Petrus Peregrinus (1240–?) experimented with such needles and found that the magnetic north pole of one would be attracted to the magnetic south pole of the other. On the other hand, the magnetic north poles of two magnetized needles would repel each other, as would the magnetic south poles of the two needles. In short, like magnetic poles repelled each other while unlike magnetic poles attracted each other.

In 1785, the French physicist Charles Augustin de Coulomb (1736–1806) measured the strength of the force by which a magnetic north pole attracted a magnetic south pole, or repelled another magnetic north pole. He found that the attraction or repulsion declined as the square of the distance (the inverse square law). That is, if you increased the distance to x times what it was before, the force between the poles became $1/x \times 1/x$, or $1/x^2$, what it was before. When Newton dealt with gravitation in 1687, he showed that the force of gravitational attraction followed what was to become the inverse square law.

Thus, the Moon is sixty times as far from the Earth's center as the Earth's surface is. The Earth's gravitational

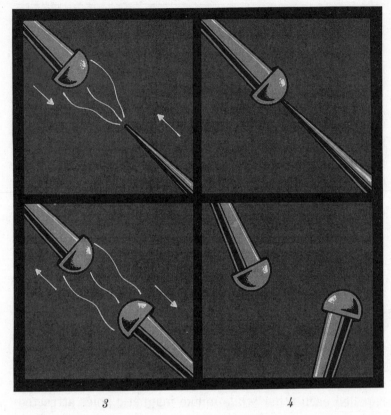

1. and 2. An unmagnetized iron needle will be attracted to either the north or the south pole of a magnet. Once magnetized, however, one end will be repelled, while the other will attract.
3. Unlike poles (N, S) attract.
4. Like poles (N, N or S, S) repel.

pull at the distance of the Moon is only $\frac{1}{60} \times \frac{1}{60}$, or $\frac{1}{3600}$ of what it is on the Earth's surface. Nevertheless, this pull is proportional to the product of the two masses involved, and the Earth and Moon are so massive that the Earth's gravitational pull is still large enough at the distance of the Moon to hold the Moon in orbit.

For that matter, the Sun can hold the Earth in orbit across a distance nearly 400 times that between the Earth and the Moon. Indeed, huge clusters of galaxies, stretching across millions of light-years of space are held together by gravitational pulls.

Yet, as it turns out, the magnetic attraction between two magnetized needles is trillions of trillions of trillions of times as strong as the gravitational attraction between those same two magnetized needles. Why is it, then, that we are so aware of gravitational pulls and hardly at all aware of magnetic pulls? Why are astronomical bodies held together by gravitation, while we never hear of two bodies held together by some magnetic force?

The answer is that magnetism involves both an attraction and a repulsion, both of equal intensity. Gravitation involves *only* an attraction. There is no such thing as gravitational repulsion.

The world is full of magnets. As we shall see, every atom is a tiny magnet. The magnets of the Universe are turned every which way, however, and there is as much chance of repulsions here and there as attractions. On the whole, the two cancel each other and, by and large, we are left with a Universe in which there is not much magnetic attraction or repulsion overall.

Gravitation, however, involving only an attraction, simply piles up, so to speak. Although the effect of gravitational pull is so small as to be unnoticeable for ordinary objects, or even for mountains, by the time you have objects the size of the Earth, or the Sun, the gravitational pull is enormous.

Still, magnetism does play its part. Suppose you place a piece of stiff paper over a magnetized steel bar. Scatter some iron filings upon the paper and tap it. The tapping allows the filings to move and to take up some natural position with respect to the magnet. When this is done, the filings arrange themselves in a group of curved lines extending from one pole of the magnet to the other. Peregrinus had noticed this, and, in 1831, the English scientist Michael Faraday (1791–1867) considered the subject.

To Faraday it seemed that the influence of the magnet stretched out through space in all directions in a magnetic field that weakened with distance according to the inverse square law. Through the field one could draw a vast number of lines (magnetic lines of force) that marked out regions where the strength of the magnetic field was the same. Such lines were followed by the iron filings and were thus made visible.

That is why magnetic needles in a compass point north and south. The Earth itself is a magnet, and the needles line up with the magnetic lines of force that go from one of Earth's magnetic poles to the other. (Earth's magnetic poles are located in the far north and the far south, but are a considerable distance from the geographic poles of rotation.) A great many other facts about magnets can be explained by the concept of the magnetic field and its lines of force, and Faraday's notion has remained valid ever since. (There are also gravitational and electric fields, and lines of force there, too.)

What about electricity, by the way? The English physicist William Gilbert (ca. 1544–1603) extended Thales's work on electrified substances. He explained in a book he published in 1600 that substances other than amber would also attract light objects if rubbed. Gilbert called all such objects electrics.

In 1733, the French chemist Charles François de Cisternay Dufay (1698–1739) experimented with rods of glass and of resin, both of which could be "electrified," when rubbed, and made capable of attracting light objects. Both would then attract small bits of cork, which were, in turn, electrified.

A piece of cork electrified by glass would attract a piece of cork electrified by resin. Two pieces of cork, each electrified by glass, would, however, repel each other; as would two pieces of cork, each electrified by resin. Dufay concluded, therefore, that there were two types of electricity. Each repelled itself, but attracted the other, as in the case of the two types of magnetic poles.

The American scholar Benjamin Franklin (1706–1790) took this one step further. He suggested, in 1747, that there was but one type of electricity, which all matter contained a normal amount of and which was undetectable. If certain objects were rubbed, however, then some of the electricity was removed; while if other objects were rubbed, some was added. Those objects that had an excess might be considered positively charged, those with a deficiency were negatively charged.

In such a case a positively charged object would attract a negatively charged one because contact would allow the excess charge in the first to flow into the second and make up the deficiency there. The two would cancel each other and leave two uncharged objects behind. (This was actually observed by Franklin in his experiments.) On the other hand, two positively charged objects would repel each other as would two negatively charged objects because in neither case was there a chance of charge flowing from one to the other.

It only remained for Franklin to decide which of the two types of electrically charged bodies had the excess and which the deficiency. There was no way of telling at the time, so Franklin chose arbitrarily. He decided that rubbed glass had an excess and should be considered positive (+), while rubbed resins had a deficiency and should be considered negative (−).

Ever since, people working with electric currents have assumed that current flows from positive to negative. Unfortunately, Franklin had had a fifty-fifty chance of guessing right, and had lost. It was the rod of resin that actually had the excess, so that the current really flows from negative to positive. That doesn't matter in electrical engineering, however. The results are the same whichever direction you imagine the current flowing, provided you stick by your decision and don't change your mind in midcourse.

Combining the Phenomena

There are, then, four phenomena that can make themselves felt across a vacuum: light, electricity, magnetism, and gravitation. All four might be pictured as making use of ether, but are they making use of the same ether, or does each one have an ether of its own? There was no way of telling, but sometimes light was pictured as waves in the luminiferous ether, from a Latin expression meaning "light carrying." Might there also turn out to be an "electriferous," a "magnetiferous," and a "gravitiferous" ether?

To be sure, the differences among the four were not equally great. Light did not seem either to attract or repel. Gravitation only attracted. Electricity and magnetism, however, each both attracted and repelled, and did so in much the same way with likes repelling and unlikes attracting. Of these last two, one seemed to arise out of the other.

In 1819, the Danish physicist Hans Christian Oersted (1777–1851) was lecturing on the electric current and, as a demonstration (it is not clear what he was trying to show), he brought a compass near a wire through which an electric current was flowing. To his own profound surprise, the compass needle reacted at once, pointing in a direction at right angles to the flow of current. When Oersted reversed the flow of current, the compass needle veered and pointed in the opposite direction, still at right angles to the current flow.

Oersted was the first to demonstrate an intimate connection between electricity and magnetism, but did not proceed in his investigations. Others, who heard of the demonstration, did, however, and at once.

In 1820, the French physicist Dominique François Arago (1786–1853) showed that a wire carrying an electric current acted as a magnet and could attract iron filings; something it would no longer do when the current ceased flowing. Because the wire was copper, this showed that magnetism was not necessarily an attribute of iron only, but might exist in any matter. Scientists began to speak of electromagnetism.

That same year, another French physicist, André Marie Ampère (1775–1836), showed that if two parallel wires had current flowing through each in the same direction, they attracted each other; if in opposite directions, they repelled each other.

If you twist a wire into a helix (the shape of a bedspring) and send a current through it, the current travels through each curve in the helix in the same direction. All of the curves attract one another and each sets up a magnetic field, every one reinforcing all of the others. The solenoid (coil of wire) then acts like a bar magnet with a north magnetic pole at one end and a south magnetic pole at the other.

In 1823, the British physicist William Sturgeon (1783–1850) wrapped wire about a U-shaped iron bar. The iron tended to intensify the magnetic field and, when the electric current was turned on, it became a surprisingly strong electromagnet.

In 1829, the American physicist Joseph Henry (1797–1878) used insulated wire (to prevent short-circuits), wrapping hundreds of turns of it about an iron bar, to produce an electromagnet that could lift phenomenal weights of iron when a current was passed through it.

Faraday then considered the reverse. If electricity could create magnetism, might not magnetism create electricity? He inserted an ordinary bar magnet into a helix of wire that was *not* connected to any battery that could start an electric current flowing in it. The magnet, nevertheless, experienced such a current when the magnet was pushed in, or pulled out. There was no current when the magnet was motionless at any point inside the helix. Apparently, the current flowed through the wire only when the wire cut across the magnetic lines of force, flowing in one direction as the magnet went in and in the other direction as it came out.

In 1831, Faraday worked out a system whereby a copper disc was turned between the poles of a magnet. An electric current was set up in the disc and flowed continually as long as the disc turned. It was an effort to keep it turning because it took work to push the disc across the magnetic lines of force. As long as this was done, however, by human or animal muscle, or by falling water, or by the force of steam produced by burning fuel, mechanical work was turned into electricity.

This time, it was Henry who reversed the situation. That same year, he invented the electric motor, in which the flow of electricity caused a wheel to turn.

All of these discoveries served to electrify the world (in a literal as well as a figurative sense) and to alter human society enormously. To scientists, however, the importance of these discoveries was that they increasingly demonstrated the close relationship between electricity and magnetism.

Indeed, there were those who began to think that there was a single electromagnetic field, one that, at times, showed its electrical face to the world, and, at times, its magnetic face. This reached its climax with the work of Maxwell. Between 1864 and 1873, he worked out the mathematical implications of Faraday's notions of fields and lines of force, and of the apparent connection of electric and magnetic fields. Maxwell ended by devising four comparatively simple equations (simple to mathematicians, at any rate) that described all known electrical and magnetic behavior. They have been known ever since as Maxwell's equations.

Maxwell's equations (whose validity is confirmed by all observations made since) show that electric fields and magnetic fields cannot exist separately. There is, indeed, only a combined electromagnetic field with an electric component and a magnetic component at right angles to each other.

If electric behavior and magnetic behavior were similar in all respects, the four equations would be symmetrical; they would exist in two mirror-image pairs. In one respect, however, the two phenomena do not match each other. In electrical phenomena, positive charges and negative charges can exist independently of each other. An object can be either positively charged or negatively charged. In magnetic phenomena, on the other hand, the magnetic poles do not exist separately. Every object that shows magnetic properties has a north magnetic pole at one location and a south magnetic pole at another location. If a long magnetized needle, with a north magnetic pole at one end and a south magnetic pole at the other, is broken in the middle, the poles are *not* isolated. The end with the north magnetic pole instantly develops a south magnetic pole at the break, while the end with the south magnetic pole develops a north magnetic pole at the break.

Maxwell included this fact in his equations, which introduced a note of asymmetry. This has always bothered scientists, in whom there is a strong drive for simplicity and symmetry. This "flaw" in Maxwell's equations is something we'll return to later.

Maxwell showed that from his equations you can demonstrate that an oscillating electric field will produce, inevitably, an oscillating magnetic field, which will in turn produce another oscillating electric field, and so on indefinitely. This is the equivalent of an electromagnetic radiation moving outward, in wave form, at a constant speed. The speed of this radiation can be calculated by taking the ratio of certain units expressing magnetic phenomena to other units expressing electrical phenomena. This ratio works out to nearly 300,000 kilometers (186,290 miles) per second, which is the speed of light.

This could not be a coincidence. Light, it appeared, was an electromagnetic radiation. Maxwell's equations thus served to unify three of the four phenomena known to pass through a vacuum: electricity, magnetism, and light.

Only gravitation remained outside this unification. It seemed to have nothing to do with the unified three. Albert Einstein, in 1916, worked out his general theory of relativity, which improved on Newton's concept of gravitation. In Einstein's interpretation of gravity, which is now widely accepted as essentially correct, there should be gravitational radiation in the form of waves, analogous to electromagnetic radiation. Such gravitational waves, however, are much more subtle and feeble, and much more difficult to detect, than are electromagnetic waves. Despite some false alarms, they have not yet been detected at this moment of writing, although virtually no scientist in the field doubts that they exist.

Extending the Spectrum

Maxwell's equations set no limitations on the period of oscillations of the field. There could be one oscillation per second or less, so that each wave would be 300,000 kilometers long, or more. There could also be a decillion oscillations per second or more, so that each wave would be a trillionth of a trillionth of a centimeter long. And there could be anything in between.

Light waves, however, represent only a tiny fraction of these possibilities. The longest wavelengths of visible light are 0.0007 millimeters long, and the shortest wavelengths of visible light are just about half this length. Does this mean there is electromagnetic radiation we don't see?

Through most of human history, the question as to whether light existed that could not be seen would have been considered a contradiction in terms. Light, by definition, was something that could be seen.

The German-British astronomer William Herschel (1738–1822) was, in 1800, the first to show this was not a contradiction after all. At that time, it was thought that the light and heat one obtained from the Sun might be two separate phenomena. Herschel wondered if heat might be spread out in a spectrum just as light was.

Instead, then, of studying the spectrum by eye, which noted only the light, Herschel studied it by thermometer, which measured the heat. He placed the thermometer at various places in the spectrum and noted the temperature. He expected that the temperature would be highest in the middle of the spectrum and that it would fall off at either end.

That did not happen. The temperature rose steadily as one progressed away from the violet, and reached its highest point at the extreme red. Astonished, Herschel wondered what would happen if he placed the thermometer bulb *beyond* the red. He found, to his even greater astonishment, that the temperature rose to a higher figure there than anywhere in the visible spectrum. Herschel thought he had detected heat waves.

In a few years, however, the wave theory of light was established and a better interpretation became possible. Sunlight has a range of wavelengths that are spread out by a prism. Our retina reacts to wavelengths of light within certain limits, but sunlight has some waves that are longer than that of the visible red, and is therefore to be found beyond the red end of the spectrum. Our retina won't respond to such long waves, so we don't see them, but they are there, anyway. They are called infrared rays, the prefix coming from a Latin word meaning "below," for you might view the spectrum as going from violet on the top to red on the bottom.

All light, when it strikes the skin, is either reflected or absorbed. When absorbed, its energy speeds up the motion of the molecules in our skin and this makes itself felt as heat. The longer the wavelength, the deeper it penetrates the skin and the more easily absorbed it is. Hence, although we can't see the infrared, we can feel it as heat, and the thermometer, for similar reasons, can record it as such.

It would help, of course, if it could be shown that infrared rays were actually made up of waves like those of light, but with longer wavelengths. One might allow two beams of infrared rays to overlap and produce interference fringes, but no one would be able to see them. Perhaps they could be detected by thermometer, with the temperature going up each time the instrument passed through a "brighter" area, and going down each time it passed through a "dimmer" one.

In 1830, the Italian physicist Leopoldo Nobili (1784–

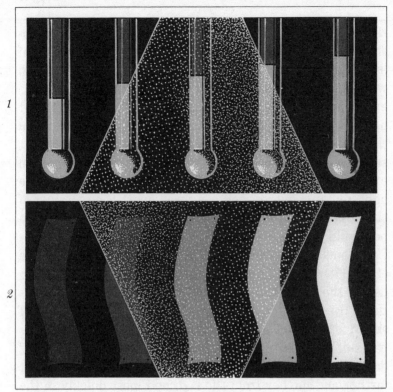

1. *Infrared Light. In 1800, when William Herschel placed his thermometer in the dark area beyond the red end of the spectrum, he was surprised to record the highest temperature.*
2. *Ultraviolet Light. In 1770, Karl Wilhelm Scheele found that paper soaked in silver nitrate solution darkened most quickly when exposed to violet light. In 1801, John Wilhelm Ritter exposed the paper in the dark area beyond the violet and the paper darkened even more rapidly.*

1835) invented a thermometer that would do the job. One of his co-workers was the Italian physicist Macedonio Melloni (1798–1854). Because glass would absorb a great deal of the infrared rays, Melloni made use of prisms formed of rock salt, which is transparent to infrared rays. As a result, interference fringes were set up and Nobili's thermometer

showed that they existed. By 1850, Melloni had demonstrated that infrared rays showed all of the properties of
light without exception—except that they could not be seen
with the naked eye.

What about the other end of the spectrum, where violet
light deepens into darkness? That story began in 1614, when
the Italian chemist Angelo Sala (1576–1637) noticed that
silver nitrate, a perfectly white compound, darkened on
exposure to sunlight. This happens, we now know, because
light contains energy and can force apart the molecule of
silver nitrate, producing finely divided silver, which appears black.

About 1770, the Swedish chemist Karl Wilhelm Scheele
(1742–1786) went into the subject in more detail, making
use of the solar spectrum, which wasn't known in Sala's
time. He soaked thin strips of white paper in solutions of
silver nitrate, let them dry, and placed them in various
parts of the spectrum. He found that the strips of paper
darkened least quickly in the red, more quickly as one went
farther and farther from the red, and most quickly in the
violet. This happens (as we now know, for reasons that will
be explained later) because light increases in energy as one
goes from red to violet.

Once Herschel discovered infrared rays in 1810, however, it occurred to the German chemist Johann Wilhelm
Ritter (1776–1810) to check the other end of the spectrum.
In 1801, he soaked strips of paper in silver nitrate solution
and repeated Scheele's experiment except that he placed
strips of paper *beyond* the violet, where no light was visible.
As he suspected they might, the strips of paper darkened
in this lightless region even more rapidly than they would
in violet light. This represented the discovery of ultraviolet
rays, where the prefix is from the Latin for "beyond."

Infrared and ultraviolet radiation existed just at the
borders of the visible spectrum. Maxwell's equations made

it seem that there could be radiation far beyond the borders. If such radiation could be found, then Maxwell's equations would be supported very strongly, for without them no one would have suspected such radiation might exist.

In 1888, the German physicist Heinrich Rudolf Hertz (1857–1894) made use of a rectangular wire, with a gap in it, as a detecting device. He set up an oscillating electric current in his laboratory. As the electric current oscillated, moving first this way, then that, it should emit electromagnetic radiation, with the radiation wave moving up while the current is going one way and then down when it is moving the other way. Such an electromagnetic wave should have a very long wavelength because even if the oscillating electric current changes direction every small fraction of a second, light can move quite far between changes.

Hertz's rectangular wire would gain an electric current if the electromagnetic wave crossed it, and there would be a spark across the gap. Hertz got his spark. In addition, as he moved his rectangular wire here and there in the room he got a spark where the wave was very high or very low, but no spark where it was in between. In this way, he could map the wave and determine its length.

Hertz had discovered what came to be called radio waves, which lay far beyond the infrared radiation and could have wavelengths of anywhere from centimeters to kilometers.

No one questioned Maxwell's equations after that. If there was a luminiferous ether, it carried electricity and magnetism also. If there was another ether, it existed only for gravitation.

In 1895, by the way, electromagnetic radiation was discovered far beyond the ultraviolet, with wavelengths exceedingly small; but we will get to that later, after we consider a few other matters.

Dividing Energy

Electricity, magnetism, light, and gravitation are all forms of energy, where energy is anything that can be made to do work. These forms of energy certainly seem different from one another, but one can be turned into another. As we have already seen, electricity can be turned into magnetism, and vice versa, and a vibrating electromagnetic field can produce light. Gravitation can cause water to fall, with the falling water turning a turbine that can force a conductor through magnetic lines of force to produce electricity. Interconversions of energy and work represent the field of thermodynamics.

Such conversions are never completely efficient. Some energy is always lost in the process. The lost energy does not, however, disappear, but makes its appearance as heat, which is still another form of energy. If heat is taken into account, then no energy is ever totally lost, nor does any energy ever appear out of nowhere. In other words, the total amount of energy in the Universe seems to be constant.

This is the law of conservation of energy, or the first law of thermodynamics, which was finally placed in compelling terms in 1847 by the German physicist Hermann Ludwig Ferdinand von Helmholtz (1821–1894).

In a way, heat is the most fundamental form of energy, for any other form of energy can be converted *completely* into heat, while heat cannot be converted completely into nonheat energy. For this reason, heat is the most convenient phenomenon through which to study thermodynamics; a word, by the way, which is from the Greek for "movement of heat."

Heat had been closely studied by scientists ever since the first truly practical steam engine had been invented, in 1769, by the British engineer James Watt (1736–1819).

Once the law of conservation of energy was understood, the study of heat became even more intense.

After the advent of the steam engine, there were two theories of the nature of heat. Some scientists thought of it as a type of subtle fluid that could travel from one piece of matter to another. Others thought of heat as a form of motion, of atoms and molecules moving or vibrating.

The latter suggestion, or the kinetic theory of heat (where kinetic is from a Greek word for "motion"), was finally established in the 1860s as the correct one when Maxwell and the Austrian physicist Ludwig Eduard Boltzmann (1844–1906) worked it out mathematically. They showed that everything that was known about heat could be interpreted satisfactorily by dealing with atoms and molecules that were moving or vibrating. As in the case of gases, the average speed (or, better, "velocity") of motion or vibration of the atoms and molecules making up *anything* is the measure of its temperature if the mass of the atoms and molecules is also taken into account. The total kinetic energy (which takes into account both mass and velocity) of all of those moving particles is the total heat of the substance.

Naturally, then, the colder an object gets, the slower the motion of its atoms and molecules. If it gets cold enough, the kinetic energy of the particles reaches a minimum. It can then get no colder, and the temperature is at absolute zero. This notion was first proposed and made clear in 1848 by the British mathematician William Thomson (1824–1907), better known by his later title of Lord Kelvin. The number of Celsius degrees above absolute zero is the absolute temperature of a substance. If absolute zero is equal to $-273.15°$ C, $0°$ C is equal to $273.15°$ K (for Kelvin) or $273.15°$ A (for absolute).

Any body at a temperature higher than that of its surroundings tends to lose heat as electromagnetic radiation. The higher the temperature, the more intense the radiation. In 1879, the Austrian physicist Joseph Stefan (1835–1893) worked this out exactly. He showed that the total radiation increased as the fourth power of the absolute temperature. Thus, if the absolute temperature was increased two times, say from 300° K to 600° K (that is, from 27° C to 327° C), then the total radiation would be increased $2 \times 2 \times 2 \times 2$, or 16 times.

Formerly, about 1860, the German physicist Gustav Robert Kirchhoff (1824–1887) had established the fact that any substance at a temperature lower than that of its surroundings would absorb light of particular wavelengths, and would then emit those same wavelengths when its temperature rose above that of its surroundings. It follows that if a substance absorbs *all* wavelengths of light (a "black body," in that it reflects none of them), it will emit all wavelengths when heated.

No object actually absorbs all wavelengths of light, in the usual sense of the word, but an object with a small hole in it does so after a fashion. Any radiation that finds its way into the hole is not likely to find its way out again and is finally absorbed in the interior. Therefore, when such an object is heated, black-body radiation—all of the wavelengths—should come pouring out of the hole.

This notion was first advanced by the German physicist Wilhelm Wien (1864–1928) in the 1890s. When he studied such black-body radiation, he found that a wide range of wavelengths was emitted, as was to be expected, and that the very long and very short wavelengths were low in quantity, with a peak somewhere in between. As the temperature rose, Wien found that the peak moved steadily in the direction of shorter wavelength. He announced this in 1895.

Stefan's law and Wien's law fit our experience. Suppose an object is at a temperature a little higher than that of our own body. If we put our hands near that object, we can feel a little warmth radiating from it. As the temperature of the object rises, the radiation becomes more noticeable and the peak radiation is at a shorter wavelength. A kettle of boiling water will deliver considerable warmth if our hand is placed near it. If temperature is raised still higher, an object will eventually give off perceptible radiation at wavelengths short enough to be recognized by our retina as light. We first see red light because that is the light with the longest wavelength, and is the first to be emitted. The object is then red-hot. Naturally, most of the radiation is still in the infrared, but the tiny fraction that comes off in the visible portion of the spectrum is what we notice.

As the object continues to rise in temperature, it glows more and more brightly. The color changes, too, as more and more of the shorter-wave light is emitted. As the object continues to grow still hotter, it becomes even brighter and the color undergoes another change as more, and shorter, wavelengths of light are emitted. The glow becomes more orange, and then yellow. Eventually, when something is as hot as the Sun's surface, it is white-hot, and the peak of the radiation is actually in the visible light region. If it grows still hotter, it becomes blue-white, and, eventually, although it is brighter than ever (assuming we can look at it without destroying our eyes in the same instant), the peak is in the ultraviolet.

This heat/light progression created a problem for nineteenth-century scientists because it was difficult to make sense out of the pattern of black-body radiation. Toward the end of the 1890s, the British physicist John William Strutt, Lord Rayleigh (1842–1919), assumed that every wavelength had an equal chance of being radiated in black-body radiation. On that assumption, he worked out an equation that showed quite well how the radiation would increase in intensity as one went from very long wavelengths to shorter wavelengths. This equation, however, didn't provide for a peak wavelength, to be followed by a decline, as one approached still shorter wavelengths.

Instead, the equation implied that the intensity would continue going up without limit as the wavelengths got shorter. This meant that any body should radiate chiefly in the short wavelengths, getting rid of all of its heat in a blast of violet, ultraviolet, and beyond. This is sometimes called the violet catastrophe. But the violet catastrophe does not take place, so there must be something wrong with Rayleigh's reasoning. Wien himself worked out an equation that would fit the distribution of short wavelengths of black-body radiation, but it wouldn't fit the long wavelengths. It seemed as though physicists could explain either half of the radiation range, but not the whole.

The problem was taken up by the German physicist Max Karl Ernst Ludwig Planck (1858–1947). He thought there might be something wrong with Rayleigh's assumption that every wavelength had an equal chance of being radiated in black-body radiation. What if the shorter the wavelength, the less the chance of its being radiated?

One way of making this seem plausible is to suppose that energy is not continuous and can't be broken up into smaller and smaller pieces forever. (Until Planck's time, the continuity of energy had been taken for granted by physicists. No one had wondered if energy, like matter, might consist of tiny particles that couldn't be divided further.)

Planck assumed that the fundamental bit of energy was larger and larger as the wavelength grew smaller and smaller. This meant that for a given temperature, the radiation would rise in intensity as wavelengths grew shorter, just as the Rayleigh equation indicated. Eventually, though, for wavelengths shorter still, the mounting size of the energy unit would increase the difficulty of getting enough energy into one place in order to radiate it. There would be a peak, and as the wavelengths continued to decrease, the radiation would actually decline.

As the temperature went up and the heat grew more intense, it would be easier to radiate the larger energy units and the peak would move in the direction of shorter wavelengths, just as Wien's law would require. In short, the use of the energy units that Planck postulated completely solved the problem of black-body radiation.

Planck called these energy units quanta (quantum in the singular), which is a Latin word meaning "how much?" What counted, after all, in the answer to the black-body radiation puzzle was how much energy there is in the quanta of different wavelengths of radiation.

Planck advanced his quantum theory, and the equation it made possible for black-body radiation (which agreed with the actual observations both for long wavelengths and short wavelengths), in 1900. This theory proved so important—far more important than Planck, at the time, could possibly imagine—that all of physics prior to 1900 is called classical physics, and all of physics after 1900 is called modern physics. For his work on black-body radiation, Wien received a Nobel prize in 1911, and Planck received one in 1918.

3

ELECTRONS

Dividing Electricity

Early experiments on electricity dealt with objects that carried rather small electric charges. In 1746, however, the Dutch physicist Pieter van Musschenbroek (1692–1761), working at the University of Leyden, invented something called the Leyden jar, in which a great deal of electric charge could be pumped.

The greater the charge, the greater the pressure for discharge. If a Leyden jar is touched to some object, the electricity flows into the object and the jar is discharged. (If touched to a human being, that human can receive a flow of electricity that will surely be painful.)

If a Leyden jar carries a sufficiently large electric

charge, direct contact need not be made. Under such conditions, if the Leyden jar merely approaches an object that will discharge it, the electric charge can force its way from the Leyden jar to the other object through the intervening air.

The result is a flash of light and a crackle. The light is not the electricity itself; rather, the electricity, whatever it is, heats the air as it passes through, and the air grows momentarily hot enough to radiate light. The heat also expands the air and, after the discharge is complete, the expanded air contracts again and the crackling sound results.

Some people saw a similarity between the light and crackle of a discharging Leyden jar and the lightning and thunder in the clouds during a storm. Could lightning and thunder be the discharge of a gigantic Leyden jar arrangement in the clouds?

In 1752, Benjamin Franklin proved this was so by flying a kite in a thunderstorm, leading the charge of the lightning down a cord and into an uncharged Leyden jar. The resultant charged Leyden jar showed that electricity from the sky had the same properties as electricity produced on Earth.

But what about the electricity itself that hides inside the charged body, or inside the light produced by heated air? One answer would be to discharge electricity through a vacuum in order to see what the bare electricity looked like. As early as 1706, an English physicist, working with charged objects far less intense than a Leyden jar, managed to get a discharge across an evacuated vessel, obtaining light as he did so.

In those days, however, evacuating a vessel was still only imperfectly possible. A remnant of air would be left inside, and that would be enough to glow as a result of the passage of the electricity. It was not the electricity itself. In order to get bare electricity, two things were needful.

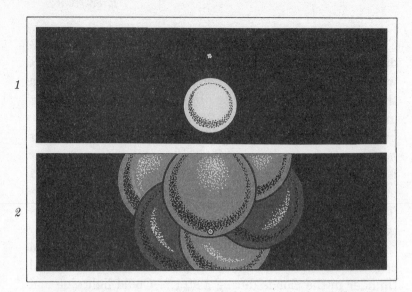

1. *If you could see an atomic nucleus from the perspective of an orbiting electron, the nucleus might appear as a tiny dot at a great distance.*

2. *Even though they have charges of equal but opposite magnitude, an electron has only $1/1,837$ the mass of a single proton.*

One was a vessel that was so well evacuated that the traces of air inside would not be enough to interfere with the electricity. The second was a way of forcing electricity in quantity through a good vacuum. A Leyden jar might do here, but its discharge lasted for only a moment. Was there a way of keeping the electricity going for a considerable period of time?

The second problem was solved in 1800 by the Italian physicist Alessandro Giuseppe Volta (1745–1827). He showed that electricity could be produced when two different metals were both dipped into a salt solution. This was accompanied by a chemical reaction, and as long as the chemical reaction proceeded, electricity continued to be produced. If some of the electricity was drawn off through a wire, the electricity would flow through the wire as long as it was being produced by the chemical reactions.

As a result, it became possible to have an electric current, instead of merely a stationary electric charge. To produce a large electric current, Volta made use of a series of two-metal combinations with salt water between. Any of a series of similar objects is called a battery. Volta had invented an electric battery.

As soon as Volta announced his discovery, scientists began constructing bigger and better batteries, and within a generation Faraday worked out a method for producing an electric current, much more cheaply, by burning fuel. There was then no problem of setting up electric currents with enough force to move across a good vacuum—if a good vacuum could be supplied.

The supplier was the German inventor Johann Heinrich Wilhelm Geissler (1814–1879), who, in 1855, invented an air pump that was a marked improvement on all of the pumps used previously. Instead of using mechanical devices involving moving parts, Geissler used only rising and falling levels of mercury. The mercury trapped a bit of air with each change in level and removed it. It was a slow process, but by the time this mercury air pump was done, over 99.9 percent of the air in a vessel had been removed.

Geissler, who was an expert glass blower, blew vessels that had two pieces of metal sealed into opposite ends, and these vessels were evacuated. Such vessels were named Geissler tubes by Geissler's friend and co-worker, the German physicist Julius Plücker (1801–1868). Plücker connected the two pieces of metal sealed in the tube to opposite poles of an electricity-generating device. One of these metal pieces, therefore, became positively charged and was called the anode, while the other became negatively charged and was called the cathode.

These words were first used by Michael Faraday. The positively charged anode is from Greek words meaning "upper way," and the negatively charged cathode from Greek words meaning "lower way." Since Benjamin Franklin's time, electricity was thought to flow from positive to negative; that is, from anode (upper) to cathode (lower), like water flowing from an upper level to a lower one.

Plücker forced electricity through the vacuum of a Geissler tube and now there was simply not enough air to make a visible glow—but there was a glow anyway. It was a greenish glow in the neighborhood of the cathode, *always* the cathode. Plücker reported his observations in 1858, and this was the first indication that Franklin might have made a wrong guess, and that electricity flowed not from anode to cathode but from cathode to anode.

Could that greenish glow represent the bare electric current itself? Plücker wasn't sure. He thought it might be pieces of the metal broken off and glowing, or that it might be the tiny wisps of gas still left in the vessel.

The German physicist Eugen Goldstein (1850–1930) studied the phenomenon carefully and found that it didn't matter what gas was in the vessel before it was evacuated. It also didn't matter what metal the anode and cathode were made of. The only thing that was the same in all cases was the electric current, so Goldstein maintained that the glow *was* associated with the current itself. In 1876, he called the vacuum-crossing material cathode rays.

This name implied that the current was emitted by the cathode and traveled to the anode. Indeed, the glass glowed on the anode side of the tube as though the cathode rays were striking and energizing it.

In 1869, the German physicist Johann Wilhelm Hittorf (1824–1914), who had been a student of Plücker, showed that if a solid object were sealed into the tube in front of the cathode, there would be a shadow of the object against the glow at the anode end. Clearly, something was traveling from the cathode, and some of it was stopped by the solid object.

The British physicist William Crookes (1832–1919) de-

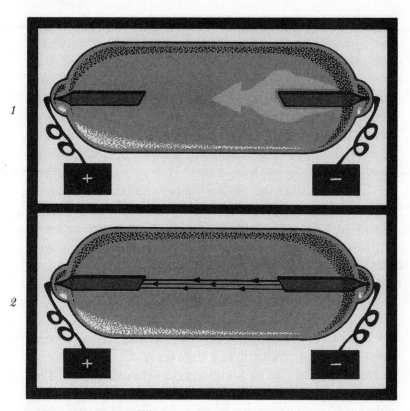

1. In 1858, Julius Plücker reported a greenish glow around the negatively charged cathode in the near vacuum of a Geissler tube. This was the first indication that Ben Franklin's guess about the direction of current flow might be wrong.

2. In 1876, Eugen Goldstein maintained that the glow was associated with the current itself. He called the vacuum-crossing material cathode rays. Current flows from the cathode to the anode.

vised still better devices for making a vacuum and, in 1878, produced a Crookes tube, in which the remaining air was only ¹⁄₇₅,₀₀₀ that in Geissler tubes. (All such tubes came to be grouped together as cathode-ray tubes.) The cathode rays now showed up more clearly, and Crookes could demonstrate that they moved in straight lines, and that they could even be made to turn a little wheel.

But what was it that flowed out of the cathodes? Did the cathode rays consist of particles or of waves? Both possibilities found support among scientists in a reprise of the type of argument about light between Newton and Huygens. The arguments in favor of particles for the cathode rays were somewhat the same as those in the case of light, the chief one being that the cathode rays cast sharp shadows.

The fact that the particle notion had been decisively defeated in the case of light made a number of scientists hesitate, however, to be caught on what might again be the losing side. (Generals are often accused of always being prepared for the previous war. Scientists are human, and they remember past battles, too, and sometimes have a tendency to bring old experiences to bear on new phenomena.)

The strongest voice in favor of the cathode rays being waves was that of Hertz, the discoverer of radio waves. In 1892, he showed that cathode rays could penetrate thin films of metals. It didn't seem to him that particles ought to be able to do so, but waves might, of course, because if the metal films were thin enough, even light waves could penetrate them.

Hertz's pupil Philipp Eduard Anton von Lenard (1862–1947) even prepared a cathode-ray tube with a thin, aluminum "window." The cathode rays were able to spray through the window and emerge into the open air. If the cathode rays were waves of very short length, they would travel in straight lines and cast sharp shadows, just as light waves do. For a while in the early 1890s, the notion of cathode rays as waves was therefore riding high.

And yet if the cathode rays were emerging from the negatively charged cathode, might they not be carrying a negative electric charge? If so, that might well indicate that cathode rays were not waves, for no wave known at that time carried any electric charge, however small. And if cathode rays did carry an electric charge, they ought to be affected by an electric field.

In 1883, Hertz tested the hypothesis. He passed cathode rays between two parallel metal plates, one of which was positively charged and one negatively charged. If the cathode rays were charged, they should veer out of their straight-line paths, but they did not. Hertz concluded that they were uncharged, and that, too, was a point in favor of waves.

What Hertz didn't realize, however, was that the cathode rays were traveling far more quickly than he expected, so that they got past the plates before they had a chance to veer perceptibly. This could have been taken care of if the plates were sufficiently highly charged, but they weren't. The combination of speedy cathode rays and weakly charged plates made the deviation from straight-line path imperceptible, so that Hertz's conclusion wasn't valid. (Scientific experiments are not always the last word. A particular experiment, however honestly and intelligently conducted, can yield the wrong answer for any of a variety of reasons. That is why it is important that experiments be checked by other scientists, using other instruments, other conditions, and, if possible, other ideas.)

Thus, in 1895, Perrin (who, in the following decade, was to demonstrate the reality of atoms) showed that cathode rays could impart a large negative charge to a cylinder on which they fell. It was hard to see how cathode rays could carry a negative charge from the cathode to the cylinder without themselves possessing the negative charge while they traveled. This seriously weakened Hertz's finding.

Then the British physicist Joseph John Thomson (1856–1940) decided to try repeating Hertz's experiment with electrically charged plates. Thomson had the advantage of knowing how rapidly cathode rays moved. In 1894, he had estimated that they moved at a speed of 200 kilometers (125 miles) per second. What's more, he had a more effectively evacuated vacuum tube than Hertz had had, and he used electrically charged plates with a considerably stronger charge.

In 1897, Thomson allowed cathode rays to speed between his charged plates and found that the electric field induced a distinct curvature away from the negatively charged plate and toward the positively charged one. That convinced him, and it convinced physicists, generally, that cathode rays consisted of speeding cathode-ray particles, each carrying a negative electric charge.

The verdict here was the reverse of that in the case of light. Where light was concerned, waves had won out over particles. With cathode rays, particles had won out over waves. (As we shall see, however, neither victory was absolute. It often happens in science that the choice between alternatives is not as clearcut as it might at first seem.)

Cathode-Ray Particles

The degree to which a charged particle is deflected by an electric field depends on three things: the size of the electric charge carried by the particle, the speed at which the particle travels, and the mass of the particle. The deflection of a charged particle by a *magnetic* field depends on the same three factors, but in a different fashion from the deflection by an electric field. If Thomson measured both types of deflection, it would be possible from the two measurements, taken together, to work out the ratio of the charge to the mass of the particles. Given that, if you knew the size of the electric charge of the particles, you could then work out their mass.

The electric charge was also not entirely impossible to work out. Faraday had made extensive studies of the way in which electrical currents induced chemical reactions, and he had established the laws of electrochemistry in 1832. From these laws, and from careful measurements of the amount of electricity required to bring about the deposition of a known mass of metal from a solution of its compounds, it was possible to calculate the size of the electric charge required to deposit a single atom of the metal.

There seemed no great risk in deciding that the electric charge involved in the chemical change of a single atom might be the smallest electric charge that could exist. It was reasonable, therefore, to suppose that a cathode-ray particle would carry this smallest electric charge. In other words, a cathode-ray particle would be to electricity what an atom is to matter—or, as was about to be discovered, what a quantum is to energy.

Taking this assumption and the measured deflections of cathode rays by an electric field and by a magnetic field of known strength, Thomson could calculate the mass of a single cathode-ray particle, and did so. He was awarded a Nobel prize in 1906 for this accomplishment.

The results were surprising. As far as atoms of matter are concerned, the smallest atom known in Thomson's day (and in our own) was the hydrogen atom. Indeed, we are now quite certain that an ordinary hydrogen atom is the smallest atom that can possibly exist. The cathode-ray particle, however, turned out to have a mass far smaller than that of the hydrogen atom. It has a mass only $\frac{1}{1837}$ that of the smallest atoms.

For a century, scientists had been quite certain atoms were the smallest things that could exist, and that the smallest atom would therefore be the smallest *anything* that had mass. Now that thought was shattered; or, at least, it had to be modified, but the modification might not have to be very great. It was possible to argue, after Thomson's experiments, that atoms were still the smallest pieces of *matter* that could exist. Electricity, it might be said, was not matter, but a form of energy that was much more subtle than matter. It should not be surprising, from that point of view, that these cathode-ray particles, which might be viewed as "atoms of electricity," were much smaller than atoms of matter.

It was the smallness of the cathode-ray particle that might account for the fact that an electric current could flow through matter, or that cathode-ray particles could themselves pass through thin films of metals. The passage of these particles through metal had been taken as strong evidence that they could not be particles, but at the time of the first discovery of such passage, there had been no idea of how *small* those particles were. (Experiments can mislead even the best scientists if some key bit of knowledge is missing.)

Because the cathode-ray particle is far smaller than any atom, it is termed a subatomic particle. It was the first subatomic particle to be discovered, and was to be the first of a flood of them that would completely change our minds about the structure of matter. Their discovery increased our knowledge, revolutionized our technology, and utterly changed our way of life. (The topic of technology and our way of life is outside the scope of this book, but the fact is worth mentioning. No matter how ivory-towerish scientific discoveries might seem, there is always a good chance that they will affect us in many crucial ways.)

What does one call a cathode-ray particle? Naming something does not increase our knowledge concerning it, but it makes it easier to refer to it and to discuss it. In 1891, the Irish physicist George Johnstone Stoney (1826–

1911) had suggested that the minimum electric charge that one could deduce from Faraday's laws be called an electron. Thomson liked the name and applied it to the particle rather than to the electric charge it carried. The name has stuck and become very familiar even to the nonscientific public (think of all the electronic devices, such as television sets and record players, that we deal with). We might say, then, that Thomson discovered the electron in 1897.

X Rays

In the previous chapter I mentioned that electromagnetic radiation lying far beyond the ultraviolet in the shortwave direction was eventually discovered. I did not go into detail then, but the time has now come when we can discuss it.

In the 1890s, the German physicist Wilhelm Konrad Roentgen (1845–1923) was working on cathode rays in his own unique way. He did not concern himself, as Hertz and Thomson had, with their nature, but with their effect on certain chemicals. Cathode rays, impinging on those chemicals, caused them to luminesce. That is, the chemicals gained energy from the cathode rays and then lost that energy again in the form of the radiation of visible light.

One of the chemicals that luminesced upon the impingement of cathode rays was a compound called barium platinocyanide. Roentgen had sheets of paper coated with that compound in his laboratory.

The luminescence was quite faint and, in order to observe it as well as possible, Roentgen darkened the room and enclosed the experimental apparatus within sheets of black cardboard. He could then peer into an enclosure that was totally dark, and when he turned on the electric current, the cathode rays would pass along the tube, penetrate the thin, far wall, fall upon a chemically coated paper, and initiate luminescence that he could see and study.

On November 5, 1895, Roentgen turned on the current and, as he did so, a dim flash of light that was *not* inside the apparatus caught the corner of his eye. He looked up and there, quite a distance from the apparatus, was one of the sheets, covered with barium platinocyanide, luminescing briskly.

Roentgen turned off the current and the coated paper darkened. He turned it on and the coated paper glowed again. He took the paper into another room and pulled down the blinds in order to darken the room. When he turned on the cathode-ray tube, the coated paper glowed in this room.

Roentgen decided that the cathode-ray tube was producing radiation that was not cathode rays—a radiation that could penetrate cardboard, and even the wall between two rooms, as cathode rays could not. He published his first report on this new radiation on December 28, 1895, and because he had no idea of the nature of the radiation, he called it X rays. The name has clung to the radiation ever since. For this discovery, Roentgen received a Nobel prize in 1901, the first year in which such prizes were given out.

Now the same problem and uncertainty arose over X rays that had previously arisen over light and over cathode rays. Some physicists thought X rays were streams of particles, some thought they were waves. Of those who thought they were waves, some (like Roentgen himself) thought they were longitudinal waves, like sound waves. Others thought they were transverse waves, like light waves. If they were transverse waves, they might be a type of electromagnetic radiation with wavelengths far shorter than ultraviolet, just as the recently discovered radio waves had wavelengths far longer than the infrared.

The problem was how to decide among the alternatives. Light had been shown to be waves because it displayed interference. In order to demonstrate the interference, light had been passed through two closely spaced slots. The interference could be made more pronounced by using diffraction gratings, glass plates on which very finely spaced parallel scratches were made. Light passing through the intervals between the scratches produced clearly visible interference phenomena, allowing wavelengths to be measured with great precision.

The shorter the wavelength, however, the finer the spacing between the scratches must be. Diffraction gratings wouldn't work on X rays if they were transverse waves with extremely short wavelengths. It then occurred to the German physicist Max Theodor Felix von Laue (1879–1960) that it was not necessary to try to manufacture a diffraction grating with impossibly closely spaced scratches. Nature had already done the job.

Crystals consist of atoms and molecules of a substance arranged in an all but endless even array. This can be inferred from the shapes of crystals and from their tendency to break in certain planes in such a way as to retain their shapes. It is as though they break "with the grain" along a plane that lies between two adjacent layers of atoms of molecules. Why should not X rays, Laue reasoned, penetrate the crystals between the layers? A crystal might serve as a diffraction grating with scratches no wider apart than the layers of its atoms, and this might show interference effects for X rays.

If X rays were to go through an object in which atoms and molecules were scattered in random disorder, the X rays would be scattered this way and that, in random fashion. There would be a uniform shadowing effect, darkest at the center and growing lighter as one moved outward in all directions.

If X rays were to go through a crystal with orderly layers of atoms and molecules, the X rays' diffraction patterns would be set up and the photographic plate would show distinct spots of light and shadow forming a symmetrical pattern about the center.

In 1912, Laue tried the experiment of passing X rays through a crystal of zinc sulfide. It worked perfectly, the X rays behaving exactly as they would be expected to if they were very short transverse waves. That settled the issue and, in 1914, Laue received a Nobel prize for his work.

The British physicist William Henry Bragg (1862–1942), together with his son, William Lawrence Bragg (1890–1971), a physics student at Cambridge, saw that X-ray diffraction could be used to determine the actual wavelength of X rays if the distance between the layers of atoms in the crystal diffracting the rays were known. This they accomplished in 1913, showing that the wavelength of X rays was anywhere from $\frac{1}{50}$ to $\frac{1}{50,000}$ the wavelength of visible light. For this they shared a Nobel prize in 1915.

Electrons and Atoms

It is clear, when one stops to think of it, that electrons might exist in matter. Suppose we consider the early studies of electricity, when one simply built up an electric charge by rubbing a glass rod or a piece of amber. Might this not be because electrons travel from the object being rubbed to the object doing the rubbing, or vice versa? Any substance that gets extra electrons forced into it will accumulate a negative charge, and any substance that loses some of its electrons will accumulate a positive charge. And if so, the electrons have to be in the matter to begin with if they are going to be transferred one way or the other.

Again, an electric current might consist of electrons moving through the material in which the current exists.

Thus, in a cathode-ray tube, when the electric current reaches the cathode, electrons accumulate there (giving it a negative charge, which is what makes it a cathode) and are forced into the vacuum as a stream of cathode-ray particles.

The electrical *impulse* travels at the speed of light, so that if you have wires strung from a telephone in New York to one in Los Angeles, a voice can modulate the electric flow in New York, which will then reproduce the voice in Los Angeles about 1/60 of a second later. The electrons themselves, however, bumping from atom to atom, travel much more slowly.

This is analogous to what happens when you flick a checker against a long line of similar checkers. As the checker you flick strikes the first in the long line, the last, at the other end of the line, flies away almost at once. The checkers in the middle barely move, but the impulse of compression and expansion moves along the line of checkers at the speed of sound and ejects the last checker.

Still, though it seemed quite likely that electrons might well exist in matter, it was somehow taken for granted that these particles of electricity existed quite apart from, and independently of, atoms, which were pictured as featureless and indivisible.

Information gathered from chemical experiments during the 1800s certainly made it seem that atoms were indivisible, but that they were featureless was a mere assumption. Nevertheless, scientists are human, and in science, as in other facets of human thought, an assumption that has been held long enough sometimes takes on the force of cosmic law. People forget that it is only an assumption and find it difficult to consider the possibility that it might be wrong.

In this connection, consider the manner in which an electric current can pass through some solutions and not others. This phenomenon was first studied systematically by Michael Faraday.

Thus, a solution of table salt (sodium chloride) will conduct electricity, as Volta found when he constructed the first electric battery. Sodium chloride is therefore an electrolyte. An electric current will *not* pass through a solution of sugar; therefore, sugar is a nonelectrolyte.

From his experiments, Faraday decided that something in the solution carried negative charges in one direction and positive charges in the other. He didn't know exactly what it was that carried the charge, but he could give it a name. He called the charge carriers ions, from a Greek word meaning "wanderers."

In the 1880s, a young Swedish chemical student, Svante August Arrhenius (1859–1927), tackled the problem in a novel way. Pure water has a certain fixed freezing point at 0° C. Water that has a nonelectrolyte dissolved in it (say, sugar) freezes at slightly below 0° C. The more sugar dissolved in the water, the lower the freezing point. In fact, the lowering of the freezing point is proportional to the number of molecules of sugar dissolved in it. This holds true for other nonelectrolytes, too. The same number of molecules of any nonelectrolyte in solution will lower the freezing point by the same amount.

The situation is different with electrolytes. If sodium chloride is dissolved in water, then the freezing point is lowered just twice as much as it ought to be, considering the number of molecules in solution. Why should that be?

Sodium chloride has a molecule made up of one atom of sodium (Na) and one of chlorine (Cl), so that its formula is NaCl. When sodium chloride is dissolved in water, Arrhenius suggested, it breaks up, or dissociates, into those two atoms, Na and Cl. For every molecule of NaCl outside of solution, you have two half molecules, Na and Cl, so to speak, in solution. There would be twice as many particles in solution as was thought, and there would be twice the lowering of the freezing point. (Molecules made up of more than two atoms might break up into three or even four parts and produce three, or even four, times the expected lowering of the freezing point.)

A molecule of ordinary sugar has a molecule consisting of 12 carbon atoms, 22 hydrogen atoms, and 11 oxygen atoms, or 45 atoms altogether. When it dissolves in water, however, it does not dissociate, but remains in full molecular form. Therefore, there are only the expected number of molecules in solution and the freezing point is lowered only by the expected amount.

When sodium chloride dissociates, however, it can't possibly break up into ordinary sodium and chlorine atoms. The properties of sodium and chlorine atoms are known, and are not to be found in a salt solution. Something must happen that makes the sodium and chlorine of dissociated sodium chloride different from ordinary sodium and chlorine.

To Arrhenius, it seemed the answer was that each dissociated fragment of the sodium chloride molecule carried an electric charge and that they were the ions that Faraday had spoken of. From the results of the experiment involving an electric current passing through a sodium chloride solution, it was easy to argue that each sodium particle formed through dissociation carried a positive charge and was a sodium ion that could be symbolized as Na^+, while each chlorine particle carried a negative charge and was a chloride ion, symbolized as Cl^-. It was because electrolytes tended to dissociate into such charged fragments that they *were* electrolytes and could conduct an electric current.

Sodium ions and chloride ions had properties far different from uncharged sodium atoms and chlorine atoms. That is why a salt solution is a mild substance, while sodium and chlorine, themselves, are both dangerous to life. Non-

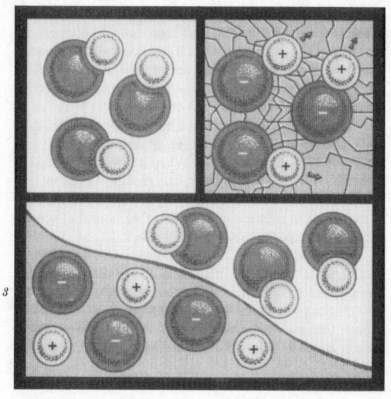

1. and 2. When ordinary table salt, sodium chloride (NaCl), dissolves in water, it dissociates into positively charged sodium ions, Na⁺, and negatively charged chloride ions, Cl⁻.
3. Atoms in and out of solution.

electrolytes, such as sugar, not being dissociated, have no charged fragments that can carry an electric charge, and therefore do not conduct an electric current.

In 1884, Arrhenius prepared his theory of ionic dissociation as a thesis for the degree of Doctor of Philosophy. The examining committee greeted the thesis with coldness, for they were not prepared to accept any theory that spoke of atoms that carried an electric charge. How could atoms carry an electric charge when atoms were featureless and incapable of modification? (They were helpless in the grip of the assumption.)

The committee could not actually reject the thesis because it was perfectly argued, and because it explained so many things that couldn't be explained any other way. Nevertheless, they passed it with the lowest possible passing grade.

When, thirteen years later, J. J. Thomson discovered the electron, it suddenly became obvious that atoms might, just possibly, carry one or two excess electrons or lose one or two of the normal quantity they might contain. With each passing year, there were new discoveries that made that possibility seem more certain, and in 1903, Arrhenius received a Nobel prize for the very thesis that, nineteen years before, had barely earned a passing grade.

Of course, it wasn't entirely satisfactory to deduce the presence of electrons in atoms merely from the behavior of electrolytes. Was there any way of observing electrons in atoms directly? Could one, for instance, knock electrons out of atoms and detect them?

In 1887, when Hertz was experimenting with the detecting device with which he was to prove the existence of radio waves the following year, a spark appeared across the gap of his detecting device whenever electricity jumped the gap. He observed something curious though, for a spark appeared more easily when light shone upon the gap.

Apparently, light had some effect on electric discharge, so this came to be called the photoelectric effect, the prefix photo- coming from the Greek word for "light."

In the very next year, 1888, another German physicist, Wilhelm Hallwachs (1859–1922), found that the photoelectric effect did not treat the two types of electric charge equally. A piece of the metal zinc, carrying a negative charge, lost that charge when it was exposed to ultraviolet rays. That same piece of zinc, carrying a positive charge, was not affected at all by ultraviolet radiation, and retained its charge. There was no ready explanation for this until Thomson discovered the electron and it began to appear that electrons might exist in matter.

In that case, a spark was formed across a gap because electrons were forced out of one of the metal points at the gap. If light somehow caused electrons to be ejected, the spark would form more easily. Again, zinc that had a negative charge would be carrying an excess supply of electrons, and if light caused those electrons to be ejected, the zinc would lose its charge. Zinc that had a positive charge would have a deficiency of electrons, and because light could not be expected to supply electrons to make up that deficiency the positive charge would remain unaffected.

At least, that was the easy explanation of the early observations of the photoelectric effect. It is, however, an advisable caution on the part of scientists not to rush toward the easy explanation too precipitously. Sometimes, one can fall into a trap that way (as when one decides that cathode rays can't consist of particles because they pass through thin films of metal).

Thus, just because electrons are knocked out of matter does not mean they necessarily exist in matter to begin with. Einstein, in 1905, showed as part of his special theory of relativity that mass was a form of energy. Mass could be turned into energy, and energy could be turned into mass.

Light contained energy. Might it be, then, that light-energy on striking metal under certain conditions would be converted into a tiny fragment of mass—an electron—that would carry off a bit of the negative charge possessed by the metal? In this way, electrons would appear that had never been part of the metal.

Einstein's theory, however, did not merely state that mass and energy were interchangeable. It presented a simple equation that showed just how much mass would be converted into how much energy and vice versa. It turns out that even a small quantity of mass could be turned into a great deal of energy; and, conversely, that it took a great deal of energy to form even a small amount of mass.

The electron is a particularly small bit of mass, but even so, the quantity of energy it would take to form it is simply not present in ultraviolet rays, as was soon to be determined. The photoelectric effect cannot, therefore, be the result of the creation of electrons out of energy; it must be the result of the ejection of electrons already present in the metal atoms.

It takes far less energy to eject an electron that already exists than to form one from scratch. In this case, then, the simpler explanation turned out to be correct (and it is a pleasant thing, indeed, that this sometimes happens in science).

Of course, it was still possible that what emerged from the metal might not be electrons. They might be some other type of particle carrying a negative charge. In 1899, however, Thomson applied magnetic and electric fields to the emerging particles and found that they had the same mass as electrons and the same negative charge. With those two properties matching, it seemed clear that photoelectric particles were electrons, and there has been nothing to disturb that view since.

Electrons and Quanta

Philipp Lenard studied the photoelectric effect in 1902 and was able to show that the electrons ejected from various metals always matched each other in properties. In other words, although there were many different atoms, they were all associated with but one type of electron. This was a hopeful bit of information considering that scientists love simplicity.

On the other hand, Lenard found that not all light was equal when it came to inducing a photoelectric effect. It often happened that red light did not produce the ejection of electrons, and that making the light more intense didn't help. No electrons would appear no matter how intense the light was.

However, if one exposed a particular metal to light of shorter and shorter wavelengths, there came a point at which electrons began to be ejected. The wavelength at which this happens is called the threshold value.

At the threshold value, the electrons that are ejected move at a very slow speed, as though the light has just barely enough energy to eject them and no more. If the light at the threshold value is made more intense, more electrons are ejected—but they still move at a very slow speed.

If the metal is exposed to light with wavelengths smaller and still smaller than that of the threshold value, the electrons are ejected with greater and greater speed. The speed of the electrons depends on the wavelength, while the number of electrons ejected depends on the intensity of the light. Different metals have different threshold values, as though some metals hold electrons more loosely than other metals.

Lenard couldn't explain this, and neither could J. J. Thomson when he tried. Ordinary nineteenth-century physics didn't work. When the solution did come, it came by way of quantum theory, which had been devised by Planck five years earlier.

Planck had supposed that electromagnetic radiation came in quanta of a certain size. The shorter the wavelength, the larger the energy content of the quantum.

It is also true that the shorter the wavelength, the greater the number of waves the radiation can produce in one second. The number of waves of radiation per second is called the frequency. The shorter the wavelength, then, the higher the frequency. We can therefore say that the size of a quantum is proportional to its frequency.

Until 1905, the notion of quanta had only been used in connection with black-body radiation. Might it not be a mathematical trick that explained that one phenomenon and nothing more? Did quanta *really* exist?

Einstein, whose theoretical work in 1905 made it possible to show, a few years later, that atoms really existed, tackled this new question concerning reality in that same year.

Einstein was the first to take the quantum theory seriously, and to consider it more than a mere convenience in solving the one problem of black-body radiation. He was willing to suppose that energy came in quanta at all times and under all conditions, so that problems that involved energy, other than black-body radiation, must also take quanta into consideration.

This meant that radiation existed in quanta form when it struck matter. It struck as quanta and, if absorbed, was absorbed as quanta. At any one moment in any one place, an entire quantum is absorbed; nothing more, nothing less.

If light that strikes is of long wavelength and low frequency, the quanta are of low energy. Such a quantum, when absorbed, simply might not contain enough energy to break an electron loose from a particular atom. In such a case, the quantum is absorbed as heat, and the electron might vibrate faster but it doesn't break away. Given enough quanta of this sort, a substance might absorb enough heat to melt, but at no moment in time is enough heat absorbed by any single atom to shake an electron loose.

As the wavelength decreases and the frequency increases, the quantum contains more energy and, at the threshold value, there is just enough energy to break an electron loose. There is no excess energy to appear as energy of motion, so the electron moves very slowly.

With still shorter wavelengths and still more energetic quanta, there is enough additional energy to eject an electron with considerable speed. The shorter the wavelength and the more energetic the quanta, the faster that motion.

Depending on the nature of the atom, electrons are held more tightly or more loosely, to begin with, and would then require larger quanta, or smaller quanta, to bring about an ejection. The threshold value will therefore be different for each element.

The quantum theory neatly explained all of the observed facts about the photoelectric effect, and this was very impressive. When a theory that has been worked out to explain one phenomenon turns out to explain another phenomenon, apparently unrelated to the first, it becomes very tempting to accept the theory as representing reality. (Here you see an example of the use of a theory; it explains widely different categories of observations. Without quantum theory, no one could see the connection between blackbody radiation and the photoelectric effect—to say nothing of many other phenomena.) It was for his work in this connection that Einstein received a Nobel prize in 1921.

Waves and Particles

If light occurs in quanta, and if each quantum goes speeding separately through space, the quantum behaves, in that way, like a particle. The quantum even received a name in its particle aspect. Because of the electron, most particles have received an -on ending, and, in 1928, the American physicist Arthur Holly Compton (1892–1962) named such a speeding quantum a photon, from the Greek word for "light."

It was fitting that Compton invented the name, for in 1923, he showed that radiation *did* act as particles, not just by being separate pieces of something, but by *behaving* as particles did. The shorter the wavelength and the more energetic the quanta, the more likely it was that they would demonstrate properties usually considered characteristic of particles rather than of waves.

Compton studied the manner in which X rays were scattered by crystals and found that some X rays, in the process of being scattered, increased their wavelengths. This meant that some of the energy of the X-ray quantum was lost to an electron in the crystal. Compton thought that the effect might be particulate in nature, like that of one billiard ball hitting another, with one losing energy and the other gaining it. He found, when he worked out a mathematical relationship that accurately described what had happened, that this seemed to be so in actual fact. This is now called the Compton effect.

It turned out, then, that both Newton and Huygens had hold of part of the truth two and a half centuries before. Light consisted of something that was both wave and particle. This can be confusing. In the ordinary world around us, there are waves, such as water waves; and there are particles, such as sand particles; and there are no confusions about it. Waves are waves and particles are particles.

The point is that light does not resemble the ordinary objects around us, and can't be forced into categories defined according to the same rules. Light, when studied in certain ways, shows interference phenomena, as water waves do. Studied in other ways, however, they show energy transfers, as colliding billiard balls do. No observation, however, can show light acting both as a wave and a particle

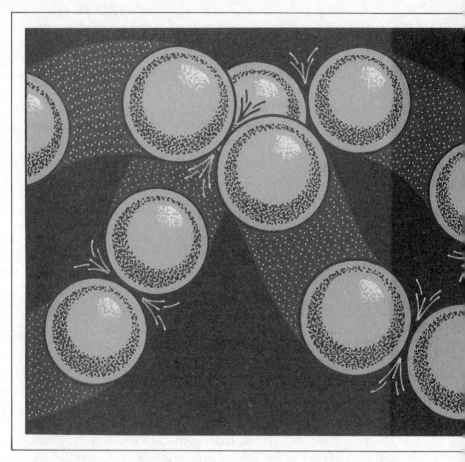

In 1923, Arthur Holly Compton found that X rays behaved as particles, losing energy when they were scattered by crystals. An X ray behaved like one billiard ball hitting another, with one losing energy

simultaneously. You can study light as either one or the other, never both at once.

This is not really such a mystery. Imagine you are looking at an empty ice-cream cone from the side, so that the wide part is at the top and the point is at the bottom. The outline is that of a triangle. Imagine next that you are looking at it with the wide opening facing you directly, and the point directed away from you. Now the outline is that

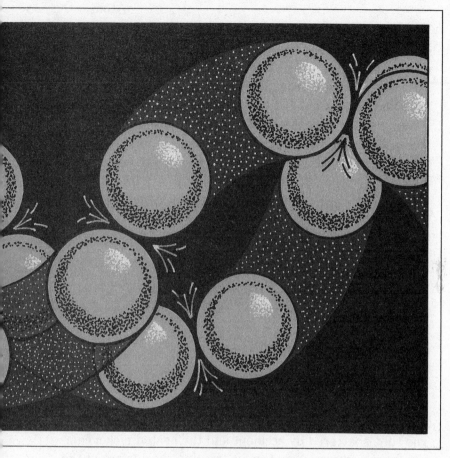

and the other gaining it. The shorter the wavelength, and the more energetic the quanta, the more likely that it would demonstrate properties considered characteristic of particles rather than waves.

of a circle. If those are the only two ways in which you are allowed to view the cone, then you can see it as *either* a circle *or* a triangle, but you can never see it as both simultaneously.

You might ask what the *real* two-dimensional outline of the cone was, but the answer would have to be "It depends on how you look at it." In the same way, you might ask whether light was really a wave, or really a particle,

85

and you would have to answer, "It depends on the particular way in which you are observing it."

One great side effect of the particle nature of light was that it made the luminiferous ether unnecessary. After a century of existence in scientists' minds, during which time the necessity of explaining it wrought more and more confusion, it disappeared as though it had never been—and, in fact, it never *had* been.

If something that had seemed to be a wave turned out to have a particle aspect, might it not be that something that had seemed to be a particle would turn out to have a wave aspect?

The French physicist Louis Victor de Broglie (1892–1987) suggested, in 1924, that this might be so. He made use of Einstein's equation relating mass to energy and Planck's equation relating the size of a quantum to frequency to show that every particle ought also to act like a wave of a certain length.

In 1925, the American physicist Clinton Joseph Davisson (1881–1958) was studying the reflection of electrons from a metallic nickel target enclosed in a vacuum tube. The tube shattered by accident and the heated nickel combined with oxygen from the air, rusting the surface of the target. To remove the film, Davisson had to heat the nickel for an extended period. Once this was done, it turned out that the electron-reflecting properties of the nickel surface had changed. The surface had consisted of many tiny crystals before the accident, but of just a few large ones afterward.

Davisson, who knew of de Broglie's suggestion, thought it would be useful to go still further and to prepare a nickel surface made of a single crystal. This might be able to show any wave aspect an electron might have. He aimed a stream of electrons at the single-crystal surface and found that electrons were not only reflected, but were diffracted and showed interference phenomena. Electrons *did* have a wave aspect.

Also in 1925, the British physicist George Paget Thomson (1892–1975), only son of J. J. Thomson, was forcing fast electrons through very thin gold films and he, too, noticed diffraction effects. As a result, de Broglie received a Nobel prize in 1929 for working out the theory of electron waves, while Davisson and G. P. Thomson shared one in 1937 for demonstrating it. Electron waves are *not* electromagnetic waves, by the way, but are "matter waves."

Physicists are now convinced that, indeed, everything has both a particle and a wave aspect, but not necessarily in equal measure. The more massive a particle is, the more prominent the particle aspect is and the more difficult it is to observe the wave aspect. A billiard ball (or the Earth itself) has a wave aspect, but this has so short a wavelength that it is quite likely it might never be observed. We know it is there in theory, but that is all. Even a grain of sand has a wave aspect too subtle to be observed, in all likelihood. An electron, however, has such a small mass that its wave aspect can be observed easily, once the proper experiment is carried out.

In the same way, the less energetic a wave is, the more prominent its wave aspect, and the more difficult it is to observe its particle aspect. A water wave is so weak (if only a single molecule of water is taken into account) that it is not at all likely its particle aspect can be observed. The same is true of a sound wave, although physicists speak of the particle aspect of sound waves as phonons, from the Greek word for "sound."

Even electromagnetic radiation is hard to observe in its particle aspect when the quanta are very small, as in radio waves. It is only when the quanta grow large and the wavelengths tiny, as in X rays, that the particle aspect can be easily observed.

Einstein showed that the gravitational field should radiate waves just as an electromagnetic field does. The gravitational field is so much less intense than the electromagnetic that gravitational waves are exceedingly weak, and their particle aspect almost hopelessly undetectable. Nevertheless, physicists speak of gravitational waves as consisting of speeding gravitons.

It is only because in the ordinary world about us particles are so massive and waves so lacking in energy that we think of the two phenomena as mutually exclusive. In the world of the atom and of subatomic particles, this exclusivity disappears.

Sometimes science is said to produce paradoxical results and to go against common sense. It is important to remember that common sense is often based on the very limited observations we make in the world about us. To go against common sense sometimes means that we are taking a broader and more accurate view of the Universe. (Remember that "common sense" once told us that the Earth is flat and that the Sun moves around it.)

4

NUCLEI

Probing the Atom

Once scientists began to suspect that electrons might be
associated with atoms, a problem arose. Electrons carried
a negative electric charge, but atoms were electrically neutral. That meant there had to be positive electric charges
located somewhere in the atom that served to neutralize
the charges of the electron.

If this were so, then if electrons were removed from
an atom, what remained would carry a positive charge. If
electrons were added to an atom, the atom plus the excess
electrons would carry a negative charge. This would account for the positive and negative ions of Faraday and
Arrhenius.

J. J. Thomson, in 1898, was the first to suggest an atomic structure that took electric charge into account. He maintained that the atom was a tiny, featureless sphere, as had been assumed for a century, but that it carried a positive electric charge. In that positively charged atom, enough electrons were embedded (like raisins in a cake) to neutralize the charge.

Thomson's notion of atomic structure still left the atom a solid affair, and if many atoms were lined up in contact, right and left, up and down, inward and outward, then the solid that was formed in this way must be exactly what the name indicated—*solid*.

Yet that could not be so. Lenard had noted in 1903 that the speeding electrons making up the cathode rays slipped through thin films of metal, which seemed to imply that an atom must consist, at least to some degree, of empty space. Lenard suggested that an atom consisted of a cloud of small particles; some of them being electrons and some of them being positively charged particles of similar size. A positive and a negative particle would revolve about each other, making up a pair that was, overall, neutral. A large cluster of such pairs might make up an atom, but between and within these pairs there would be empty space through which a small object such as a speeding electron could easily pass.

But if that were so, then an atom ought to lose either type of particle with equal ease. If exposure to light caused the ejection of negatively charged electrons from metal, why were not positively charged particles also ejected—at least now and then? Again, if speeding electrons left a cathode under the force of an electric current, why did not speeding positively charged particles emerge from the anode? Clearly, if the positively charged particles existed, they would have to be quite different in nature from electrons. The positive particles would have to be, for some reason, much less mobile than electrons.

In 1904, the Japanese physicist Hantaro Nagaoka (1865–1950) suggested that the positive portion of the atom did not take up the whole of the volume, as Thomson had suggested, and that it did not take up as much volume as the electrons, as Lenard had suggested. Nagaoka offered a compromise. Nagaoka believed that the positively charged portion of the atom was located in the atomic center, and was smaller than the atom as a whole. It was surrounded by circling electrons, held by electromagnetic attraction, as the planets circled the Sun, held by gravitational attraction.

Nagaoka's suggestion provided a neutral atom under ordinary conditions, allowed for the production of positive and negative ions, and left empty space for speeding electrons to pass through. In addition, it explained why electrons were so easily removed from atoms, while positively charged particles were not. The electrons, after all, were on the outskirts of the atom, while the positively charged portion was in the protected center.

Still, none of these suggestions really won acceptance. They were all speculative and uncompelling. What was needed was direct evidence concerning the internal structure of the atom. Such direct evidence did not seem as though it would be easy to come by. After all, how could anyone possibly probe the interior of so small an object as an atom? And yet, even as Thomson, Lenard, and Nagaoka were advancing their suggestions, such an atom-probing device already existed. The story of its discovery goes as follows.

As soon as X rays were discovered by Roentgen, other physicists hastened to study the new radiation, and many wondered if it could be found elsewhere, in places where it hadn't been noticed only because no one had thought of looking for it there.

The French physicist Antoine Henri Becquerel (1852–1908) was particularly interested in fluorescent compounds, substances that absorb sunlight (or other energetic radiation) and then give up the energy by emitting light of just a few restricted wavelengths. Fluorescence is very similar to phosphorescence, except that fluorescent substances cease to give off light as soon as they are no longer exposed to energetic radiation, while phosphorescent substances continue to give off light for a period of time after exposure ceases.

Becquerel wondered if fluorescent substances emitted X rays along with visible light. In order to test this, he planned to wrap photographic plates in black paper and put the package in sunlight with a crystal of a fluorescent chemical upon it. The sunlight would not penetrate the black paper, and neither would any fluorescent light the crystal gave off. If, however, the crystal gave off X rays, these would penetrate the black paper and fog the photographic film.

The crystals he used were of potassium uranyl sulfate, a well-known fluorescent material. Each molecule of that compound contained one atom of the metal uranium.

On February 25, 1896, Becquerel performed his experiment and, sure enough, the photographic film was fogged. He decided that the crystal was indeed giving off X rays, and he prepared to repeat the experiment with new film in order to make sure. However, there followed several cloudy days. Becquerel put the photographic film, with its black paper wrapping and the crystal on top, in a drawer and waited for sunlight.

By the first of March Becquerel was restless. To give himself something to do, he decided to develop the film just to make sure that nothing was getting through the dark paper in the absence of fluorescence. To his amazement, something *was* getting through, and a lot of it. The plate was strongly fogged. The crystals must be giving off radiation that did not depend on sunlight and did not involve fluorescence. Forgetting the Sun, Becquerel began to study the radiation instead.

He quickly realized that the radiation given off by the potassium uranyl sulfate originated in the uranium atom, for other compounds containing uranium atoms gave off similar radiation, even when they were not fluorescent. In 1898, the Polish-French physicist Marie Curie (1867–1934) showed that another metal, thorium, also gave off radiation. She termed such behavior on the part of uranium and thorium radioactivity. Both Becquerel and Curie suspected that more than one type of radiation was involved.

In 1899, the New Zealand-born physicist Ernest Rutherford (1871–1937) studied the manner in which radioactive radiations penetrated sheets of aluminum. He found that some of the radiation could be stopped by $\frac{1}{500}$ of a centimeter of aluminum, while the rest required a considerably thicker sheet to be stopped. Rutherford called the first type of radiation alpha rays, from the first letter of the Greek alphabet, and the second type beta rays, from the second letter. A third type of radiation, which was the most penetrating of all, was discovered in 1900 by the French physicist Paul Ulrich Villard (1860–1934), and was called gamma rays, from the third letter of the Greek alphabet.

It was not long before these various radiations were quantified. The beta rays were deflected by a magnetic field in such a way that it was clear they consisted of negatively charged particles. In 1900, Becquerel determined the mass and the size of the charge of these particles, and it turned out that beta rays, like cathode rays, were made up of speeding electrons. A speeding electron is, therefore, sometimes called a beta particle.

Gamma rays were not deflected by a magnetic field, and this made it appear that they did not carry an electric charge. Rutherford suspected that gamma rays might be electromagnetic in nature, and passed some through a crystal. The existence of a diffraction pattern showed that they were very much like X rays, except that they possessed even shorter wavelengths.

As for alpha rays, they were deflected by a magnet in such a way as to show that they consisted of positively charged particles. Might these be the positively charged particles that, along with electrons, Lenard thought made up atoms?

No. Lenard had imagined that the positively charged particles were, rather, like electrons in their properties except for the nature of the charge. The alpha particles, however, were very different from electrons in other ways than their electric charge. In 1906, Rutherford showed that the alpha particle was much more massive than an electron. We now know that it is about 7,344 times as massive as an electron.

As soon as Rutherford found the alpha particles to be particularly massive, it seemed to him that they would be the very thing with which to probe the atom. A stream of alpha particles striking a thin film of metal would penetrate it, and the manner of its penetration might yield useful information.

Rutherford placed a piece of radioactive substance in a lead box that had a hole in it. The radiations could not penetrate the lead, but a thin stream of radiation would emerge from the hole, and, traveling outward, would strike a thin film of gold. Behind the gold was a photographic plate, which would be fogged by any alpha particles that passed through the gold.

The gold sheet was so thin as to be semitransparent, but, just the same, so tiny are atoms that that same sheet was about 20,000 gold atoms thick. Even so, the alpha particles smashed through as though those 20,000 atoms simply weren't there. They fogged the photographic plate in precisely the spot that would have been fogged if the gold film had not been there.

Yet not entirely. Rutherford noticed that a few alpha particles were deflected. There was a faint haze of clouding around the dark central spot on the photographic plate. The haze faded off quickly with distance, but didn't entirely disappear. About 1 in 8,000 alpha particles were deflected ninety degrees or more. In fact, an occasional alpha particle seemed to hit something and bounce directly backward.

To explain this, Rutherford advanced his idea of atomic structure in 1911. The atom, he said, had almost all its mass concentrated in a small, positively charged body at its very center. At the outskirts of the atom, spread over a volume that took up almost all of the atom, were nothing but electrons. This was something like Nagaoka's atom, except that the positively charged body at the center of the atom was much smaller and more massive in Rutherford's atom.

Furthermore, Rutherford had experimental observations, which Nagaoka did not. The alpha particles penetrated the electron portion of the atom as though it were empty because the alpha particle was so much more massive than the electron. If the alpha particle neared the massive, positively charged central body, the alpha particle (itself positively charged) was deflected. From the proportion of deflections, Rutherford could calculate the size of the nucleus. Nagaoka had no evidence of this type.

It is Rutherford, then, who rightly gets the credit for the advance. The central body is called the nucleus (plural nuclei) of the atom, from a Latin word meaning "little nut" in that it resembles a tiny nut inside the comparatively roomy atomic shell. Because, in biology, living cells also have central bodies called nuclei, that of the atom is sometimes specified as the atomic nucleus. For the purpose of this book, however, the qualifying word is not used.

Rutherford's picture of the nuclear atom proved entirely satisfactory, although many details have been filled in, as we shall see, in the three-fourths of a century since. For this and other work, Rutherford received a Nobel prize in 1908. (He received it in the category of chemistry, which displeased him, for he viewed himself as a physicist, of course.)

Positively Charged Particles

The nucleus has anywhere from 99.945 to 99.975 percent of the mass of the atom of which it is part. For this reason it became very important to study the nucleus. Indeed, you might almost say that the nucleus was the "real" atom. What had been thought of as the atom in the nineteenth century was mostly empty space; or at least, space filled with the very insubstantial particle/waves of the electrons. It was the nucleus that might have been the tiny, spherical, solid, and ultimate bit of matter that was first envisioned by Leucippus and Democritus.

Despite its mass, the nucleus is tiny in size, with a diameter only $\frac{1}{100,000}$ that of the atom. For that reason, it is considered as much a subatomic particle as the electron.

The nucleus must carry a positive electric charge; one of sufficient size to neutralize the charge of all of the electrons that are ordinarily to be found in a particular atom. Nevertheless, the history of such positively charged subatomic particles does not begin with Rutherford.

Goldstein, who had invented the name cathode ray, was interested in trying to find signs of any radiation traveling in the opposite direction. He could detect no such radiation emanating from an anode. In 1886, however, it occurred to him to devise a cathode that would itself allow radiation to travel in the other direction. This he tried to

The positively charged nucleus contains almost all of the mass of the atom, but it is only $1/100,000$ of its diameter.

achieve by making use of a cathode that was perforated and had little holes (or "channels") in it. When such a cathode was enclosed in the middle of an evacuated tube, and an electric current forced through it, cathode rays were formed. However, positively charged radiation, originating near the cathode, could pass through the channels, moving in the opposite direction.

This is precisely what Goldstein observed, and he called this new radiation *Kanalstrahlen*, which is German for "channel rays." However, this was incorrectly translated as "canal rays" in English.

In 1895, Perrin collected some of these canal rays on an object he placed in their path and showed that the object gained, in this way, a positive electric charge. In 1907, therefore, J. J. Thomson suggested they be called positive rays.

In 1898, Wien subjected these rays to magnetic and electric fields. He found that the particles of which positive rays were composed were much more massive than electrons. They were, indeed, as massive as atoms. In addition, the mass of the positive-ray particles depended on the traces of gas present in an evacuated tube. If it was hydrogen, the positive-ray particles had the mass of a hydrogen atom; if oxygen, they had the mass of an oxygen atom, and so on.

Once the basic theory of Rutherford's nuclear atom was accepted, it was immediately understood what the positive-ray particles were. The speeding electrons that made up the cathode rays collided with the stray atoms in the cathode-ray tube—hydrogen, oxygen, nitrogen, or whatever. The electrons were insufficiently massive to disturb the atomic nuclei and, in any case, struck them very rarely. They did, however, strike electrons and knock them out of the atom. The atoms, minus their electrons, would be nuclei carrying a positive electric charge, and would move off in the direction opposite to that taken by the cathode-ray particles.

As early as 1903, Rutherford had recognized that alpha particles were very similar to positive-ray particles in their properties. By 1908, he was quite certain that an alpha particle was just about equal in mass to the helium atom. It seemed to him there had to be some connection between alpha particles and helium because uranium minerals, which constantly produced alpha particles, also seemed, just as constantly, to contain small quantities of helium.

In 1909, Rutherford placed some radioactive material in a double-walled glass vessel. The inner glass wall was quite thin, but the outer glass wall was considerably thicker. In between the two walls was a vacuum.

The alpha particles ejected by the radioactive material could pass through the thin inner wall, but not through the thick outer wall. The alpha particles thus tended to be trapped in the space between the walls. After several days, the particles between the walls had accumulated to a volume at which they could be tested; and when this was done, helium was detected. It was clear, then, that alpha particles were helium nuclei. Other positive rays were nuclei of other types of atoms.

One of the ways in which positive-ray particles differed from electrons was that whereas all electrons had the same mass and the same electric charge, positive-ray particles had different masses and electric charges. Naturally, physicists wondered if they could somehow break up the positive-ray particles into smaller pieces and perhaps locate a very small positive particle no bigger than the electron.

Rutherford was among those who searched for such a tiny "positive electron," but didn't find it. The smallest positively charged particle he could find weighed as much as a hydrogen atom, and must be a hydrogen nucleus. In 1914, Rutherford decided that this particle must be the smallest positively charged particle that could exist. It has an electric charge exactly equal to that of the electron (although positive rather than negative), but has a mass, as we now know, 1836.11 times that of the electron.

Rutherford called this smallest positive-ray particle a proton, from the Greek word for "first," because when these particles are listed in order of increasing mass, the proton is first.

Atomic Numbers

Nuclei, which might be viewed as the essential cores of atoms, differ among themselves, as I have said, in two ways: in their mass and in the size of the positive charge they carry. This represents a significant advance over an earlier level of knowledge. Through the nineteenth century, nothing was known of the electric charges within the atom, and the only known difference among atoms was mass. Mass alone was not entirely satisfactory.

Earlier in the book, I mentioned that when the elements are arranged in the order of the mass of their atoms (atomic weight), a periodic table can be established. In this table, elements are so arranged that those with similar properties fall into the same row.

Such a table, based on atomic weight alone, has its faults. The size of the difference in mass varies as one goes up the scale. Sometimes the mass difference from one atom to the next is very small and sometimes quite large. In three cases, the atomic weight of a particular element is actually a bit *greater* than that of the element next higher in line.

In actual fact, if mass were all-important, the position of the two elements in these three cases ought to be reversed. They are not reversed, however, because, if they were, each of the elements involved would then be placed with a group that did not share its properties. Mendeleev, who first devised the periodic table, felt that keeping the elements with their own families was more important than strictly following the order of increasing atomic mass, and later chemists agreed.

Then, too, with only mass as a distinguishing characteristic, one could never be sure when an element might be discovered with an atomic weight in between two already known elements. As late as the 1890s, an entire family of elements, hitherto unknown, had been discovered and a new column had to be added to the periodic table. But it was possible that the confusing aspects of the periodic table might be done away with if the newly discovered second differentiating characteristic of atoms, the size of the positive charge on the nucleus, could be dealt with.

The possibility of doing so came by way of X rays. (There was no way of predicting, when X rays were first discovered, that they would be useful in connection with the periodic table. However, all knowledge is one. When a light brightens and illuminates a corner of a room, it adds to the general illumination of the entire room. Over and over again, scientific discoveries have provided answers to problems that had no apparent connection with the phenomena that gave rise to the discovery.)

The X rays first detected by Roentgen were produced when cathode rays impinged on the glass of a vacuum tube. The speeding electrons were suddenly slowed, and kinetic energy was lost. Such energy cannot be truly lost, but can only be converted into another form of energy; into electromagnetic radiation, in this case. The energy lost in a given moment was so great that unusually energetic photons were formed and the radiation was emitted in the form of X rays.

Once this was understood, it was quickly seen that if something denser than glass, and made up of more massive atoms, was put in the way of speeding electrons, they would be decelerated even more sharply. X rays would then be formed of still shorter wavelengths and higher energies. The obvious thing to use were various metal plates. These were placed at the opposite end of the tube from the cathode, where the speeding electrons would impinge upon them. Such metal plates were called anticathodes, where anti- is from a Greek word meaning "opposite." (Ordinarily, anodes are placed opposite the cathode, but to make room for an anticathode, the anodes are placed at the side of the tube.)

In 1911, the British physicist Charles Glover Barkla (1877–1944) noticed that when X rays were produced by an anticathode of a particular metal, they tended to penetrate substances just so far. Each metal produced X rays of penetrating power specific to the metal. Later on, when X rays were recognized as electromagnetic radiation, this was interpreted as meaning that each metal would produce X rays of a particular wavelength. Barkla called these the characteristic X rays of a particular metal.

Barkla also found that sometimes two types of X rays were produced by an anticathode of a particular metal, each with its own penetration, but with nothing much in between. He called the more penetrating beam *K* X rays, and the less penetrating one *L* X rays. Later on, still less penetrating beams were found to be produced in some cases, and the letters designating them continued through the alphabet, *M* X rays, *N* X rays, and so on. For his work, Barkla received a Nobel prize in 1917.

Barkla's work was carried on by one of Rutherford's students, Henry Gwyn-Jeffreys Moseley (1887–1915). In 1913, he studied the characteristic X rays very carefully, making use of the X-ray diffraction of crystals, which had just been discovered by the Braggs.

Moseley found that if he went up the list of elements in the periodic table, the wavelength of the X rays produced decreased regularly. The greater the atomic weight of the atoms in the anticathode, the shorter the wavelength of the X rays. Moreover, the change in wavelength was much more regular than the change in atomic weight.

Physicists were sure that the deceleration of electrons was brought about chiefly by the size of the positive charge on the atomic nucleus, which was an indication that the size of the charge as one went up the periodic table increased more regularly than did the mass of the atomic nucleus.

Moseley suggested, in fact, that the size of the charge increased by one with each step up the table. Thus, hydrogen, the first element, had as its nucleus the proton, which had a charge of $+1$. Helium, the second element, had a nucleus (the alpha particle) with a charge of $+2$. Lithium, the third element, had a nuclear charge of $+3$, and so on up to uranium, with the most massive atom then known, which had a nuclear charge of $+92$.

Moseley called the size of the nuclear charge the atomic number of the element, and this proved to be more fundamental than the atomic weight. Indeed, the atomic number solved many of the problems of the periodic table as Moseley's concepts were refined and extended by later physicists.

Thus, in those cases, in going up the periodic table, for which you have an element with a slightly higher atomic weight than the element following, this does not happen if atomic numbers are considered instead. An element that would seem out of place because it has a higher atomic weight than the one that follows turns out to have a lower atomic number. If all atoms are arranged by atomic number, every one, without exception, turns out to be with its proper family, and there need be no reversals. Then, too, when two neighboring elements have atomic numbers that differ by one, there can be no hitherto unknown element in between.

It soon became clear that all negative electric charges are exact multiples of the charge on an electron, while all positive electric charges are exact multiples of the charge on a proton. You can have nuclear charges of $+16$ and $+17$, but you can't have $+16.4$ or $+16.837$.

Where there is a missing element in the periodic table, the change in wavelength of the characteristic X rays, in going from one element to the next over the gap, is twice as great as expected, and that is a sure sign of an element in between.

At the time Moseley worked out the concept of the atomic number, there were seven gaps in the table, each gap representing an as yet unknown element. By 1948, the gaps were all filled. Physicists were able to form atoms with atomic numbers higher than 92 by methods to be explained later. At the present time, all of the elements are known from atomic number 1 to atomic number 106. (Moseley would almost certainly have been awarded a Nobel prize for his work in this regard within a few years but, in 1915, he was killed in action in World War I at Gallipoli in Turkey.)

The atomic number tells us the size of the positive charge on the nucleus. Because the normal atom, as a whole, is electrically neutral, there must be one electron in the outer reaches of the atom for every positive charge on the nucleus. Thus, because hydrogen has a charge of $+1$ on the nucleus, the normal atom must possess one electron. Helium, with a nuclear charge of $+2$, must have two electrons in each atom; oxygen, with a nuclear charge of $+8$ must have eight electrons; uranium, with a nuclear charge of $+92$ must have ninety-two electrons, and so on. In short, the atomic number reflects not only the size of the nuclear charge, but the number of electrons in a normal atom.

It seems to make sense that chemical reactions take place when atoms, either independently, or as part of molecules, collide with each other. If so, the collisions are made basically between the electrons of one atom and the electrons of another. The nuclei of the two atoms are far off in the center of the atom, hidden behind the electrons, and are not at all likely to take part in chemical reactions or even to influence them in any crucial way.

This not only makes sense (things are not necessarily

so, just because they make sense), but seems to follow from such things as Arrhenius's theory of ionic dissociation. The formation of ions seems to be the result of the transfer of one or more electrons from one atom to another.

In the case of a molecule such as that of sugar, no ions seem to be formed. Instead, the atoms within the molecule simply cling together, perhaps because they share electrons and therefore cannot separate easily and remain intact atoms. However, there would seem to be some cases in which the transferring of electrons is the more stable situation, and others in which the sharing of electrons is—but why?

A hint comes from a group of six elements that are made up of atoms that do not tend to transfer or share electrons, but remain as single atoms at all times. The three lightest atoms in the group—of helium, neon, and argon— never transfer or share electrons at all, at least as far as chemists have ever observed. The three heaviest—of krypton, xenon, and radon—do share electrons under some extreme circumstances, but not very firmly.

These six elements are called the noble gases ("noble" because they are standoffish and do not tend to deal with the common herd). The reason for the "nobility" of these elements is best understood if we imagine that electrons are arranged about an atom in concentric shells, one outside the other. Naturally, as one moves outward from the nucleus, each successive electron shell is larger than the one before, and holds more electrons. Thus, the helium atom has two electrons, which seem to fill the innermost shell. This is not surprising, in that as the shell nearest the nucleus it should be the smallest of the shells, capable of holding the fewest electrons.

The American chemists Gilbert Newton Lewis (1875–1946) and Irving Langmuir (1881–1957), beginning in 1916, independently worked out notions of shells and electron transfer, or sharing, because these phenomena seemed to explain chemical behavior so well. (Actually, the subject was greatly refined in later decades, but we'll get to that later.)

Individual shells are associated with the series of characteristic X rays first discovered by Barkla. The K series of X rays are the most penetrating and seem to originate from the electron shell nearest the nucleus. This first electron shell is therefore called the K shell.

Following this same reasoning, the shell just beyond the K shell is called the L shell because it seems to be the origin of the less-penetrating X rays of the L series. Beyond the L shell is the M shell, the N shell, and so on.

It might be that helium atoms are noble and will neither transfer nor share electrons (and therefore engage in no chemical reactions) because a filled shell is particularly stable. Either sharing or transferring an electron would lessen the stability of the situation, and stabilities are never lessened spontaneously. (It always takes energy to force something to destabilize, but stabilization takes place all by itself. These are properties associated with what is called the second law of thermodynamics.)

The next noble gas is neon, which has ten electrons in its atoms. The first two fill the K shell, and the next eight fill the L shell, which is larger and can hold more electrons. The electron pattern of neon is, therefore, 2, 8. With an L shell filled and stable, neon is a noble gas.

After neon is argon, with eighteen electrons in its atoms. These are arranged thus: two in the K shell, eight in the L shell, and eight in the M shell; or 2, 8, 8. The M shell, being larger than the L shell can hold more than eight electrons. Indeed, it can hold eighteen. However, eight electrons in the outermost shell (however many it can hold altogether) is a particularly stable configuration, making argon a noble gas.

After argon comes krypton, with thirty-six electrons,

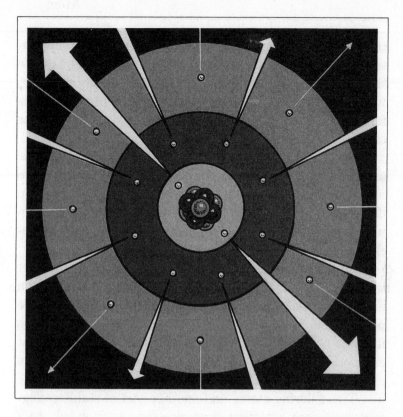

The innermost K shell, the middle L shell, and the outermost M shell of electrons in the noble gas argon, showing the relative strengths of X rays produced from each shell when struck by cathode rays.

arranged 2, 8, 18, 8; then xenon with fifty-four electrons, arranged 2, 8, 18, 18, 8; and finally, radon, with eighty-six electrons, arranged 2, 8, 18, 32, 18, 8.

Apparently, atoms interact with one another, when possible, so as to attain a noble gas configuration of electrons. Sodium, with eleven electrons, has them arranged 2, 8, 1. The eleventh electron is the only one in the M shell and is easily lost. When this happens, the sodium atom becomes a sodium ion, with one positive charge, because

the charge of $+11$ on the nucleus is not completely neu-
tralized with only ten electrons in the outer regions. (Notice
that losing an electron does *not* change sodium into neon,
which also has ten electrons in its atom. What counts, as
far as the identity of the atom is concerned, is the nuclear
charge, not the electron number.)

On the other hand, chlorine has seventeen electrons
arranged 2, 8, 7. It needs one more electron to attain the
noble gas configuration. It therefore has a tendency to gain
one electron and become a negatively charged chloride ion
with eighteen electrons in its atom, overbalancing the nuclear charge of $+17$.

For this reason, sodium and chlorine atoms easily react
with each other. A transfer of one electron forms sodium
ions and chloride ions, which cling to each other because
positive and negative charges attract. When salt dissolves
in water, the ions are less tightly bound and can slip past
each other. In this way, such a solution will conduct an
electric current.

Two chlorine atoms find another sort of stable configuration if each contributes an electron, so to speak, to a
shared pool. Each atom has six electrons in its outermost
shell that are entirely its own, and two electrons it shares
with the other atom. Each outermost shell is full and stable
provided the two atoms remain in contact so that they can
maintain the two-electron sharing between them. The result is that the two chlorine atoms produce a two-atom
chlorine molecule (Cl_2), which is more stable than two single
chlorine atoms would be.

By thus dealing with electron arrangements in atoms,
chemists found that they could understand why the periodic
table is arranged as it is—based on chemical reactions that,
in turn, depend on the electron arrangements in the outermost shell. Again, chemists found that they could use the
electron arrangements to explain many chemical reactions

that they had earlier simply accepted without knowing why.

It is, so far, sufficient to accept electrons as tiny, solid particles existing in geometric arrangements. However, such a view is insufficient in explaining the spectral lines that also distinguished each element from all others.

Spectral Lines

After Newton had demonstrated the existence of the light spectrum, it was studied closely by many scientists. If sunlight, for instance, is passed through a thin slit before being passed through a prism, each different wavelength casts an image of that slit in some characteristic color. The wavelengths line up very closely, seeming to form one smooth band of changing color (as in a rainbow). But what if some wavelengths happen to be missing for some reason? In that case, there will be places in the spectrum that produce no color image through the slit. That will produce a dark line across the spectrum.

In 1802, the British chemist William Hyde Wollaston (1766–1828) observed such dark lines, but he did not pursue the subject, and neither did anyone else for a while.

In 1814, however, the German optician Joseph von Fraunhofer (1787–1826) produced excellent prisms and other optical equipment, and was able to produce sharper spectra than anything produced before. At once he observed hundreds of dark lines in the spectrum. He carefully mapped their positions, and their prominence, and showed that the same lines fell in the same position whether their source was sunlight, moonlight, or light from the planets. (Of course, the light given off by the Moon and planets is reflected sunlight, so this is not, perhaps, surprising.)

From then on, the Fraunhofer lines, as they were frequently called, were studied carefully, but were regarded as little more than curiosities until an important break-through was made in 1859 by Kirchhoff.

Kirchhoff found that if particular elements were heated, they did not produce a continuous spectrum as the Sun did. Instead, they radiated light in separate wave-lengths, so that the spectrum consisted of a number of bright lines, separated by stretches of darkness. If sunlight was sent through the relatively cool vapors of a particular element, the vapors would absorb just those wavelengths that they would emit when radiating. Moreover, each element emitted, when hot, or absorbed, when cool, its own characteristic wavelengths. In this way, the elements present in a particular mineral could be identified by the wave-lengths emitted when the mineral was strongly heated. Elements hitherto unknown could be detected by the presence of wavelengths not given off by any known element. The elements present in the Sun and in other stars could be identified by the dark lines in their spectra.

All of this knowledge about spectral lines made them extremely important to chemists and to astronomers, but no one knew why different elements radiated or absorbed different wavelengths. One step forward in solving this puzzle was taken by the Swiss physicist Johann Jakob Balmer (1825–1898). He was particularly interested in the spectrum of glowing hydrogen, which seemed simpler than those of other elements (and why not, since hydrogen was the light-est and, presumably, the simplest of the elements).

The hydrogen spectrum consisted of a series of lines, spaced more and more closely with decreasing wavelength. In 1885, Balmer worked out a formula for the wavelengths of these lines. The formula contained a symbol that could be replaced by successive square numbers: 1, 4, 9, 16, and so on. As a result, the successive wavelengths of the lines in the hydrogen spectrum could be calculated. That still didn't explain why the lines were where they were, but at least it showed that there was a deep regularity in the lines that must somehow be reflected in the structure of the atom. There was no way of going further until more was known about the structure of the atom. Let's see how that worked out.

Once physicists accepted the nuclear atom, they had to consider what kept electrons in place. After all, if electrons are negatively charged and the nucleus is positively charged, and if opposite charges attract, why don't the electrons fall into the nucleus? The question might also be asked about the Earth—why doesn't it fall into the Sun, in that the two attract each other. In Earth's case, the answer is that it is in orbit. It *is* falling toward the Sun, but its additional motion at right angles to that fall keeps it forever in orbit.

There was a tendency to think, therefore, that the atom was a sort of miniature solar system, with the electrons whipping about the nucleus. There is a catch to that, however. From electromagnetic theory, it was known (and observed) that when an electrically charged object revolved in this fashion, it would give off electromagnetic radiation, losing energy in the process. As it lost energy, it would spiral inward and eventually fall into the nucleus.

In the same way, the Earth, in revolving about the Sun, gives off gravitational radiation, losing energy in the process, so that it is spiraling into the Sun. However, gravitation is so much weaker than electromagnetism that the amount of energy the Earth loses in this way is excessively small, allowing it to revolve about the Sun for billions of years without spiraling in appreciably closer.

An electron, however, subject to the much more intense electromagnetic field, loses so much energy in the form of radiation that its collapse into the nucleus, so it would seem, cannot be long delayed—yet this is not the case. Atoms remain stable for indefinite periods, their electrons remaining in the outer portions.

This problem was tackled by the Danish physicist Niels Henrik David Bohr (1885–1962). He decided there was no use in saying that an electron radiated energy when it orbited an atom, when it clearly didn't. He insisted that as long as an electron remained in orbit, it *didn't* radiate energy.

Yet hydrogen, when heated, did radiate energy—and when cool, absorbed it. It emitted certain wavelengths that fit Balmer's equation, and absorbed those same wavelengths. To explain this, Bohr supposed, in 1913, that the electron in the hydrogen atom could take on any of a number of different orbits at different distances from the nucleus. Whenever it was in a particular orbit, whatever its size, it didn't either gain or lose energy. When the electron changed orbits, however, it either absorbed energy, if it moved farther from the nucleus, or emitted energy, if it moved closer to the nucleus.

But why should an electron be in a particular orbit and then, absorbing energy, suddenly shoot outward into the next larger orbit—never, by any chance, being in an orbit halfway between the two. Bohr saw that this had to have some connection with the quantum theory. If the atom could only handle quanta of a certain size, it could only absorb light of a certain wavelength, and that would automatically send it outward to the next orbit.

Bohr worked out a series of calculations that showed how one could map a series of permitted orbits that would result in the absorption or emission of quanta of fixed size (and hence, radiation of fixed wavelength) that would perfectly account for the particular wavelengths of the lines in the hydrogen spectrum.

What Bohr did was to show that one could not work out the structure of the atom solely according to classical

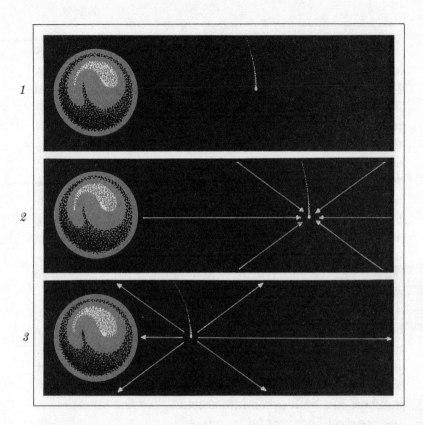

1. Whenever an electron stays in a particular orbit, it doesn't gain or lose energy.
2. When it changes to a higher orbit, it absorbs energy.
3. It emits energy when it moves closer to the nucleus.

physics; one had to make use of quantum theory. For this he received a Nobel prize in 1922.

Bohr had to use whole numbers for one of the terms of his formula, each number representing a set of different spectral lines. The need for whole numbers arose out of the fact that a whole number of quanta were involved. You couldn't have fractions of a quantum. The number that was inserted in the formula was called a quantum number for that reason.

Although Bohr's formula gave the figures for the wavelengths of the spectral lines, it didn't explain everything. If the spectral lines were studied with increasingly refined instruments, it turned out that each line had a "fine structure"—a number of very closely spaced thinner lines. It was as though each of Bohr's orbits consisted of a family of orbits with very small differences among themselves.

In 1916, the German physicist Arnold Johannes Wilhelm Sommerfeld (1868–1951) pointed out that Bohr's orbits were all circular. The orbits might be elliptical, too, and to different degrees. In order to take these new orbits into account, a second quantum number had to be introduced. This could have any whole number from zero up to one less than Bohr's quantum number.

If Bohr's quantum number (or principal quantum number) is 1, the lowest it can be, Sommerfeld's quantum number (or orbital quantum number) can only be 0. If the principal quantum number is 2, then the orbital quantum number could be either 0 or 1, and so on. If the two quantum numbers are both taken into account, the fine structure of the spectral lines can be expressed.

Complications continued, however. If atoms are in a magnetic field, lines that seemed absolutely single split into finer parts. Both Bohr and Sommerfeld pictured the orbits (whether circular or elliptical) as all being in a single plane, so that the nucleus and all possible orbits formed a system that was as flat as a sheet of paper. It was possible, though, for orbits to be tipped, so that all of the orbits taken together might be distributed symmetrically through three-dimensional space, and the atom given a spherical outline. This makes sense, because atoms do, in many ways, act as though they were tiny spheres.

In order to take the three-dimensional system of orbits into account, a third quantum number had to be added—the magnetic quantum number. This could have any positive whole value from 0 to whatever the principal quantum number might be, and any negative integer value of the same sort. If the principal quantum number were 3, the magnetic quantum number might be -3, -2, -1, 0, $+1$, $+2$, or $+3$.

With three dimensions taken care of, it might seem there was nothing further that needed to be done. Nevertheless, there were still certain features of the spectral lines that were puzzling, and one more quantum number was added by the Austro-Swiss physicist Wolfgang Pauli (1900–1958). This was thought to represent the spin of the electron on its axis. This spin could be in one direction or the other, clockwise or counterclockwise. In order to make the calculations fit the observed facts about the spectral lines, this spin quantum number had to be either $+\frac{1}{2}$ or $-\frac{1}{2}$.

Pauli went on to show that within an atom no two electrons could exist with all four quantum numbers identical. This is called the exclusion principle because once an electron has its four quantum numbers, any other electron is excluded from the particular orbit represented by those numbers. For this, Pauli received a Nobel prize in 1945. (Sometimes, a scientist must wait twenty years, or, in rare cases, even as long as fifty years, for a Nobel prize. It takes time, now and then, to see that a discovery is truly significant. If prizes were awarded immediately for something that *seemed* important, many would be given for discoveries that would turn out to be trivial or even wrong.)

A detailed system of mathematics that uses the four quantum numbers and the exclusion principle to describe how electrons are distributed in an atom was worked out by the Italian physicist Enrico Fermi (1901–1954) in 1926, and by the British physicist Paul Adrien Maurice Dirac (1902–1984) in 1927. This system is called Fermi-Dirac statistics, which applies to any particle that has a spin of $+\frac{1}{2}$

or $-\frac{1}{2}$. Such particles are lumped together as fermions, for Fermi. The electron is an example of a fermion. So is the proton.

There are particles that have spins of 0, 1, or 2. (The photon has a spin of 1, for instance, and the graviton has a spin of 2.) The exclusion principle does not hold for such particles, whose manner of distribution was worked out by the Indian physicist Satyendra Nath Bose (1894–1974) in 1924. Einstein praised Bose's work and, in 1925, added to it. This system is called Bose-Einstein statistics, and any particle with an integral spin, or one of zero, is called a boson, for Bose.

Bohr's electron orbits, although an enormous advance, were nevertheless not entirely satisfactory. There was still the image of electrons as particles in motion, racing about their orbits. If this is so, there is still no clear explanation of why the electron isn't emitting radiation and spiraling into the nucleus. It is all very well to say that while an electron is in orbit it doesn't radiate; but why doesn't it? It is also a compelling argument to say that it can only emit quanta of a certain size; but why? Something is missing.

The German physicist Werner Karl Heisenberg (1901–1976) thought there would always be trouble if one attempted to picture the structure of the atom in terms of ordinary everyday life. What we are used to—planets circling the Sun, or billiard balls striking each other—involves masses so large compared with the atom that the tiny quanta out of which energy is built are too small to have any noticeable effect on such objects. All of our mental images, therefore, are of a nonquantum world. In dealing with atoms, electrons, and radiation, however, we are dealing with a world in which quantum effects are noticeable, so our images fail. (The quantum theory is, in a way, a system of saying that the Universe is grainy, not totally smooth. It is like a newspaper photograph that looks smooth

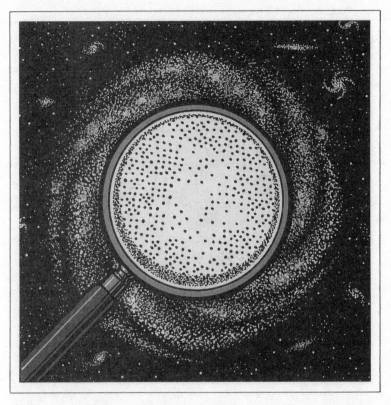

Taken to the highest level of magnification, the substance of the Universe is grainy, not smooth.

because the little dots of black and white of which it is made up are too small for us to see. If we magnify the picture sufficiently, our ordinary-world images fail. All we see are the dots, which no longer form a discernible picture.)

It seemed to Heisenberg that one should use only the figures taken from the spectra and manipulate them in some way that would enable physicists to work out numerical answers useful in connection with atomic behavior, rather than to attempt to interpret the behavior in terms of orbits, ellipses, tilts, spin, and all of the rest of it. In 1925, Heisenberg worked out what was called matrix mechanics, for

In dealing with the world in which quantum effects are noticeable, all our conventional images fail. If we consider the electron to be a wave rather than a particle, it might appear to form a wavy hoop about the nucleus.

the purpose, because it made use of a mathematical device called a matrix.

In that same year, however, Davisson had proved the existence of electron waves, and it occurred to the Austrian physicist Erwin Schrödinger (1887–1961) that these waves could explain the nature of electron orbits.

If we consider the electron to be a wave, rather than a particle, we can picture an orbit about the nucleus as having to consist of whole-number wavelengths. Then, if

118

we imagine ourselves tracing the wave around the nucleus, it doubles back on itself to form a "path" that looks like a wavy hoop about the nucleus. The smallest orbit would be that consisting of a single wavelength up and down. The electron couldn't spiral into the proton because it couldn't take up an orbit with a length less than a single wave. All of the other orbits would be at distances and with shapes such that a whole number of waves would fit around the orbit. That is why orbits can only be at certain distances, in certain ellipses, at certain tilts, with certain spins, and so on.

Schrödinger worked out a mathematical treatment that could solve problems by taking electron waves into account, and announced it in 1926. His system is called wave mechanics. Dirac also contributed to the working out of the treatment, and he and Schrödinger shared a Nobel prize in 1933 for this work.

It turned out eventually that wave mechanics and matrix mechanics are equivalents; they deliver the same results. The mathematical system is, therefore, simply called quantum mechanics. With improvements and refinements added, the system has proven completely satisfactory in dealing with electrons, and with subatomic phenomena generally.

In 1939, the American chemist Linus Carl Pauling (b. 1901) applied the principles of quantum mechanics to the manner in which atoms transferred and shared electrons. This replaced the older particle system of Langmuir and Lewis; it was subtler and explained many things the older system could not. For this, Pauling received a Nobel prize in 1954.

Earlier, in 1927, Heisenberg had demonstrated that it was impossible, even in principle, to work out certain types of measurement with perfect accuracy because of the graininess of the Universe implied by quantum mechanics. For example, suppose you tried to determine the exact position of a particle, as well as its exact momentum (which is its mass multiplied by its velocity). Any device you might use to determine position would change the particle's velocity and, therefore, its momentum. Any device you might use to determine momentum would change the particle's position. The best you could do would be to get the combined position and momentum with a tiny fuzziness; with a small, unavoidable inaccuracy. The uncertainty of the position multiplied by the uncertainty of the momentum, if both are taken at the absolute minimum that can be obtained, comes to an amount closely related to a fundamental constant of quantum theory.

Heisenberg's uncertainty principle also says that one can't determine time and energy content simultaneously and exactly. For this work, Heisenberg received a Nobel prize in 1932. The uncertainty principle is a very important discovery, explaining a great deal that would otherwise be mysterious in subatomic physics. Nevertheless, many scientists shied away from it because it made it seem as though there was a certain element of randomness in the Universe that could never be wiped away. Einstein, for instance, never accepted the uncertainty principle, and always thought that quantum mechanics was an incomplete theory because of it.

Still, not wholly accepting the uncertainty principle doesn't remove it. Moreover, the uncertainty principle seems to describe the Universe as it *is*, and there's no point in fighting it.

Bohr's picture of electron orbits seemed to describe an electron particle whose position and motion could, and should, be known at every moment. Schrödinger's use of waves, which proves to work much better, does *not* do this. An electron wave goes up and down, and somewhere in it is the electron in its particle aspect. We can't tell exactly where the particle is, however. In a way, it's everywhere along the wave. The height of the wave tells us the probability that it would happen to be there at any given moment, but it doesn't *have* to be there. In this way, quantum mechanics features probability and uncertainty—and that, indeed, seems to be the way the Universe is.

Because quantum theory deals with things so far removed from what we are used to in ordinary life, scientists speak of "quantum weirdness." There are aspects about it that seem so paradoxical that scientists have simply not managed to agree on what it all means. Perhaps someday, new discoveries, new concepts, new thoughts will clarify what seems now to be hopelessly mysterious.

The game of science never stops, you see, for new problems arise whenever old problems are solved; but who would want it any other way? To solve everything would stop the game, and nothing else that life could offer would, in my opinion, make up for the intellectual loss.

5

ISOTOPES

Nuclear Energy

Working out the details concerning the electron distribution in atoms was, in a way, a rather simple problem. All electrons are alike, and no differences among them, whether they are in one type of atom or another, or exist independently, have been discovered to this day. Atoms differ among themselves in the number of electrons they possess, but not in the type of electrons.

What about atomic nuclei, though? They differ from atom to atom in both mass and electric charge. Are they single particles of many different types, one for each different element, or do they have internal structure? Are they built out of different numbers of simpler particles, and are these simpler particles the same in all nuclei of all elements? Moreover, are these questions we can practically hope to answer? After all, nuclei are tiny objects hidden at the very center of the atom, sometimes behind layers and layers of electrons. How would one reach and study them?

The first hints concerning nuclear structure came with the discovery of radioactivity, fifteen years before the existence of the atomic nucleus was demonstrated. One question that naturally arose at once in connection with radioactivity was where all of that energy might be coming from. Uranium seemed to ceaselessly emit alpha rays, which were streams of helium nuclei, and beta rays, which were streams of electrons. All of these particles traveled at very high speeds, alpha particles at about a tenth the speed of light, and beta particles at about nine-tenths the speed of light. It takes considerable energy to make them move that quickly from a standing start. (After all, the uranium atoms aren't moving to begin with.) Then, too, there were gamma rays, which were far more energetic than even X rays.

Uranium's radiation isn't just a brief spurt of energy. A sample of uranium metal will continue to radiate indefinitely at an apparently constant rate, and this was a serious problem. By the law of conservation of energy, it would seem that energy could not be created out of nothing, and yet energy *seemed* to be created out of nothing in connection with radioactivity.

Of course, it might have been that the law of conservation of energy was wrong, or was limited to only certain conditions. Scientists, however, found the law so useful in all aspects of science that they hated to scrap it. There was the definite feeling that a search must be made to explain radioactivity *without* giving up the law of conservation of energy; that the law must be given up only as an absolutely last resort. (This is an example of intelligent conservatism among scientists. A theory or a law that has proved itself over and over should not lightly be discarded. It should be, and would be, discarded if there were no alternative, but one must be sure that there were indeed no alternatives.)

The situation grew quickly worse in the years immediately following the discovery of radioactivity. Marie Curie and her husband, Pierre Curie (1859–1906), started with pitchblende, a uranium-containing rock, from which they hoped to obtain samples of pure uranium for study. To their astonishment, they found that pitchblende was more radioactive than it would be even if it were made of pure uranium. Possibly, it contained elements still more radioactive than uranium. There was no sign of such elements in ordinary analysis, so they must exist in very small quantities, and, if so, must be *very* radioactive.

In 1898—after long, tedious, and painstaking work that began with tons of pitchblende and ended with tiny pinches of radioactive powder—the Curies isolated two elements: polonium (named for Marie Curie's native Poland) and radium (named for its radioactivity). Each was far more radioactive than uranium.

If one wondered at uranium's energy emission, how much more one must wonder at radium, which gave off energy at nearly three million times the rate. In 1901, Pierre Curie measured the energy given off by radium and found that 1 gram of radium gave off energy at a rate amounting to 140 calories per hour. That wasn't much in itself, but it continued for hour after hour indefinitely. Where did all that energy come from?

Some scientists wondered if radioactive atoms might not absorb energy from the surrounding environment and convert it into the energy of radiation. This hypothesis, however, would break the second law of thermodynamics, and scientists were as reluctant to do this as to break the first law (conservation of energy).

In 1903, Rutherford suggested that all atoms possessed large volumes of energy within their structure. Ordinarily, this energy was never tapped, so that people remained unaware of its existence. Radioactivity was, however, a spontaneous outpouring of a little of this energy. This was a daring suggestion, but it caught the imagination of the public and people began to speak of atomic energy, as a newer and far more concentrated form of energy than had ever before been known. (The English writer H.G. Wells even wrote of "atomic bombs" in his science fiction stories forty years before such a thing existed in reality.)

And yet Rutherford's suggestion might have seemed to be a case of pulling a rabbit out of a hat. Just *saying* that the atom contained energy explained nothing. But then, in 1905, Einstein showed, convincingly, that mass was a very concentrated form of energy. If radioactive substances were to turn even a small fraction of their mass into energy, then all of the energy liberated in radioactivity could easily be accounted for.

Once the nuclear atom was discovered, it was clear that because almost all the mass of an atom was concentrated in the nucleus, the necessary loss of mass must take place there. It was within the nucleus, then, that the energy source of radioactivity lay, and, eventually, people began to speak of nuclear energy rather than atomic energy.

Nuclear Varieties

If the energy of radioactivity arises out of the loss of mass of the atomic nucleus, what happens to that nucleus as a result? The beginnings of an answer came even before it was understood that the atomic nucleus was the source of energy, or even that there was an atomic nucleus at all. (It

often happens that scientific observations provide the beginnings of an answer to a question even before the establishment of a comprehensive theory that provides order and
reason to a section of science. Such early observations are
difficult to understand, and advance knowledge slowly without the theory. Once the theory is established, however,
the earlier observations quickly fall into place. Advance is
then rapid until it is slowed by the absence of some deeper
and broader understanding of another aspect of science.)

In 1900, Crookes, working with uranium metal, decided to purify it as much as he could, and subjected it to
chemical proceedings that separated out apparent impurities. He found, to his astonishment, that the purified uranium was hardly radioactive at all, while the impurities
were markedly radioactive. He suggested that it was not
uranium that was radioactive, but something else that existed in uranium as an impurity.

Becquerel, however, having discovered the radioactivity of uranium, was not ready to let go of it that easily.
(Scientists often treat their discoveries as their babies, and
defend them vigorously against any attempt to wipe them
out. This is a very human reaction, even if, sometimes, in
hindsight, wrong. In this case, though, Becquerel was
proved to be right.) Becquerel showed that uranium, when
purified in Crookes's fashion, did indeed show little radioactivity, but if this purified uranium were allowed to stand,
it would regain its radioactivity after some time.

In 1902, Rutherford and a co-worker, the British chemist Frederick Soddy (1877–1956), showed that this was also
true of thorium. If the metal were purified, most of its
radioactivity was lost, but was then regained on standing.
Rutherford and Soddy suggested, therefore, that when an
atom of uranium gave off radioactive radiations, its nature
was changed and it became an atom of another element that
was more radioactive. This new element, being radioactive,

also changed. Uranium was not very radioactive in itself, but its daughter elements were. When uranium was purified so that the daughter elements and their radioactivity were removed, uranium seemed far less radioactive than it had been, but it slowly formed additional quantities of daughter elements and its radioactivity returned and became as before. Atoms, it seemed, in undergoing radioactivity, experienced something that could be described as radioactive disintegration.

As it turned out, this was a correct view. Both uranium and thorium broke down into other elements, which, in turn, broke down until finally an element was reached that was not radioactive. In this way, one had a radioactive series. Scientists began to search for these intermediate elements in disintegrating uranium and thorium. Polonium and radium, earlier detected by the Curies, were two of them—along with others. It was also discovered that both uranium and thorium, after undergoing many changes, become nonradioactive lead.

The notion of radioactive disintegration came as a shock to scientists. After all, from the time of Leucippus and Democritus on it had been assumed that atoms were unchangeable—but that had only been an assumption. To be sure, atoms are unchangeable as far as chemical changes are concerned, but radioactivity is not a chemical change. Chemical changes involve only the outermost electrons of an atom. The atom might gain an electric charge, or form a bond with another atom as a result, but its essential identity, which depends upon its nucleus, remains intact. Radioactivity, however, *does* involve the nucleus. It is a nuclear change and, if the nucleus undergoes a change, it is quite likely that, in the process, one type of atom changes into another.

(A change in point of view, such as this, does not mean that all chemical textbooks need to be torn up and thrown

127

out as though all of the information they contain is now worthless. The new point of view merely broadens and extends knowledge, and supplies explanations that are fuller and more useful. Thus, twentieth-century textbooks must take into account the existence of nuclear change, but they can, if they wish, still discuss *chemical* change, as before, and treat atoms as changeless—as, indeed, they are, where chemical change is concerned.)

The search for the radioactive intermediates between uranium and lead, and between thorium and lead, was successful—too successful, actually. Far too many were found.

The atomic number of uranium is 92, and that of thorium 90. The atomic number of lead is 82, and another known element, bismuth, has an atomic number of 83. The as yet undiscovered elements lying at the end of the periodic table were elements numbered 84, 85, 86, 87, 88, 89, and 91. That's seven altogether, minus the newly discovered elements polonium (84) and radium (88), which left five. There were no other elements beyond the remaining five to be discovered between uranium and lead. None! That was quite certain after Moseley's work in 1914.

And yet by Moseley's time, more than thirty intermediates had been discovered. Each one of these was distinctly different, at least as far as its radioactive properties were concerned. Some would expel an alpha particle and some a beta particle. Some would expel a gamma ray with or without an alpha particle or a beta particle. Even when each of two intermediates emitted an alpha particle, say, one would do so with greater energy than the other, and more rapidly.

Soddy tackled this problem. Already in 1912 and 1913, before the concept of the atomic number had been worked out, he had found that certain intermediates had the same chemical properties, and, if mixed, could not be separated by ordinary chemical procedures. They were the same elements, and that meant (as was later understood) that their electron configurations were the same, as was the positive charge on their nuclei. Because they differed in their radioactive properties, however, something about the nucleus, other than the charge, was *not* the same.

The periodic table was based on the chemical properties of the elements. It followed that if two different atoms were identical in chemical properties and different only in radioactive properties, they were the same element (chemically) and must both fit into the same place in the periodic table.

Soddy announced his findings in 1913, naming these different types of atoms—which were of the same element and belonging in the same place in the periodic table—*isotopes*, from Greek words meaning "same place." He received a Nobel prize for this work in 1921.

This was yet another blow to long-held views about atoms. Leucippus, Democritus, and Dalton had all assumed that all of the atoms of a particular element were identical. There hadn't seemed to be any observations that pointed to the contrary—until now. Scientists working with radioactive intermediates found as many as five or six varieties of an element, each with distinct radioactive properties.

Once the concept of atomic number became clear in 1914, it was possible to see the details of how one type of atom changed into another. Thus, the uranium atom has a nucleus with an atomic weight of 238, and an atomic number of 92. We call it U-238. However, in its radioactive transformation it gives off an alpha particle, which has an atomic weight of 4 and an atomic number of 2. The atomic weight and atomic number of the alpha particle must be subtracted from that of the uranium nucleus. What is left, then, is a nucleus with an atomic weight of 234, and an atomic number of +90. (An alpha particle, when emitted, always reduces the atomic weight of the emitting nucleus by 4 and the atomic number by 2.)

When this disintegration of the uranium nucleus was

discovered by Crookes, he called the product uranium X, which was just a way of saying he hadn't the slightest idea what it could be. But now, with the cold figures of the change, it could be seen that the new atom was thorium, all atoms of which, after all, have an atomic number of 90.

Ordinary, well-known thorium has an atomic weight of 232 and is, therefore, Th-232. The product of uranium disintegration has an atomic weight of 234 and is Th-234. Here we have an example of two isotopes. Both possess an atomic number of 90 and have a nuclear charge, therefore, of +90. There is, however, a difference in mass. Th-234 is two units more massive than Th-232.

Does this really make a difference? Chemically, it doesn't. Both Th-232 and Th-234, having a nuclear charge of +90, have 90 electrons in their atoms, arranged in the same way in each case, so that all chemical properties are the same. However, from the radioactive standpoint, it does make a difference. Thorium-232, the ordinary thorium we find in minerals, gives off alpha particles, while thorium-234, the product of uranium disintegration, gives off beta particles. Moreover, the atoms of thorium-234 disintegrate some 200 billion times as rapidly as the atoms of thorium-232. That's quite a difference.

There are other isotopes of thorium that turn up as part of one radioactive series or another. They include Th-227, Th-228, Th-229, Th-230, and Th-231. They all break down in different ways and at different rates, and all do so much more quickly than Th-232. But let's go back to thorium-234, as it gives off a beta particle. Does it change as a result?

A beta particle is an electron. It has a charge of -1, so it can be considered to have an atomic number of -1. Its mass is $1/1,837$ that of a hydrogen atom, or about 0.00054. This is so small a figure that we won't go far wrong if we consider it as just about 0. This means that if a nucleus emits a beta particle, one must subtract 0 from the nucleus's atomic weight—leaving it unchanged. We must also subtract -1 from the nucleus's atomic number. Subtracting -1 is equivalent to adding $+1$, so the atomic number goes *up* by 1. Therefore, the Th-234 nucleus with an atomic number of 90 and an atomic weight of 234, emitting a beta particle, is changed to a nucleus having an atomic number of 91 and an atomic weight of 234. The element of atomic number 91 is protactinium, which was first isolated and identified in 1917 by the German chemist Otto Hahn (1879–1968) and his co-worker, the Austrian chemist Lise Meitner (1878–1968). What we have, then, is Th-234 changed to Pa-234.

The emission of a gamma ray by an atomic nucleus does not change the nucleus. The gamma ray has an atomic number of 0 because it has no charge, and an atomic weight of 0 because it has no mass. A nucleus, in emitting a gamma ray, merely loses energy.

Once scientists knew how each of the radioactive radiations changed an atomic nucleus, they were able to work out the precise identity of all intermediates in a radioactive series.

The concept of isotopes left the periodic table intact. Each place contained only one type of atom as far as atomic number was concerned. That isotopes differed in atomic weight did not matter where chemical properties were concerned. What it meant in connection with nuclear structure and properties we will come to later.

Half-Lives

The various intermediates in a radioactive series break down quite quickly. If a given quantity of one of these intermediates is observed, it will be seen that the number of breakdowns declines with time. The reason is clear. As the atoms break down, fewer and fewer of the original variety are left to break down, and the fewer further breakdowns there are to be observed.

The manner in which the rate of breakdown declines is precisely what is to be expected of something that chemists were familiar with in the case of many chemical reactions. It is what is called a first-order reaction. This means that each radioactive atom of a particular variety has a certain chance of breaking down, and that this chance doesn't change with time. It might have one chance in two of breaking down on a particular day, but if a hundred days pass without its having broken down, it still has only one chance in two of breaking down on the 101st day. (This is analogous to the situation with respect to tossing a coin. You have one chance in two of tossing heads. Still, if you toss the coin a hundred times and get tails each time, the chance of tossing a head on the 101st time is still only one in two—assuming, of course, that it is an honest coin. It is often erroneously believed that the more times one tosses a tail, the greater the chance of a head on the next occasion.)

You can't tell when an individual atom will break down, but, if you are dealing with a great many atoms, you can calculate how many will break down in the course of a day, or a minute. You won't know *which* atoms will break down, but you will know the number. This is similar to the way in which statisticians can predict how many motorists are likely to die on a holiday weekend, even though they can't possibly tell which particular motorists will die.

This means that you can calculate how long it will take for half of all atoms present to break down. It turns out, in the case of first-order reactions, that it always takes the same time for half of any quantity to break down. Thus, if you start with 120 grams of a given isotope, and if it takes

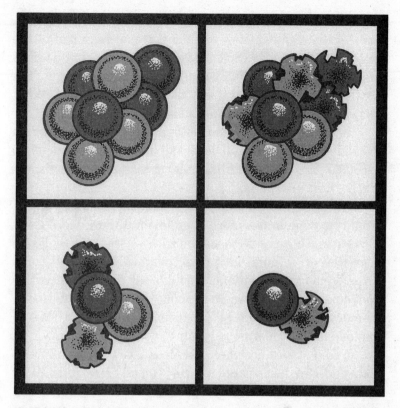

Half-life is the average time required for one-half of the atoms of a radioactive material to undergo radioactive decay.

a year for half of it to break down, it will take another year for half of the remaining half to break down. Put another way, starting with 120 grams, you will have 60 grams at the end of one year, 30 at the end of two, 15 at the end of three, 7.5 at the end of four, and so on. Theoretically, you will never get to zero, but, eventually you will have a single atom and, after some unpredictable period of time, it, too, will break down and your radioactive isotope will be gone.

In many cases, scientists can count the number of actual alpha particles or beta particles given off per unit of time. From the way in which that number falls off, they can calculate the amount of time it would take half of the isotope to be gone. (Scientists have worked out various ways of detecting individual alpha and beta particles, but I will discuss devices in this book only when I must. What I am concentrating on is ideas and concepts.)

Thus, the protactinium isotope obtained from uranium breakdown, Pa-234, will lose half its atoms in about 70 seconds. This is its half-life, a term introduced by Ernest Rutherford in 1904.

Naturally, protactinium-234, if it existed all by itself, would be gone in not much time, even if there were enormous quantities of it to begin with. If the entire Earth consisted of nothing but protactinium-234, and if it could be imagined that the atoms would break down quietly, this vast amount would be gone in about three hours. (Actually, so much energy would be produced that the Earth would explode like an enormous bomb.)

And yet protactinium-234 does exist in the Earth's soil, and can be isolated in very small quantities. Why isn't it *all* gone? The answer is that any such atoms that existed when the Earth was formed were gone a few minutes later; however, more are constantly being formed from uranium.

Other isotopes have longer half-lives. Radium-226 (the isotope isolated from pitchblende by the Curies), which emits alpha particles, has a half-life that is quite long, so that over short periods of time the decline in breakdown rate is too small to notice. If one waits long enough, however, the decline can be measured, and it turns out that the half-life is 1,620 years. But even this is not long enough for radium to endure for the lifetime of Earth. Radium exists only because it is constantly being formed from uranium.

Because uranium has a very slow rate of breakdown, radium forms very slowly. Radium, however, breaks down as it forms, but quite slowly at first because there is so little of it. As more and more accumulates, however, it breaks down faster and faster (characteristic of a first-order reaction), eventually breaking down as rapidly as it forms and reaching radioactive equilibrium.

It turns out that in any mineral that contains uranium, there will also be radium, with the amount of radium much smaller than uranium because of radium's shorter half-life. This is so even if uranium doesn't produce radium directly, but only through several other intermediate stages.

As it happens, the concentration of uranium in uranium ore is 2,780,000 times greater than that of radium, so the half-life of uranium-238 is 2,780,000 times longer than that of radium-226. This means that the half-life of uranium-238 is about 4.51 billion years.

This is why there is still primordial uranium on Earth. The Earth was first formed about 4.6 billion years ago, and included a certain amount of uranium in its makeup to begin with. In all that vast length of time, only about half of that primordial uranium has broken down. The other half is still here. It will take another 4.51 billion years for half of what is now left to break down. It is because uranium has been around all this time that the intermediates of its breakdown are in existence as well, although in much smaller quantities, of course.

Thorium-232 has an even longer half-life than uranium—13.9 billion years. Only about ⅕ of the original supply of thorium on Earth has yet had a chance to break down.

There is the uranium isotope, uranium-235, discovered in 1935 by the Canadian-American physicist Arthur Jeffrey Dempster (1886–1950). It is not nearly as long-lived as uranium-238 or thorium-232. The half-life of uranium-235 is only 710 million years. That is still long enough, however, to allow a little over ¹⁄₇₀ of the original quantity present at Earth's beginning to exist today.

Stable Nuclear Varieties

Soddy's discovery of isotopes involved only radioactive atoms, yet his discovery immediately cast suspicion on nonradioactive atoms. As early as 1905, the American chemist Bertram Borden Boltwood (1870–1927), noting that uranium minerals always seemed to contain lead, wondered if lead might not be the final product of radioactive disintegration. As investigation proceeded, this proved to be so. This meant that lead, although a nonradioactive element, was intimately involved with radioactivity.

The one way a radioactive atom ordinarily changes its atomic weight is by emitting an alpha particle. A beta particle affects the atomic weight insignificantly, and a gamma ray does so not at all. Every time an alpha particle is emitted, the atomic weight decreases by four. This means that if the original radioactive atom had an atomic weight evenly divisible by four, all of the intermediate products, without exception, would also have to have an atomic weight evenly divisible by four—as would the final lead atom. Thus, thorium-232, the only long-lived thorium isotope, has an atomic weight divisible by 4 ($232 = 58 \times 4$). As its breakdown proceeds, it loses a total of 6 alpha particles with a total atomic weight of 24, leaving an atomic weight of 208 for what remains of the nucleus. These six alpha particles also cause thorium-232 to lose a total of 12 positive charges; however, 4 beta particles are emitted, which restores 4 positive charges. The net loss in positive charges is, therefore, 8.

Thorium has an atomic number of 90. Losing 8 positive charges produces an atom with an atomic number of 82, which is that of lead. Consider the atomic weight loss of 24, and you see that the final product of the disintegration of thorium-232 is lead-208, which is not radioactive but sta-

ble, and of which there is a reasonable quantity on Earth—
and always has been, and always will be.

That's fine, so far, but consider uranium-238. Its atomic
weight, when divided by 4, leaves a remainder of 2 ($238 =
59 \times 4 + 2$). If it loses atomic weight by emitting alpha
particles, all of its intermediate products as well as its final
product will have atomic weights that when divided by 4
will leave a remainder of 2. An atom of uranium-238, in its
disintegration, loses 8 alpha particles and 6 beta particles,
which makes it lead-206.

Finally, uranium-235 has an atomic weight that, when
divided by 4, leaves a remainder of 3 ($235 = 58 \times 4 + 3$),
as do all of its intermediates and its final product. Each
uranium-235 atom gives off 7 alpha particles and 4 beta
particles, ending up as lead-207. (There is a fourth series
in which all of the atomic weights, when divided by 4, leave
a remainder of 1. We will have occasion to mention it later.)

We are left, then, with three different lead isotopes:
lead-206, lead-207, and lead-208. Each is stable, and each
possesses the usual properties of lead. Which of these, then,
if any, exists in nature independent of radioactivity?

Suppose we consider the atomic weight of lead. As
found in nature, in rocks that have no suspicion of radioactivity about them, lead has an atomic weight of 207.19.
In that the various stable isotopes are always present in a
fixed proportion, could this number simply be an average
atomic weight? (Because all of the various geological processes depend on the chemical properties of the various
minerals, they cannot separate isotopes as they separate
the various elements according to their chemical properties,
but leave the isotopes thoroughly mixed in the same proportions at all times.)

Let's test the proposition. Suppose you had a rock rich
in uranium. In addition to the original supply of lead, if
any, you would have a slow, constant addition of lead-206

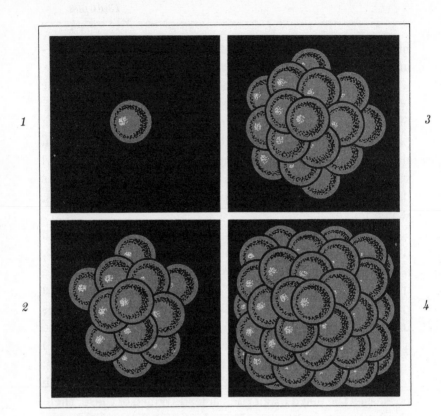

In 1815, William Prout suggested that all atoms are combinations of hydrogen atoms.
(1) Hydrogen 1.   (2) Carbon 12.   (3) Oxygen 16.   (4) Sulfur 32.

and lead-207, making the atomic weight of lead in the rock measurably less than it would be in nonradioactive rock. A rock rich in thorium would be undergoing slow, constant addition of lead-208, making the atomic weight of its lead content higher than it would be in nonradioactive rock.

In 1914, Richards measured the molecular weight of lead taken from various radioactive ores. He found that thorium minerals gave values as high as 207.9 for lead, while uranium minerals gave values as low as 206.01.

In the same year, then, that atomic numbers had replaced atomic weights as fundamental to the periodic table, it suddenly began to appear that atomic weights were not fundamental at all. They might merely be averages of isotopic weights (mass number), which might themselves be much more significant.

But, of course, lead isotopes were prepared through radioactive breakdown. Perhaps that is a special case. What about elements that have nothing whatever to do with radioactivity? There were hints on this point even before the existence of lead isotopes were virtually proved by Richards's findings.

Suppose we consider positive rays, which are streams of positively charged atoms possessing less than their normal number of electrons. (Sometimes positive rays contain no electrons, consisting simply of bare nuclei.) If these positive-ray particles are placed in electromagnetic fields, their paths curve away from the straight lines they normally move in. The extent of curvature depends on both the electric charge of the particles and their mass. If we're dealing with an element whose atoms all have the same number of electrons removed, then all of the particles that make up the ray have the same positive electric charge. Therefore, if we witness any deviation in the curvature of the ray path, it must be because of differences in particle mass—that is, in their atomic weights.

Suppose the gas in a tube were neon, all of whose atoms carried the same positive charge. If all of the atoms also had the same atomic weight (as had always been taken for granted since the atomic theory had been established), they would all curve along the same path. If photographic film were placed in the way of the speeding particles, they would all strike the film at the same place, forming a small, fogged spot.

Thomson tried the experiment in 1912 and found that the neon ions did fog the photographic plate nearly in the expected position; however, very close to it there was a second, considerably less prominent spot of fogging. The position at which the second fogging took place was about that expected of atoms with an atomic weight of 22. No atomic weight of this size was expected, but Thomson suggested that if, out of every ten neon atoms, nine had a mass number of 20 and one of 22, the weighted average of the two would come out to 20.2, which was close to the measured atomic weight of neon as it occurs on Earth. In other words, neon, which had nothing to do with radioactive processes, might be made up of two isotopes: neon-20 and neon-22. This possibility suddenly opened up a new view of nuclear structure.

In 1815, in the infancy of the atomic theory, the British chemist William Prout (1785–1850) had suggested (anonymously because the idea was too far-out for him to dare attach his name to it) that all atoms are combinations of hydrogen atoms. Atomic weights were just being determined, and they seemed to be whole numbers. That is, hydrogen was 1, carbon was 12, oxygen was 16, sulfur was 32, and so on. Prout suggested that the carbon atom was made up of 12 hydrogen atoms in close association, the oxygen atom of 16 hydrogen atoms, the sulfur atom of 32 hydrogen atoms, and so on.

This suggestion was called Prout's hypothesis when the authorship was revealed. It didn't stand up, however, for as atomic weights were determined more and more accurately, it turned out that they were by no means all integers, or even close to them. Chlorine, for instance, was 35.456; copper was 63.54; iron was 55.85; magnesium was 24.31; mercury was 200.59; and so on.

Prout's hypothesis, if true, would have made the atomic theory much more elegant; that is, simpler and neater. However, observations had forced its abandonment for a century. Now it was suddenly back in the forefront of thinking.

If a positively charged ionic stream of an element is placed in an electromagnetic field, the less massive isotopes will be more easily deflected from their usual straight-line path. In a mass spectrometer, this effect will produce closely spaced dark lines on a photographic plate. The position of the line allows the mass number of the isotope to be calculated, and the darkness of the line indicates the relative amount of that isotope.

What if all atomic weights that were not integers were simply the averages of the mass numbers of various isotopes—mass numbers that *were* integers. If so, atomic weights might be useful in chemical calculations, but it would be the isotopic mass numbers that would be useful in considering nuclear structure.

In 1919, the British chemist Francis William Aston (1877–1945), who had been a student of J. J. Thomson's, devised what he called a mass spectrograph. This caused charged ions with the same charge and mass to be concentrated in a fine line on a photographic plate. In this way, the presence of isotopes could be seen as closely spaced dark lines. The position of a line allowed the mass number of an isotope to be calculated, and the darkness of the line the relative amount of that isotope. The results would be much more precise than those obtained by Thomson's groundbreaking but crude instrument.

Using such a mass spectrograph, the lines for neon-20 and neon-22 were clearly detected—as was, eventually, a very faint line for neon-21. We now know that out of every 1,000 neon atoms, about 909 are neon-20, 88 are neon-22, and 3 are neon-21. All three isotopes are stable, with the weighted average of their mass numbers giving neon, as found in nature, an atomic weight of 20.18. For his work with the mass spectrograph, Aston received a Nobel prize in 1922.

Other elements were, of course, tested, and a majority of them found to consist of several isotopes. Chlorine, for instance, is made up of two isotopes: chlorine-35 and chlorine-37. Of every 1,000 chlorine atoms, 755 are chlorine-35 and 245 are chlorine-37, making the weighted average of their mass numbers just about the atomic weight of chlorine as it exists in nature. (The weighted average doesn't come out *exactly* to the measured atomic weight because the mass numbers, as we shall see, aren't *quite* integers, either.)

Sometimes one isotope is present in an overwhelming majority. Out of every 1,000 atoms of carbon, for instance, 989 are carbon-12 and 11 are carbon-13. Out of every 1,000 nitrogen atoms, 996 are nitrogen-14 and 4 are nitrogen-15. Out of every 10,000 hydrogen atoms, 9,999 are hydrogen-1 and 1 is hydrogen-2. Out of every million helium atoms, all but one are helium-4, the odd one being helium-3. In all of these cases, the atomic weight is close to an integer.

In 1919, the American chemist William Francis

Giauque (1895–1982) discovered that out of every 10,000 oxygen atoms, 9,976 are oxygen-16, 20 are oxygen-18, and 4 are oxygen-17. What made the finding significant was that oxygen had been used as the standard for atomic weight since Berzelius's time, its atomic weight being set exactly equal to 16.0000. However, that was now seen to be only an average that could vary slightly from sample to sample. In 1961, therefore, physicists and chemists officially agreed to tie the standard to a mass number and not to an atomic weight. The mass number of carbon-12 was set to 12.0000 exactly, which shifted the old atomic weights only very slightly. For instance, the atomic weight of oxygen is no longer 16.0000, but 15.9994.

Some atoms come in only one variety in nature. Thus, all fluorine atoms in nature have a mass number of 19, all sodium atoms 23, all aluminum atoms 27, all cobalt atoms 59, all gold atoms 197, and so on. For these cases, many physicists believed the word *isotope* to be inappropriate. Isotope implies that at least two atomic varieties are packed into the same place in the periodic table. To say that an element has one isotope is like saying that a parent has one twin offspring. In 1947, therefore, the American chemist Truman Paul Kohman (b. 1916) suggested that the term *nuclide* be used instead. It is a perfectly good term, but isotope has become too firmly established to be displaced.

There are eighty-one elements, each of which has at least one stable isotope. Of these, the most complicated is bismuth, which has an atomic number of 83. All of its atoms have a mass number of 209. The most massive stable atom, then, is that of bismuth-209.

There is no stable atom with an atomic number greater than 83 or an atomic weight greater than 209. More massive atoms exist on Earth only because uranium-238, uranium-232, and thorium-232, although radioactive, are very long-lived.

The total number of stable isotopes distributed among

the eighty-one elements is 272, which is enough to supply three or four per element, if they were distributed evenly. They are not, of course. The elements with even atomic number generally have a greater than average supply. Tin, with an atomic number of 50, holds the record with ten stable isotopes, with mass numbers of 112, 114, 115, 116, 117, 118, 119, 120, 122, and 124.

Elements with odd atomic number generally have either one or two stable isotopes. There are nineteen elements (all but one with an odd atomic number) that are made up of a single stable isotope. The one exception, with an even atomic number, is beryllium (atomic number 4), which has a single stable isotope, beryllium-9.

You might wonder why it is that there are only eighty-one elements with stable isotopes, if elements numbers 82 (lead) and 83 (bismuth) have them. Clearly, there must be two elements without them in the list of elements between 1 and 83, and this is so. Elements 43 and 61 (both odd atomic numbers) do not have any stable or nearly stable isotopes. They were sought for diligently in the 1920s and, occasionally, they were reported to have been isolated from one ore or another, but all such reports turned out to be mistaken. Neither element was actually isolated until scientists learned to form nuclei in the laboratory that don't exist in measurable quantities on Earth itself. (We'll get to such matters later on.)

Another peculiarity is that potassium (atomic number 19) is the only element of odd atomic number to have more than two isotopes occurring in nature. It has three, with mass numbers 39, 40, and 41. Of these three, however, potassium-40 makes up only about 1 out of every 10,000 potassium atoms.

As early as 1912, Otto Hahn had noted that potassium seemed to be weakly radioactive, and, eventually, this was narrowed down to potassium-40. Potassium-40 is long-lived, with a half-life of 1.3 billion years. This is a longer half-life than that of uranium-235. Less than a tenth of the quantity present in the Earth when it first formed still exists today. Potassium, however, is so common an element in Earth's rocks that even though only 1 out of 10,000 potassium atoms is potassium-40, there is more potassium-40 in rocks than uranium-238 and uranium-235 combined.

If this is so, why wasn't radioactivity discovered in potassium sooner than it was in uranium? The answer is that, first, uranium emits energetic alpha particles, and potassium-40 only rather feeble beta particles. Second, uranium breaks down into a long series of intermediates, each of which is more strongly radioactive than uranium itself. Potassium-40, on the other hand, breaks down directly into a stable isotope, argon-40.

Potassium-40 is not the only nearly stable isotope among those we have listed as stable. There are about a dozen others, all of which have half-lives much longer than potassium-40, or even than thorium-232. They are so long-lived that their radioactivity can barely be detected. Vanadium-50, for instance, has a half-life of about 600 trillion years, which is some 130,000 times as long as that of uranium-238; neodymium-154 has a half-life of about 5,000 trillion years, and so on. None of these nearly stable isotopes with atomic numbers less than that of thorium (90) gives rise to a series of breakdown products. All but one give up a single beta particle and become a stable isotope. Samarium-147 is the exception, giving up a single alpha particle to become the stable neodymium-143.

The fact that the mass numbers of isotopes are all very close to integers made it very tempting to think that nuclei are (as Prout had suggested) made up of smaller particles, and that the various particles to be found within nuclei might be very few. The chance of simplifying nature in this way was very enticing, and in the 1920s physicists worked hard at puzzling out the structure of the atomic nucleus.

6

NEUTRONS

Protons and Electrons

The desire for simplicity was not the only force driving physicists to look into the realm of nuclear structure. From actual observations of radioactive materials, it seemed clear that at least some nuclei *must* have a structure; that is, be a collection of still simpler particles. Thus, some radioactive nuclei gave off beta particles (electrons) and others gave off alpha particles (helium nuclei). The simplest explanation for these emissions was that nuclei contained within themselves simpler nuclei and electrons, which were, for some reason, occasionally released.

If we are satisfied that some nuclei are made up of smaller nuclei plus electrons, then it is an easy jump to speculate that all nuclei might possess this structure. For simplicity's sake, we might also suppose that nuclei, when composed of simpler nuclei, should be composed of the simplest nuclei possible.

The smallest nucleus known was that of hydrogen-1, a nucleus with a mass number of 1 and an electric charge of +1. Rutherford had called the hydrogen-1 nucleus a proton, and there was the general feeling in the 1920s that the proton was the smallest and simplest particle capable of carrying a positive charge. There arose the theory, then, that atomic nuclei might be made up of protons and electrons crushed together into a tiny volume.

The alpha particle, which emerges from some radioactive atoms, has a mass number of 4, so it could be made up of 4 protons, each with a mass number of 1. However, the alpha particle also has an electric charge of +2, and 4 protons have a total charge of +4. It would seem, then, that in addition to the 4 protons in the alpha particle, there must be 2 electrons, canceling two of the positive charges while adding nothing significant to the mass. A 4-proton/2-electron alpha particle would then have, as observed, a mass of 4 and a charge of +2.

This sort of thing could be worked out for other nuclei, too. It could be used to explain isotopes. For instance, the oxygen-16 nucleus has a mass of 16 and a charge of +8, so it should consist of 16 protons and 8 electrons. The oxygen-17 nucleus could be viewed as having an additional proton-electron pair, which increases the mass by 1, without changing the charge. A total of 17 protons and 9 electrons yields a mass of 17 and a charge of +8. Again, the oxygen-18 nucleus can be viewed as having still another proton-neutron pair, so that it would be made up of 18 protons and 10 electrons, with a mass of 18 and a charge of +8.

For a while, physicists rode high on this proton-electron theory of nucleus structure, particularly because it

Protons and neutrons are both present in the nucleus.

reduced the universe to extraordinary simplicity. All material objects, so the theory stated, in the universe are made up of about 100 types of atoms, and every atom, in this view, is made up of equal numbers of two types of subatomic particles: protons and electrons. All protons are located in the nucleus, while some electrons are located in the nucleus and others outside the nucleus.

Furthermore, it seemed that the entire universe was held together by two fields. The nucleus was held together by the electromagnetic attraction between protons and electrons; the atom, as a whole, was held together by the electromagnetic attraction between nuclei and electrons. Various atoms were combined to form molecules, or crystals, or solid objects as large as planets by the transfer of electrons from one atom to another, or by the sharing of electrons. Was there anything not held together by electromagnetic fields? Yes, of course.

The molecules of gases are scattered far apart and are subjected to only very feeble electromagnetic forces. But if this were the only force exerted on them, the molecules would dissipate and distribute themselves throughout the vastness of space. Gases, however, are held to a large body through the influence of something else: gravitational attraction. That is why our atmosphere clings to the Earth.

The gravitational field, however, is so weak that it takes a very large body to hold gases. For instance, low-boiling liquids on Earth would tend to evaporate and their molecules flee into space if gravitation were not strong enough. It is because of Earth's gravitational pull that we have an ocean, while the Moon is not large enough to have free water on its surface.

Bodies separated by considerable distances in space are also held together by gravitational fields: satellites to planets, planets to stars, stars to one another to form galaxies, and galaxies to one another to form clusters. The Universe as a whole is, indeed, held together by gravitational attraction.

Add to this the fact that the electromagnetic field is associated with a radiation of photons, and the gravitational field with a radiation of gravitons, and it would seem that the entire Universe consists of but four types of particles: protons, electrons, photons, and gravitons. Protons have a mass number of 1, a charge of $+1$, and a spin of $+\frac{1}{2}$ or $-\frac{1}{2}$. Electrons have a mass number of 0.00055, a charge of -1, and a spin of $+\frac{1}{2}$ or $-\frac{1}{2}$. Photons have a mass number of 0, a charge of 0, and a spin of $+1$ or -1. Gravitons have a mass number of 0, a charge of 0, and a spin of $+2$ or -2.

How simple that is! It is even simpler than the Greek

notion of four elements that applied to Earth and a fifth element assigned to the heavenly bodies. As a matter of fact, the Universe was never again to seem quite as simple as it seemed for a few years in the 1920s.

Indeed, there was a mighty attempt to make it simpler still. Why should there be two fields: electromagnetic and gravitational? Might these not be two aspects of the same phenomenon? Might not a single set of equations describe both?

To be sure, the electromagnetic field and the gravitational field seem utterly different. The electromagnetic field involves only electrically charged particles, while the gravitational field involves all particles with mass, charged or not. The electromagnetic field involves both attraction and repulsion, while the gravitational field involves only attraction. The electromagnetic field is trillions of trillions of trillions of times as intense as the gravitational field for a given pair of particles that respond to both. Thus, in considering a proton-electron pair, we need only take into account the electromagnetic attraction between them; the gravitational attraction is, in comparison, insignificant.

Nevertheless, such differences need not be a bar to unification. Magnetism, electricity, and light seemed, at first, to be three widely different phenomena, and yet Maxwell found a set of equations that held for all three and showed them to be different aspects of the same phenomenon.

None other than Einstein spent the final decades of his life trying to complete the work of Maxwell by finding still more fundamental equations that would include the gravitational field as well, in what was called a unified field theory. He failed, but, as we shall see, that did not end the attempts.

The proton-electron system of nuclear structure itself did not hold up. It contained a fatal flaw.

Protons and Neutrons

The nucleus has a spin, just as electrons, protons, photons, and gravitons do. The amount of spin can be determined by a close study of the fine lines of the spectrum produced by given nuclei, and by other methods.

If the nucleus is made up of constituent particles such as protons and electrons, then it stands to reason that the total nuclear spin is the sum of the spins of the constituent particles. This is because the spins represent angular momentum, and for a long time physicists have found that there is a law of conservation of angular momentum. In other words, you can't create spin out of nothing, or destroy it. It can only be transferred from one body to another.

This holds for all ordinary bodies as far as the law can be tested. An ordinary spinning object (such as a coin you set spinning with your hand) might seem to get its spin from nowhere. The spin, however, comes from the motion of your hand, and when you twist the coin, your hand, the rest of your body, and whatever you are attached to—a chair, the ground, the planet Earth—gets a reverse twist. (Angular momentum can be in either of two directions, plus and minus, and the two can cancel each other out. In addition, one can be formed out of nothing if the other is simultaneously formed. It is the *net* angular momentum— what you get when all the pluses and minuses are added together—that is conserved.)

The amount of angular momentum depends not only on the speed of turning, but on the mass of the turning object. When you twist your hand to set a coin spinning, the Earth is so much more massive than the coin that its reverse twist takes place at a speed far too small to measure by any conceivable method. And when the coin slows its spin through friction with a surface and finally stops spinning, the Earth's reverse spin, incomprehensibly small, also slows and stops.

In the case of spinning particles, the spin is potentially eternal if the particles are left to themselves. Both protons and electrons have spins that can be represented by half-integers, $+\frac{1}{2}$ and $-\frac{1}{2}$. (The total spin is the same for both particles despite the difference in mass. The electrons just spin faster to make up for their lesser mass. Of course, the spin can be in either direction.)

If the spins of an even number of protons and electrons within a nucleus are added, the total spin has to be zero or an integer. Two spins, for instance, can be $+\frac{1}{2}$ and $+\frac{1}{2}$, or $+\frac{1}{2}$ and $-\frac{1}{2}$, or $-\frac{1}{2}$ and $+\frac{1}{2}$, or $-\frac{1}{2}$ and $-\frac{1}{2}$. The sums are respectively $+1$, 0, 0, and -1. If you imagine four half-integer spins, or six, or eight, or any even number and add them with whatever combination of pluses and minuses you please, you will always come out with zero, a positive integer, or a negative integer.

If you have an odd number of particles, each with a half-integer spin, you end up with a half-integer spin no matter how you shift the pluses and minuses. If there are three particles, for instance, you can have $+\frac{1}{2}$ and $+\frac{1}{2}$ and $+\frac{1}{2}$, for a total of $+1\frac{1}{2}$; or $+\frac{1}{2}$ and $+\frac{1}{2}$ and $-\frac{1}{2}$, for a total of $+\frac{1}{2}$. However you shift the pluses and minuses for three particles, or five, or seven, or any odd number, you always end either with $+\frac{1}{2}$, $-\frac{1}{2}$, $+$ some integer and $\frac{1}{2}$, or $-$ some integer and $\frac{1}{2}$.

This brings us to the nitrogen-14 nucleus, which was shown by spectroscopic studies to have a spin of either $+1$ or -1. The nitrogen-14 nucleus has a mass number of 14 and an electric charge of $+7$. By the proton-electron scheme of nuclear structure, its nucleus must be made up of 14 protons and 7 electrons, or 21 particles altogether. However, 21 particles, being an odd number, must add up to a total spin of a half-integer. They cannot add up to either $+1$ or -1.

This bothered physicists very much. They didn't want to give up the proton-electron picture of the nucleus because it was so simple and explained so much, but they didn't want to give up the law of angular conservation, either.

As early as 1920, some physicists, notably Ernest Rutherford, wondered if a proton-electron combination could be viewed as a single particle. It would have the mass of a proton (or a tiny bit more, thanks to the electron), with an electric charge of zero.

Of course, you couldn't consider such a particle to be just a proton and electron fused together because each would contribute a spin of $+\frac{1}{2}$ or $-\frac{1}{2}$ to the fusion, whereby the fused particle would have a spin of either 0, $+1$, or -1. The spin of the nitrogen nucleus would then still be a half-integer altogether, whether you count the protons and electrons separately, or in combinations with each other.

You must instead think of a particle that has a mass of 1, like a proton, a charge of 0, and a spin of $+\frac{1}{2}$ or $-\frac{1}{2}$. Only so could the requirement of the nitrogen nucleus be met. In 1921, the American chemist William Draper Harkins (1871–1951) applied the name *neutron* to such a particle, because it was electrically neutral.

Throughout the 1920s, this possibility remained in the minds of physicists, but because the hypothetical neutron was never actually detected, it was difficult to take the position seriously. The proton-electron scheme continued in use, therefore, even though it didn't fit all of the facts of reality. (Scientists don't generally abandon a notion that seems useful until they are sure they have a better one to put in its place. Replacing something useful with nothing, or with something very vague, is not a good idea in scientific procedure.)

In 1930, the German physicist Walter W.G.F. Bothe (1891–1957) reported that when he bombarded the light element beryllium with alpha particles, he got a radiation of some sort. It was very penetrating and it didn't seem to carry an electric charge. The only thing he could think of that had these characteristics were gamma rays, so he suspected that that was what he had.

In 1932, the French physicist Frédéric Joliot-Curie (1900–1958) and his wife, Irène Joliot-Curie (1897–1956), the daughter of Pierre and Marie Curie, found that Bothe's radiation, when it struck paraffin, brought about the ejection of protons from that substance. Gamma rays were not known to do that, but the Joliot-Curies could not think of any other explanation.

The British physicist James Chadwick (1891–1974), however, repeated the work of Bothe and the Joliot-Curies, in 1932, and reasoned that in order for the radiation to eject a massive particle such as a proton, it had to consist of massive particles itself. Because it certainly didn't carry an electric charge, he decided that here was the massive, neutral particle physicists were searching for—the neutron. That was indeed what it was, and in 1935 Chadwick received a Nobel prize for his discovery.

Once the neutron was discovered, Heisenberg immediately suggested that the atomic nucleus consisted of a closely packed mass of protons and neutrons. The nitrogen nucleus, for instance, was made up of 7 protons and 7 neutrons, each with a mass number of 1. The total mass number was, therefore, 14, and because only the protons had a $+1$ charge while the neutrons had a charge of 0, the total charge was $+7$, as it was supposed to be. Moreover, there were now 14 particles altogether—an even number—thus the total spin of the nucleus could be either $+1$ or -1, as measured.

It turned out that the proton-neutron structure explained, without exception, the nuclear spin of all atomic nuclei, while at the same time explaining everything the proton-electron structure had explained (with one exception that was later filled in, as I will eventually explain). Indeed, in the better than half a century since the discovery of the neutron, nothing has been found that would in the least shake the proton-neutron structure of the nucleus, although there have been refinements of the idea, which we will come to later.

Consider, for instance, how neatly the new notion explains the existence of isotopes. All of the atoms of a given element have the same number of protons in the nucleus and, therefore, the same nuclear charge. The number of neutrons, however, might differ.

Thus, the nitrogen-14 nucleus is made up of 7 protons and 7 neutrons, but one out of 3,000 nitrogen nuclei contains 7 protons and 8 neutrons and is, therefore, nitrogen-15. Although the most common oxygen nucleus contains 8 protons and 8 neutrons in its nucleus, which makes it oxygen-16, a few have 8 protons and 9 neutrons, or even 8 protons and 10 neutrons (oxygen-17 and oxygen-18, respectively).

Even hydrogen, with a nucleus made up of a single proton and nothing else (hydrogen-1), is not immune. In 1931, the American chemist Harold Clayton Urey (1893–1981) showed that 1 out of 7,000 hydrogen atoms was hydrogen-2, and he received a Nobel prize in 1934 for this work. The hydrogen-2 nucleus consists of 1 proton and 1 neutron, which is why it is often called deuterium, from the Greek word for "second."

Similarly, uranium-238 has a nucleus made up of 92 protons and 146 neutrons, while uranium-235 has one made up of 92 protons and 143 neutrons. There isn't an isotope of any type that doesn't fit in perfectly with the proton-neutron nuclear structure.

Protons and neutrons are both present in the nucleus (they are sometimes lumped together as nucleons), both are of almost equal mass, and both can, under the proper conditions, be ejected from the nucleus. Yet the proton was recognized as a particle in 1914, while the neutron had to wait an additional eighteen years for its discovery. Why did it take so long to discover the neutron? The reason is that electric charge is the most easily recognized portion of a particle, and while the proton carries an electric charge, the neutron does not.

One of the earliest ways of recognizing the existence of subatomic particles was by means of the gold-leaf electroscope. This device consists of two thin and very light sheets of gold leaf attached to a rod and enclosed in a box designed to protect the assemblage from disturbing air currents. If an electrically charged object is touched to the rod, the charge enters the gold leaf. Because both sheets of gold leaf receive the same charge, they repel each other and move apart in an inverted V.

Left to itself, the electroscope would remain with its leaves apart. Any stream of charged particles entering the electroscope, however, will knock electrons from molecules of air. This produces negatively charged electrons and positively charged ions (a stream of charged particles is an example of ionizing radiation). One or the other of these charged particles will neutralize the charge on one of the gold leaves, causing them to come slowly together. A stream of neutrons, however, is not an example of an ionizing radiation because, being uncharged, they neither attract nor repel the electrons out of atoms and molecules. Neutrons, therefore, cannot be detected by the electroscope.

The German physicist Hans Wilhelm Geiger (1882–1945) invented a device, in 1913, that consisted of a cylinder containing a gas under a high electric potential, but one that was not quite high enough to force a spark of electricity through the gas. Any bit of ionizing radiation entering the cylinder would form an ion, which would be pulled through the cylinder by the electric potential, creating more ions.

Even a single subatomic particle produced a discharge that would make a clicking sound. The Geiger counter became famous as a way of counting subatomic particles.

Even earlier, in 1911, the British physicist Charles Thomson Rees Wilson (1869–1959) invented a cloud chamber. He allowed dust-free moist air to expand in a cylinder. As it expanded, it cooled, and some of the moisture would condense into tiny droplets, provided there were dust particles present as centers about which the droplet could form. Without dust particles, water remains in vapor form. If a subatomic particle entered the cloud chamber, it would form ions all along its path, and these would act as water condensation centers. About each ion a tiny water droplet would form. In this way, not only the particle, but its pathway as well, could be detected. If the cloud chamber were placed in an electric or magnetic field, the speeding charged particle would curve in response—its path visible. Wilson received a Nobel prize in 1927 for this device.

In 1952, the American physicist Donald Arthur Glaser (b. 1926) invented a similar device. Instead of a gas out of which liquid droplets were ready to form, Glaser used a liquid raised to a temperature at which vapor bubbles were about to form. Those bubbles formed along the pathway of an entering subatomic particle. For this "bubble chamber," Glaser received a Nobel prize in 1960.

All of these devices, and many others like them, respond to the formation of ions by ionizing radiation; that is, by electrically charged particles. None of them will work for neutrons, which enter and leave the devices silently, so to speak.

The presence of neutrons can be detected only indirectly. If a neutron formed inside a detecting device travels a distance and then collides with some other particle that *can* be detected—provided the neutron alters the pathway of the other particle or forms new, detectable particles—

there will be a gap between the pathways that mark the formation of the neutron at one end and the collision of the neutron with something else at the other. That gap has to be filled with something, and from the nature of the two sets of pathways that can be seen, it is logical to deduce the presence of a neutron in between.

Physicists who work with particle-detecting devices learn to photograph complex pathways marked out in drops of water, bubbles of gas, lines of sparks, and so on, and interpret all of the details as easily as we might read this book.

It is because neutrons leave no marks in these devices that caused their discovery to be delayed for so many years. Once found, however, they proved to be of enormous importance, as we will see if we now move back a little in time.

Nuclear Reactions

All of the innumerable interactions of atoms and molecules that involve transfers and sharings of electrons are called chemical reactions. Until 1896, all of the interactions scientists knew about, either in living tissue or in inanimate nature, were chemical reactions, although their nature was not really understood, of course, until the structure of atoms came to be known.

In this respect, radioactivity is different. The changes involved in radioactivity involve the ejection of portions of the nucleus, or of changes in the nature of particles within the nucleus. Such events are called nuclear reactions, which, in general, involve much greater intensities of energy change than do chemical reactions.

Radioactivity is a spontaneous nuclear reaction. If

there weren't just a few spontaneous nuclear reactions, those that take place without any initiation or interference by human beings, it is possible we might never have discovered the existence of such things.

It is, after all, much more difficult for human beings to initiate or control nuclear reactions than it is to do the same with chemical reactions. To produce, prevent, or modify a chemical reaction, chemists need only mix chemicals, or heat them, cool them, put them under pressure, blow air through them, or carry out other easily managed procedures. After all, it is only the outer electrons that are involved, and they are so exposed that we can easily fiddle with them.

Nuclear reactions take place in the tiny nuclei at the very center of atoms—nuclei that are shielded by numbers of electrons. All of the procedures used to bring about chemical change do not reach or affect the atomic nuclei. Thus, when radioactivity was first discovered, chemists were astonished to find that the rate of breakdown was not altered by temperature change. Whether a radioactive substance was heated to melting, or placed in liquid air, the rate of radioactivity continued as before, unchanged. Subjecting a radioactive substance to chemical change did not alter the rate of radioactive breakdown, either.

Was there any way, then, of interfering with the nucleus at all? If such a method existed, it would have to involve reaching through the electronic shield and, so to speak, touching the nucleus itself. It was precisely in this way that Rutherford had discovered the existence of the nucleus. He had bombarded atoms with energetic alpha particles, which were massive enough to brush electrons aside, and small enough to bounce away from a nucleus as it approached.

In 1919, Rutherford placed a bit of radioactive material at one end of a closed cylinder, with an inner coating of

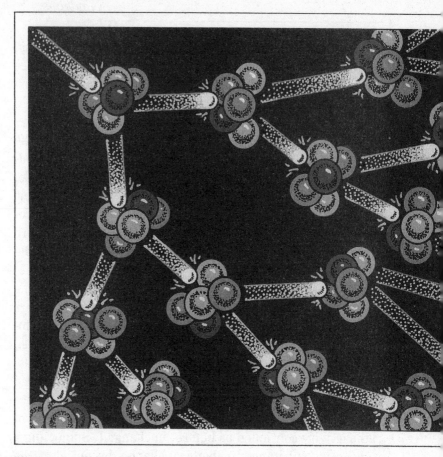

Neutrons collide with nuclei, causing more and more neutrons to be emitted in a nuclear chain reaction.

zinc sulfide at the other end. The radioactive material emitted alpha particles. Whenever an alpha particle struck and was stopped by the zinc sulfide, the alpha particle lost its kinetic energy, which was converted into a tiny flash of light that could just be seen if the room were kept very dark and the eyes were allowed to adapt to the darkness. By counting the scintillations of light, Rutherford and his co-workers could actually count the individual particle strikes. Such a device is called a scintillation counter.

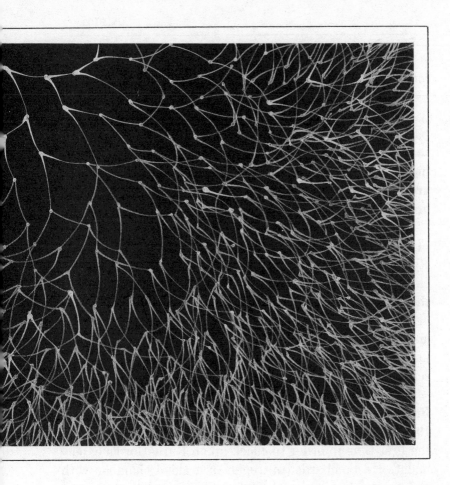

If alpha particles are passed through a vacuum, the scintillations are many and bright. If, however, some hydrogen is allowed in the cylinder, particularly bright scintillations appear. This would seem to be because the alpha particles occasionally strike the proton nuclei of the hydrogen, and the protons, being lighter than the alpha particles, can be knocked forward at a greater speed. Speed counts for more than mass where kinetic energy is concerned, so the very speedy protons produce very bright scintillations.

If oxygen or carbon dioxide are allowed into the cylinder, the scintillations grow dimmer and fewer. The comparatively massive nuclei of the oxygen and carbon atoms (respectively, four and three times the mass of the alpha particles) tend to slow down the alpha particles, sometimes to a point at which they pick up electrons and become ordinary helium atoms. The massive carbon and oxygen nuclei are knocked forward slowly, and what scintillations there are, are dim.

If, however, nitrogen is placed in the cylinder, the bright scintillations one observes with hydrogen appear. Rutherford supposed that the particles in the nitrogen nucleus were less firmly bound together than those in either the carbon or oxygen nucleus. The alpha particle could bang against the carbon or oxygen nuclei without shaking them apart, but when it collided with a nitrogen nucleus, it would knock a proton out of the nucleus, causing the usual proton scintillation.

This was only speculation at first, but, in 1925, the British physicist Patrick Maynard Stuart Blackett (1897–1974) put the Wilson cloud chamber to wholesale use for the first time in order to check up on Rutherford's experiment. He bombarded nitrogen in a cloud chamber with alpha particles, and took 20,000 photographs, catching a total of more than 400,000 alpha particle tracks. Of these, just eight involved a collision between an alpha particle and a nitrogen molecule.

By studying the tracks going into the collision and coming out of it, Blackett showed that Rutherford was right, that a proton had been knocked out of the nitrogen nucleus. The alpha particle, with a charge of $+2$, had entered the nucleus, and a proton, with a charge of $+1$, had left the nucleus. That meant the nucleus had a net increase of $+1$ in its charge. Instead of being $+7$ (nitrogen), it had become $+8$ (oxygen). Furthermore, the alpha particle had entered

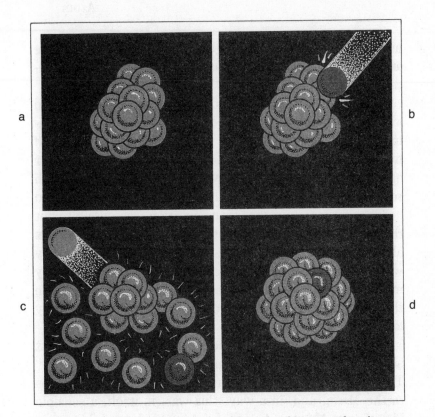

The alpha particle with a mass number of 4 (b) enters the nitrogen nucleus (a & b), and a proton with a mass number of 1 is expelled (c). The net result is that nitrogen-14 (a) combines with helium-2 (an alpha particle) (b), to throw off hydrogen-1 (a proton) (c), and produces oxygen-17 (d).

with its mass number of 4, and a proton had left with its mass number of 1. The nitrogen nucleus had gained 3 in its mass number, increasing from 14 to 17. The net result was, then, that nitrogen-14 had combined with helium-2 (an alpha particle) to yield oxygen-17 and hydrogen-1 (a proton).

Rutherford, therefore, was the first to bring about a nuclear reaction in the laboratory; that is, he was the first to bring about the change of one element into another—

nitrogen into oxygen—through human agency. For the use of the cloud chamber in this case and others, Blackett received a Nobel prize in 1948.

In a way, Rutherford had brought about what the old alchemists would have called the transmutation of elements, and there were some people who, upon hearing of his work, said, "See, the old alchemists were right, after all. Modern scientists were wrong to dismiss them with scorn." This view is wrong, however. The alchemists did not merely maintain that transmutation was possible, but thought it could be brought about by chemical means alone—by mixing, heating, distilling, and so on. In this they were wrong. Transmutation can be brought about only by nuclear reactions, which were beyond anything possible to the old alchemists and beyond anything they had any idea of.

In short, an idea is not enough. Significant details must also be correct before one can get credit for "being right." Thus, there were people before Newton who talked of trips to the Moon, and, in a way, this was, in itself, a sensible idea. It was Newton, however, who first showed that a trip to the Moon could only be achieved by means of the rocket principle. It was he, not his predecessors, who should get the credit—not for a mere dream, but for a dream that included a practical pathway to its fulfillment.

Artificial Isotopes

Rutherford had changed one isotope known in nature, nitrogen-14, into another known isotope, oxygen-17. After the possibility of such laboratory transmutations had been established, other nuclear reactions, producing other known isotopes, were brought about by bombarding various types of atoms with speeding particles.

But need the changes always produce isotopes that were already known? Might not adding and subtracting particles produce nuclei with mass numbers and charges not quite like any that occurred naturally? In 1932, the Latvian-American chemist Aristid V. Grosse (b. 1905) suggested that it might be possible to do this.

In 1934, the Joliot-Curies were continuing Rutherford's work on the bombardment of various elements by alpha particles. They were bombarding aluminum in this fashion and not only knocked protons out of the aluminum nuclei, but in some cases knocked neutrons out of them, too. When the bombardment was over, the streams of protons and neutrons that emerged from the aluminum nuclei stopped at once. To their astonishment, however, some sort of radiation (of a type that will be taken up later in the book) continued, and declined with time just as the radiation intensity of a radioactive substance might be expected to decline. They could even calculate the half-life of the radiation as being 2.6 minutes.

All aluminum atoms in nature have an atomic number of 13 and a mass number of 27. In other words, their nuclei are all composed of 13 protons and 14 neutrons. If you add an alpha particle (2 protons and 2 neutrons) and knock out a proton, the new nucleus contains 14 protons and 16 neutrons, which is silicon-30, a well-known isotope.

But what about the cases in which a neutron is knocked out of the nucleus? If, to an aluminum nucleus (13 protons and 14 neutrons) you add an alpha particle (2 protons and 2 neutrons) and knock out a neutron, you end up with a new nucleus consisting of 15 protons and 15 neutrons. This is phosphorus-30. Phosphorus-30 does not, however, occur in nature. All phosphorus atoms in nature are phosphorus-31 (15 protons and 16 neutrons), which is the only stable phosphorus isotope. Phosphorus-30 is radioactive and quickly breaks down (by a method we'll describe later) into the stable silicon-30.

Phosphorus-30 was the first "artificial" isotope produced, and it introduced the concept of artificial radioactivity. The Joliot-Curies shared a Nobel prize in 1935 for this work.

After the Joliot-Curies had shown the way, a great many different artificial isotopes were formed through nuclear reactions of one sort or another. Every one of them was radioactive, and so one spoke of radioactive isotopes or radioisotopes.

All of the stable isotopes or nearly stable isotopes that exist are to be found in Earth's rocks. None of the radioisotopes formed in the laboratory, so far, have half-lives long enough to make it possible for any measurable quantity of them to have endured since the Earth's beginnings.

Every known element has radioisotopes. Even hydrogen, the simplest, has a radioactive isotope, hydrogen-3, the nucleus of which consists of 1 proton and 2 neutrons. It is sometimes called tritium, from a Greek word for "third." It has a half-life of 12.26 years. Tritium was first prepared in the laboratory in 1934 by the Australian physicist Marcus Laurence Elwin Oliphant (b. 1901).

For a quarter century after Rutherford's pioneering work, scientists kept pounding away at atoms, with alpha particles as their projectiles. This had its advantages. For one thing, alpha particles were always available. Uranium, thorium, and several of their breakdown products (radium, for instance) produced alpha particles, therefore there would always be a supply of them.

There were disadvantages, too. Alpha particles were positively charged, as were atomic nuclei. (After all, the alpha particle is itself an atomic nucleus.) This meant that the nuclei repelled the alpha particles, and before an alpha particle could collide with and enter an atomic nucleus, that alpha particle had to overcome the repulsion. Some of its energies were consumed in doing so, and that reduced its effectiveness. Aditionally, the more massive the nucleus being bombarded, the greater the repulsion. Beyond a certain point, available alpha particles could not enter a nucleus at all.

Once the neutron was discovered, however, Enrico Fermi realized that here was a unique new projectile. If a stream of neutrons was produced, let us say, by having a stream of protons strike paraffin, these neutrons, being uncharged, were not repelled by atomic nuclei. If a neutron happened to be moving in the direction of a nucleus, it could strike it and enter it even if it had very little energy. The discovery of the neutron thus revolutionized the entire technique of atom bombardment.

Fermi found that if he sent a stream of neutrons through water or paraffin, many of the neutrons would strike nuclei but bounce off without penetrating, losing some energy in the process. Eventually, these neutrons would have only the energy expected of something jiggling along with the usual speed that particles have at a given temperature. These would become thermal neutrons or slow neutrons. Such slow neutrons, Fermi found, were actually more likely to be absorbed by nuclei than were fast neutrons.

Fermi also found that when a neutron entered a nucleus, a beta particle (an electron) was usually emitted. The addition of the neutron raised the nucleus's mass number by 1, and the ejection of a beta particle, by subtracting 1 negative charge, raised the nuclear charge (that is, the atomic number) by 1. In short, neutron bombardment of a particular element tended to produce the next higher element in the atomic number scale.

In 1934, it occurred to Fermi that it might be very

interesting to bombard uranium with neutrons. Uranium, with an atomic number of 92, had the highest atomic number known at that time. If it were bombarded with neutrons, and made to emit beta particles, would it not form element 93, which was unknown in nature?

Fermi tried the experiment, and it seemed to him that he did indeed get element 93. The experiment, however, yielded complex and confusing results, and (as we shall see later) it took some years to straighten out the findings.

The Italian physicist Emilio Segrè (1905–1989), who had worked with Fermi, decided it was not necessary to bombard uranium with neutrons in order to create an unknown element. At this time, the mid-1930s, there were four spaces in the periodic table that remained unfilled. These represented unknown elements. Of these, the one with the lowest atomic number was element 43.

In 1925, a group of German chemists, including Walter Karl Friedrich Noddack (1893–1960) and Ida Eva Tacke (b. 1896), reported the discovery of element 75, which they called rhenium after the Latin name for the Rhine in Germany. It turned out to be the last of the 81 stable elements to be discovered. The group also announced that they had found traces of element 43, which they called masurium after a region in eastern Germany.

The second announcement, however, turned out to be a false alarm, and element 43 remained undiscovered. Why not bombard molybdenum (element 42) with neutrons, thought Segrè, to see if element 43 could at least be manufactured, if not found.

In 1937, Segrè went to the United States to have molybdenum bombarded with neutrons by a new technique (which I will describe later), and he did indeed locate element 43 in the bombarded material. Segrè hesitated to name the new element, however, because he wasn't certain that an artificially produced element was the equivalent of finding one in nature. In 1947, however, the German-British chemist Friedrich Adolf Paneth (1887–1958) maintained vigorously that it *was* the equivalent, and this view was accepted. Segrè therefore named element 43 technetium, from a Greek word meaning "artificial."

Enough technetium was formed one way or another for its properties to be studied, and it was found that three of its isotopes were quite long-lived. The longest-lived is technetium-97 (with a nucleus containing 43 protons and 55 neutrons), which has a half-life of 2,600,000 years. On a human scale, a sample of this isotope would seem permanent; only a tiny bit of the material would have broken down in the course of a human lifetime. Nevertheless, there are *no* stable isotopes of technetium, and even the most nearly stable one, technetium-97, is not long-lived enough to have persisted since Earth was formed. Even if there had been large quantities in the soil in Earth's youth, none would now remain. This is especially so because no isotope of technetium is formed from any other, longer-lived radioactive element.

The remaining three vacancies in the periodic table at this time were elements 61, 85, and 87. All three had been occasionally reported as having been discovered in minerals of one sort or another, but all of the reports turned out to be mistaken.

In 1947, however, the American chemist Charles D. Coryell (b. 1912) and his co-workers located element 61 in the products of uranium breakdown after neutron bombardment (something we are going to return to). They named it promethium after the Greek god Prometheus, who brought fire from the Sun to humanity, because the element had been found in the Sun-like fire of a nuclear reaction. None of the isotopes of promethium are stable, and even the longest-lived, promethium-145 (61 protons and 84 neutrons), has a half-life of only 17.7 years.

In 1939, the French chemist Marguerite Perey (b. 1909) located tiny traces of element 87, as a very minor breakdown product of uranium-235. She named the element francium for France. The isotope she had located was francium-215 (87 protons, 128 neutrons). Its half-life is only slightly over a millionth of a second, so Perey certainly didn't detect the isotope itself. What she did detect were the very energetic alpha particles it produced (the shorter the half-life of an alpha-particle producer, the more energetic the alpha particles), and was able to reason from what was already known about the breakdown pattern the isotope that had to be responsible. Even the most long-lived francium isotope, francium-223 (87 protons, 136 neutrons) has a half-life of only 21.8 minutes.

In 1940, element 85 was produced, by Segrè and others, by the bombardment of bismuth (element 83) with alpha particles. Element 85 was named astatine, from Greek words meaning "unstable," because, like all other elements discovered since 1925, it was exactly that. Its longest-lived isotope, astatine-210 (85 protons, 125 neutrons), has a half-life of 8.1 hours.

By 1948, then, the periodic table had been filled from hydrogen (1) to uranium (92), and elements beyond uranium had been discovered. Fermi thought he had produced, in 1934, element 93 in his bombardment of uranium by neutrons, but it was not until 1940 that the element was actually isolated in bombarded uranium by the American physicists Edwin Mattison McMillan (b. 1907) and Philip Hauge Abelson (b. 1913). In that uranium had been named for the then newly discovered planet Uranus, McMillan named element 93, which lay just beyond uranium, neptunium, after Neptune, the planet that lay beyond Uranus.

Neptunium-237 is the longest-lived isotope of that element, having a half-life of 2,140,000 years. This is long, but not nearly long enough to allow any neptunium to remain in the Earth's crust, even if there had been quite a bit there to begin with. Nevertheless, neptunium-237 has its interest because it breaks down through a series of intermediate compounds, as do uranium-238, uranium-235, and thorium-232.

In fact, neptunium-237 initiates the fourth radioactive series referred to earlier. It and all of its breakdown products have mass numbers that are divisible by 4, with a remainder of 1. There are only four radioactive series possible in these upper reaches of the periodic table: thorium-232 (remainder 0), neptunium-237 (remainder 1), uranium-238 (remainder 2), and uranium-235 (remainder 3). Of these, three exist today, but neptunium-237 is extinct, for even the longest-lived member of the series didn't have a half-life long-lived enough for the isotope to be around now.

Another odd thing about the neptunium-237 series is that it is the only one that doesn't end in a stable isotope of lead. It ends with bismuth-209, the only stable isotope of bismuth.

In 1940, the American physicist Glenn Theodore Seaborg (b. 1912) joined McMillan to find that certain neptunium isotopes give off beta particles and become isotopes of the same mass number, but with an atomic number higher by 1. Thus, they discovered element 94, naming it plutonium after Pluto, the planet beyond Neptune. Its most long-lived isotope is plutonium-244 (94 protons, 150 neutrons), with a half-life of 82,000,000 years. McMillan and Seaborg shared a Nobel prize in 1951 for finding a transuranian (i.e., "beyond uranium") element.

McMillan went on to other endeavors, but Seaborg and others continued to produce additional elements. The following transplutonian elements have been isolated:

Americium (for America), with an atomic number of 95. Its longest-lived isotope is americium-243 (95 protons, 148 neutrons), with a half-life of 7,370 years.

Curium (for the Curies), with an atomic number of 96. Its longest-lived isotope is curium-247 (96 protons, 151 neutrons), with a half-life of 15,600,000 years.

Berkelium (for Berkeley, California, where it was discovered), with an atomic number of 97. Its longest-lived isotope is berkelium-247 (97 protons, 150 neutrons), with a half-life of 1,400 years.

Californium (for California, the state in which it was discovered), with an atomic number of 98. Its longest-lived isotope is californium-251 (98 protons, 153 neutrons), with a half-life of 890 years.

Einsteinium (for Albert Einstein), with an atomic number of 99. Its longest-lived isotope is einsteinium-252 (99 protons, 153 neutrons), with a half-life of 1.29 years.

Fermium (for Enrico Fermi), with an atomic number of 100. Its longest-lived isotope is fermium-257 (100 protons, 157 neutrons), with a half-life of 100.5 days.

Mendelevium (for Dmitri Mendeleev), with an atomic number of 101. Its longest-lived isotope is mendelevium-258 (101 protons, 157 neutrons), with a half-life of 56 days.

Nobelium (for Alfred Nobel, who established the Nobel prizes), with an atomic number of 102. Its longest-lived isotope, so far detected, is nobelium-259 (102 protons, 157 neutrons), with a half-life of about 58 minutes.

Lawrencium (for Ernest Lawrence, mentioned later in the book), with an atomic number of 103. Its longest-lived isotope, so far detected, is lawrencium-260 (103 protons, 157 neutrons), with a half-life of 3 minutes.

Rutherfordium (for Ernest Rutherford), with an atomic number of 104. Its longest-lived isotope, so far detected, is rutherfordium-261 (104 protons, 157 neutrons), with a half-life of 65 seconds.

Hahnium (for Otto Hahn), with an atomic number of 105. Its longest-lived isotope, so far detected, is hahnium-262 (105 protons, 157 neutrons), with a half-life of 34 seconds.

Element 106 has been found, but two groups claim the discovery. The case hasn't been settled, and until it is, there can be no official name. The longest-lived isotope so far detected has a mass number of 263 (106 protons, 157 neutrons), with a half-life of 0.8 second.

It is uncertain how much further scientists can go. As the atomic numbers go up, the elements are harder to produce and, because the half-lives tend to decrease, harder to study. There is considerable pressure to reach elements 110 and 114, however, for there are compelling arguments in favor of supposing that some isotopes of these elements are long-lived or even stable.

7

BREAKDOWNS

Mass Defect

As previously noted, the current standard used in atomic weight measurements is carbon-12. The mass number of carbon-12 is defined as 12.0000, and all other mass numbers are measured against this number. Atomic weights, which are the weighted averages of the mass numbers of the isotopes making up a particular element, are also measured against the carbon-12 standard.

Carbon-12 has twelve particles in its nucleus: 6 protons and 6 neutrons. On average, each of the particles should, then, have a mass of 1.0000 if all twelve constitute mass 12.0000. However, modern mass spectrographs, capable of measuring the mass of individual protons curving in a magnetic field of known intensity, show the mass of the proton to be not 1.0000, but 1.00734.

A neutron, being uncharged, does not curve in passing through a magnetic field, but its mass can be worked out in other ways. In 1934, Chadwick measured the exact amount of energy it took to break apart the proton and neutron in a hydrogen-2 nucleus. The mass of a hydrogen-2 nucleus is known. From that, subtract the mass of the proton and add the mass of the energy used to break it up (calculated from Einstein's equation relating mass and energy). What is left over is the mass of the neutron.

It turns out that the mass of the neutron is 1.00867. In other words, the proton and neutron are not exactly equal in mass. The neutron is about 1/7 of 1 percent more massive than the proton (which, as we shall see, is important).

If we imagine ourselves taking 6 protons and 6 neutrons, considering them as separate particles, and adding up their individual masses, the total mass is 12.096. If, however, we squeeze all 6 protons and 6 neutrons tightly together into a carbon-12 nucleus, we have a total mass of 12.0000.

The mass of the carbon-12 nucleus is 0.096 less than it ought to be if the masses of the individual particles making it up are added together. In 1927, Aston, working with his mass spectrograph, found that all nuclei had masses a trifle less than would be expected if the masses of their separate particles were added together. Aston referred to this as a *mass defect*.

What fraction of the total mass of carbon-12 is the mass defect of this nucleus? The fraction is 0.096 divided by 12, or 0.008. In order to avoid working with this small decimal, scientists multiply it by 10,000. This gives us 80, which is the *packing fraction* of carbon-12.

In general, if we start with hydrogen-1 and its single-

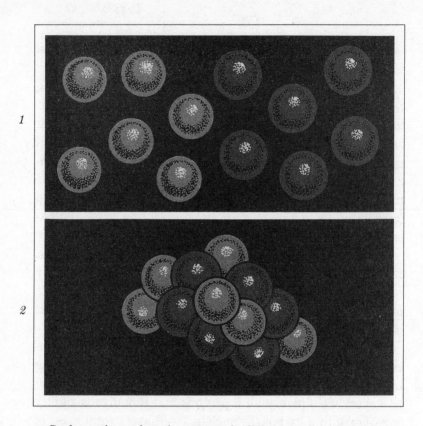

1. *Carbon 12's nucleus is composed of 6 protons and 6 neutrons. Their individual masses add up to 12.096.*
2. *Squeezed together into a nucleus, the total mass is only 12.000. This phenomenon is called the* mass defect.

proton nucleus and go up the scale of stable isotopes, we find that the packing fraction gets greater and greater, until we get to iron-56. Iron-56 has a nucleus containing 26 protons and 30 neutrons. If these particles are considered separately, the total mass is 56.4509. The mass of the iron-56 nucleus, however, is measured as 55.9349. The mass defect is 0.5260 (about half a proton). The packing fraction, 0.5260 divided by 55.9349 multiplied by 10,000, comes out to 94.0. If we continue going up the scale of stable isotopes that follow iron-56, we find that the packing fraction begins to decrease again. By the time we reach uranium-238, the packing fraction is only 79.4.

What happens to the mass that disappears when protons and neutrons are packed into a nucleus? There is only one thing that can happen: it turns into energy in accordance with Einstein's equation. In other words, if a carbon-12 nucleus is formed out of 6 protons and 6 neutrons, a small fraction of the mass of these particles is turned into energy, which dissipates into surrounding space. Any process that dissipates energy has a tendency to take place spontaneously (although not always quickly). This means that there is a tendency, under proper conditions, for protons and neutrons to combine to form nuclei.

On the other hand, in order to break up an atomic nucleus into its individual protons and neutrons, an amount of energy must be supplied that is exactly equal to the energy dissipated in the formation of that nucleus. But how can dissipated energy be collected once more and crowded into the tiny volume of a nucleus. Except under the most extraordinary conditions, this will not happen; nuclei do *not* tend to break up into their individual protons and neutrons. Once nuclei have formed, they tend to retain their identity indefinitely.

A nucleus, however, need not blow apart into single particles in losing its identity. What if it simply gains or loses a proton, or a neutron? Such an event would be sufficient to change one nucleus into another with, possibly, a further dissipation of energy. On the whole, if one nucleus changes into another nucleus with a higher packing fraction, there is a further dissipation of energy and there is a certain tendency for that change to take place.

We might, then, expect that nuclei with very low mass numbers would tend to change into those of higher mass numbers, while nuclei with very high mass numbers would tend to change into those of lower mass numbers. The nuclei at both extremes of the mass number scale would tend to converge at iron-56. Iron-56, with its maximum packing fraction, would require an energy input to become either larger or smaller in order for it to stay put.

A tendency, however, need not be observed in fact. If we are standing on sloping land, we have a tendency to slide downhill. If, however, the land is rough and uneven and we are wearing sneakers with ridged rubber soles, the amount of friction set up would prevent our sliding down, despite the tendency. On the other hand, if the slope were made steeper, or if it remained unchanged but grew icy, the friction might not suffice to keep us put and we would again tend to slide downhill.

To take another example, paper has a tendency to burn; that is, to combine with the oxygen of the atmosphere. Nevertheless, the chemical changes required to produce the effect of burning do not take place at ordinary temperatures because an energy of activation is required. (This is a type of friction that prevents a chemical change that "should" take place from actually taking place.) If the paper is heated, however, more and more energy enters it and, eventually, the value of the energy activation is topped, with the paper bursting into flame.

Once the paper flames, enough heat is produced to serve as an energy of activation for neighboring regions of the paper, which flame in their turn, spreading the flame still farther. The paper can thus continue burning indefinitely without further supplies of energy from outside. The small initial supply is all one needs. The proverbial smoldering cigarette butt that eventually burns down an entire forest is a very realistic scenario. This process, whereby a chemical reaction produces what is needed to initiate a further installment of that same chemical reaction, is called, simply, a chain reaction.

In the case of nuclei, there are factors that prevent the natural tendency of sliding toward iron-56 from turning into actuality. This is particularly true of the light nuclei, for reasons we will go into later.

For the massive nuclei, the tendency is more easily realized. Indeed, for all known nuclei that are more massive than that of bismuth-209, the tendency *is* realized. The more massive nuclei tend to emit particles in such a way that the newly formed nucleus is smaller and, therefore, has a higher packing fraction than the old. In this way, energy is produced and dissipated.

The greater the dissipation of heat in the change from one nucleus to another, the more likely is the change, the more rapidly the change takes place, and the shorter the half-life of the original nucleus. In the cases of thorium-232, uranium-235, and uranium-238, the initial change involves so small a dissipation of heat that their half-lives are very long. Nevertheless, even here the half-lives are not infinite and the change does take place, but slowly. (By way of analogy, although paper doesn't burst into flame at ordinary temperatures, it does undergo changes very slowly. Every once in a while a chemical change can take place even though the energy of activation requirement is not met. Thus, as the years pass, the pages of a book often slowly turn browner and more brittle until they eventually crumble into an ash that is the result of very slow "burning." We might say that the paper molecules have a "burn half-life" at room temperature, which might be long from the standpoint of a reader, but which is much shorter than that of uranium-238.)

To put it briefly, the type of natural radioactivity discovered in the 1890s changes nuclei with mass numbers 232 to 238 into nuclei with mass numbers 206 to 208. In the process, the packing fraction increases, whereby the protons and neutrons in the final nuclei have slightly less mass than those in the original nuclei. The missing mass is dissipated as energy, which is how we explain where the energy produced by radioactivity comes from.

Nuclear Fission

I mentioned earlier in the book that Enrico Fermi, in 1934, had bombarded uranium with slow neutrons in order to form element 93 (definitively located six years later and named neptunium). Fermi thought he had located the element—and, in a way, he had—but studies of the bombarded uranium seemed to show such a confusing melange of particles that it was difficult to precisely locate element 93 in the mixture. (Still, Fermi received a Nobel prize in 1938 for this work.)

Ida Tacke (one of the codiscoverers of the element rhenium) suspected that what had happened in Fermi's work was that the uranium nucleus was so complex and, therefore, so rickety, that on absorption of a neutron the nucleus simply fell apart into fragments. This was so different from anything that had been observed in nuclear breakdowns, however, that no one paid much attention to the suggestion.

Beginning in 1937, however, the team of Hahn and Meitner tackled the problem in Germany. (Meitner was Jewish, but she was an Austrian national, so she was momentarily safe from Adolf Hitler, 1889–1945—who then ruled Germany—and his malignant anti-Semitic policies.)

Hahn decided that what might have happened in the case of the neutron bombardment of uranium was that the uranium lost *two* alpha particles rather than one. That was as far as he dared go in the direction of Tacke's notion of fragmentation. The loss of two alpha particles would reduce the atomic number by 4, from uranium's 92 to radium's 88. If Hahn were correct, then, there ought to be slightly higher quantities of radium in the bombarded uranium than the usual traces to be expected from ordinary radioactive changes.

How could one detect these minute traces of radium and estimate the quantity present? Marie Curie had isolated traces of radium from uranium ore, but she had had tons of the ore to work with. Hahn and Meitner had only a small quantity of bombarded uranium.

As it happens, radium, in the periodic table, is placed just under the stable element barium, the two elements being very similar in chemical properties. If bombarded uranium is dissolved in acid and barium is added to the solution, the barium can be separated out again by simple chemical procedures and the radium will come out with it. (The radium will do whatever the barium does.)

If, then, Hahn and Meitner put in perfectly stable barium and got out distinctly radioactive barium, they would know that radium had been extracted with the barium. From the amount of radioactivity present (easily measured) they could determine, delicately, the amount of radium obtained. Before this experiment could be carried through, however, Nazi Germany invaded Austria, annexing it in March of 1938. Meitner then became subject to Hitler's anti-Semitism, whereupon she slipped across the border to the Netherlands, and from there to Stockholm, Sweden.

Hahn continued to work, with the German chemist Fritz Strassman (1902–1980). They added stable barium to solution and got out radioactive barium, from which they could estimate the quantity of radium they had extracted. Then, in order to produce the final step of the proof, it was necessary to separate the radium from the barium in order to produce a solution containing the radium alone.

That separation, however, didn't work. Nothing Hahn

and Strassman tried would separate the radium from the barium. Hahn decided that if nothing would separate the stable barium from the radioactive radium, the radioactive atoms were *not* radium, but barium—or more correct, a radioisotope of barium. (Hahn received a Nobel prize in 1944 for this insight.)

But how could uranium break down to yield barium? The atomic number of barium was 56. If uranium (atomic number 92) had split into barium, it would have to give off 18 alpha particles, or it would have to split in two. Either alternative seemed so unlikely that Hahn didn't quite dare suggest them publicly.

At the same time, in Stockholm, Meitner was coming to precisely the same conclusion when she received news of the failure to separate the supposed radium from barium. She *did* decide to go public. With the help of her nephew, the physicist Otto Robert Frisch (1904–1979), Meitner prepared a letter, dated January 16, 1939, and sent it to the British journal *Nature*. Frisch, who was working in Bohr's laboratory in Copenhagen, told Bohr of the contents of the letter before it was published. Bohr went to the United States to attend a physics conference in Washington, D.C., on January 26, 1939, and there he spread the word—also before the letter was published.

Concurrently, in Great Britain, the Hungarian physicist Leo Szilard (1898–1964)—who, like Meitner, had fled Germany because he was Jewish—was thinking about what H. G. Wells termed, in his science fiction, an atomic bomb. Szilard thought that such a bomb might exist if a neutron were to strike an atomic nucleus, thus bringing about a change that would cause the ejection of two neutrons, which would strike two nuclei and bring about the ejection of four neutrons, and so on. The number of breakdowns per second and the energy liberated would quickly rise to enormous proportions, thereby producing a vast explosion. Szilard was, in fact, visualizing a nuclear chain reaction.

Szilard even patented the process. He enlisted the help of another Jew, the Russian-British biochemist Chaim Weizmann (1874–1952), who tried to carry through the necessary experimentation. However, it failed. Nuclei that absorbed and ejected neutrons absorbed only fast, energetic ones, and ejected slower ones that were too lacking in energy to keep the reaction going.

But then Szilard heard of the shattering of the uranium nucleus as a result of the absorption of a neutron. (This break of a nucleus into two nearly equal parts came to be called fission, from a Latin word meaning "to split." We often speak of uranium fission, but uranium nuclei are not the only ones that undergo such a split; therefore, it is better to use the more general term nuclear fission.)

Szilard saw at once that here was the possibility of a practical version of the nuclear chain reaction he had visualized. It was a slow neutron that split the uranium nucleus and, in the process, it was soon discovered that two or three slow neutrons were liberated for each nucleus that split.

In 1940, Szilard labored to persuade American physicists to establish self-censorship over their investigations into nuclear fission, for fear that German physicists might benefit from American findings and give Adolf Hitler a devastating new type of bomb. (World War II had already started, with Germany making successful advances and the United States still neutral.)

Szilard next had to persuade the United States government to pour enormous funds into the research. He obtained the help of two other Hungarian-born refugees from Nazism: Eugene Paul Wigner (b. 1902) and Edward Teller (b. 1908). In 1942, the three visited Albert Einstein, still another such refugee. Einstein, as the only scientist whose word would carry sufficient weight, agreed to write a letter to President Franklin Delano Roosevelt (1882–1945). Roosevelt received the letter, considered it, was persuaded by it, and, on a Saturday late in the year, signed the order

that set up what came to be called the Manhattan Project—
a deliberately meaningless name designed to mask its real
purpose.

As it happened, the timing of Szilard's push for the
project was precarious. It is not often that work is done on
a weekend. If Roosevelt had delayed the signing until Monday, chances are that it would not have been signed at all—
for the Saturday he signed it was December 6, 1941, and
the next day Japan attacked Pearl Harbor. Who can tell
when Roosevelt would have been able to think of the order
again. At any rate, the project proceeded and an atomic
bomb (more properly, a nuclear fission bomb) was developed by July 1945, after Germany was crushed and Hitler
had committed suicide. It was used to finish off a virtually
helpless Japan, on August 6 and 8, 1945.

When a uranium nucleus undergoes fission, it does not
always divide in exactly the same way. The packing fraction
among nuclei of moderate size does not vary a great deal,
and the uranium nucleus may well break at one point in
one case and at a slightly different point in another. For
this reason, a mixture of a great variety of radioisotopes
are produced during uranium fission. They are lumped together under the term fission products.

The probabilities are highest that the division will be
slightly unequal, with a more massive part in the mass
region from 135 to 145, and a less massive part in the region
from 90 to 100. It was among the more massive part that
the element promethium was located in 1948.

As a result of nuclear fission, the uranium nucleus
slides farther down the packing fraction hill than it does in
its ordinary radioactive breakdown to lead. Uranium fission, therefore, releases considerably more energy than
ordinary uranium radioactivity. (Nuclear fission can also
release it much more quickly if a chain reaction is set up,
pouring forth its energy content in a fraction of a second

where ordinary uranium radioactivity would take billions of years.)

But if more energy is released in nuclear fission than in ordinary radioactivity, why doesn't the uranium nucleus break down by fission naturally, instead of giving off a series of alpha and beta particles? The answer is that there is a higher energy of activation involved in the nuclear fission process. The necessary energy of activation can be supplied if a neutron floats into a nucleus, changes its nature, and sets it vibrating, but not otherwise. At least, *almost* not otherwise. Despite the higher energy of activation, a uranium nucleus will very occasionally undergo spontaneous fission, slipping through the wall, so to speak, set up by the energy of activation. It happens very rarely, however. A uranium-238 nucleus undergoes 1 spontaneous fission for every 220 times such nuclei simply give up alpha particles. Such spontaneous fission was first detected in 1941 by the Soviet physicist Georgii Nikolaevich Flerov (b. 1913).

Just as some radioactive nuclei have much shorter half-lives than others, so do some show much greater likelihood of undergoing spontaneous fission. The transuranic isotopes, for instance, become more unstable not only with regard to ordinary radioactivity but with regard to spontaneous fission as well. Where the spontaneous fission half-life of uranium-238 is about one trillion years, that of curium-242 is 7,200,000 years, and that of californium-250 is only 15,000 years.

Uranium-238 undergoes fission with such difficulty that even neutron bombardment is insufficient to bring about much of it. It takes fast, energetic neutrons to do the trick, while only slow neutrons are ejected so that no chain reaction is possible.

It was Bohr who pointed out, soon after fission was established, that, on theoretical grounds, it was uranium-235 that should set up the chain reaction. Uranium-235 is less stable than uranium-238. Uranium-235 has a half-life only ⅙ that of uranium-238, and even a slow neutron will cause it to undergo fission. One of the more difficult aspects of developing the fission bomb, in fact, was separating uranium-235 from uranium-238, since ordinary uranium, as found in nature, simply doesn't have enough uranium-235 in it to support a nuclear chain reaction.

It is possible, however, to bombard uranium-238 with neutrons in such a way as to form first neptunium-239 and then plutonium-239. Plutonium-239 has a half-life of over 24,000 years, which is long enough to allow it to be accumulated in quantity. Plutonium-239 is fissionable with slow neutrons, as is uranium-235.

Again, thorium-232 is not fissionable with slow neutrons. When thorium-232 is bombarded with neutrons, however, it can become thorium-233, which, in turn, becomes uranium-233. Uranium-233 was first discovered by Seaborg in 1942. It has a half-life of 160,000 years and is fissionable by slow neutrons.

In other words, all of the uranium and thorium in the world can, in theory, be converted into nuclei that can be made to undergo fission, and, if controlled, to yield useful energy instead of merely a wild explosion. Thorium and uranium are not very common elements, but taken together, there is ten times as much energy available from them as from the Earth's entire supply of coal, oil, and gas.

In the 1950s, nuclear reactors yielding controlled energy began to be built, and a significant portion of the world's energy is now derived from them. To be sure, safety is a concern. (The Three-Mile Island accident in the United States in 1979 and the true disaster at Chernobyl in the Soviet Union in 1986 caused considerable alarm.) Then, too, there is a question of the disposal of the ever-accumulating fission products, which are dangerously radioactive. For these reasons, the future of fission energy seems cloudy at the moment. There is, however, another type of nuclear energy that might prove just as useful, while inherently safer to use than fission.

Nuclear Fusion

Both natural radioactivity and nuclear fission affect nuclei with large mass numbers, bringing about a change toward nuclei of intermediate mass numbers and greater stability. In the process, mass is lost and energy is produced and dissipated. It is also possible for nuclei with small mass numbers to combine with each other, or fuse (that is, melt together), forming a somewhat more massive nucleus. This, too, represents a change toward nuclei of intermediate mass numbers and greater stability. Here, too, mass is lost and energy is produced and dissipated.

Indeed, whereas the packing fraction, as one goes from large mass numbers to intermediate ones, is marked by a rather gentle rise; the packing fraction, as one goes from small mass numbers to intermediate ones rises much more steeply. This means that nuclear fusion can produce more energy from a given mass of starting material than can nuclear fission.

Let's examine how this works by taking the example of a nucleus of hydrogen-2 (1 proton plus 1 neutron) fusing with another nucleus of hydrogen-2 to form a nucleus of helium-4 (2 protons plus 2 neutrons). The mass number of hydrogen-2 is 2.0140, and two of them are 4.0280. Helium-4, however, has an unusually high packing fraction for its size, whereby its mass number is only 4.0026. The loss of mass number in going from two hydrogen-2 nuclei to a helium-4 nucleus is $4.0280 - 4.0026 = 0.0254$. The loss in mass of 0.0254 is 0.63 percent of the original mass of 4.028.

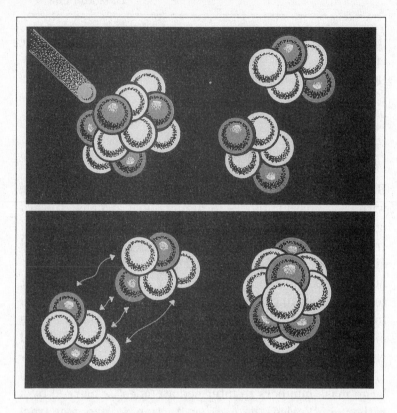

Both fission, the splitting of an atom (top), and fusion, the joining together of nuclei (bottom), liberate energy. Kilogram for kilogram, the fusion of hydrogen can produce eleven times as much energy as can uranium fission.

This doesn't sound like much (hydrogen-2 in fusing to helium-4 loses only ⅝ of 1 percent of its mass), but it is actually a great deal. The natural radioactive change of uranium-238 to lead-206 results in the loss of only 0.026 percent of the original mass, whereas the nuclear fission of uranium-235 results in the loss of only 0.056 percent of its original mass. The fusion of hydrogen can produce about twenty-four times as much energy, kilogram for kilogram, as can natural radioactivity, and 11 times as much as can uranium fission.

The energy that could be produced by nuclear fusion proved of crucial importance to an understanding of the Universe even before that other energy source, nuclear fission, was discovered. The story goes as follows.

Ever since the development of the law of conservation of energy in 1847, scientists had been wondering as to the source of the energy radiated by the Sun. This radiation had been occurring throughout human history, and geological studies made it plain that it had been occurring long before human beings had evolved.

No source of energy known in the nineteenth century could account for the Sun's having kept burning for more than a hundred million years or so, which proved to be a conservative estimate. In the first decade of the twentieth century, scientists began measuring the ages of rocks and of meteorites by the amount of radioactive change that had taken place in them. It soon appeared that our solar system (including the Earth and the Sun) was several billion years old. The best current figure for its age is 4,550,000,000 years.

By 1910, it was recognized that nuclear energy was more powerful than any other that had been considered. In 1920, the British astronomer Arthur Stanley Eddington (1882–1944) suggested that the Sun's energy might be derived from the fusion of hydrogen to helium. This suggestion looked better as the decade progressed. When Aston worked out the notion of packing fraction, it became clear that the fusion of hydrogen to helium was the only simple nuclear reaction that could possibly yield enough energy to power the Sun.

Then, in 1929, the American astronomer Henry Norris Russell (1877–1957) worked out the chemical composition of the Sun by a careful study of its light spectrum, and discovered that it was mostly hydrogen. About 90 percent of the atoms in the Sun are hydrogen, and 9 percent are helium. All of the other elements together make up the remaining 1 percent of the atoms. This meant that not only was the fusion of hydrogen to helium the only simple nuclear reaction that could yield enough energy, it was the only significant nuclear reaction that could take place at all. It was hydrogen fusion or nothing.

In 1938, the German-American physicist Hans Albrecht Bethe (b. 1906) worked out the details of what must go on in the center of the Sun, basing his theory on what was known from laboratory studies of nuclear reactions and on astronomical inferences as to conditions at the center of the Sun. He received a Nobel prize in 1967 for this work.

It is now thought that most normal stars are constantly fusing hydrogen, which can serve as an energy source for billions of years. Eventually, especially in the case of the more massive stars, conditions at the center of a star become such that helium nuclei are further fused into still more massive nuclei, such as those of carbon, oxygen, neon, silicon, through to iron (where the process stops because the packing fraction has reached its maximum).

Very massive stars that have gone as far as they can in the process of fusion find their energy source failing and can no longer support the weight of their own outer layers. The star collapses and, in the process, all of the hydrogen remaining in the outer layers (together with other atoms of mass number smaller than iron) fuses at once. The result is a vast outpouring of energy—an enormous explosion we call a supernova. Much of the material of the exploding star is spewed into surrounding space by the explosion, while what remains collapses into a tiny object called a neutron star, or into an even tinier object called a black hole.

It is now thought that at the time the Universe was first formed only hydrogen and helium nuclei were created. The more massive nuclei were formed only in the centers of stars, and it was only because some massive stars exploded that these more massive nuclei were added to the dust, gas, and debris of material in space. In fact, as a supernova explodes, so much energy becomes available that iron nuclei are driven "uphill" to form even more massive nuclei all the way up to uranium and beyond—and these, too, pour into surrounding space.

Eventually, new stars form out of the interstellar dust and gas containing the massive nuclei that had been spread outward by supernova explosions. These new stars are second-generation stars. They and their planets contain great quantities of the massive nuclei.

The Sun is such a second-generation star. The Earth, and we, ourselves, are made almost entirely of massive nuclei that were once formed at the center of a giant star and then spread through space by a giant explosion.

But if hydrogen fusion produces so much more energy than natural radioactivity does, why doesn't it take place spontaneously, and much more rapidly, than natural radioactivity does? On Earth, uranium and thorium slowly break down into less massive nuclei and even, very occasionally, undergo spontaneous fission, but hydrogen remains stable and shows no signs whatever of fusing.

The reason for this is not difficult to see. Massive nuclei such as those of uranium and thorium have all of their protons and neutrons squeezed together into one place. Any changes that might take place among them *do* take place. In the case of hydrogen fusion, however, two hydrogen-2 nuclei, or four hydrogen-1 nuclei, which exist apart from each other, must possess enough energy to break through the electron barrier, overcome the mutual repulsion of the nuclei for each other, and then collide with sufficient force to initiate fusion. At ordinary temperatures, their motions simply don't involve more than the tiniest fraction of the required energy.

In order to supply the required energy, the temperature must go very high indeed—to the millions of degrees. Even then, it helps to compress the hydrogen to very high densities so that there will be enormous numbers of collisions as hydrogen nuclei jiggle back and forth across the abnormally small distances separating them.

These conditions are satisfied in the centers of stars. In 1926, Eddington produced a convincing line of argument to show that the Sun was gaseous throughout. At the center of the Sun, the temperatures and pressures were so high that the atoms broke down. The electrons were crushed together and the nuclei approached one another freely.

We now believe that the center of the Sun has a temperature of about 15,000,000° C. and a density of about 160 grams per cubic centimeter, or about 8 times the density of gold. (Yet that center is gaseous because the atoms are broken and the nuclei move about as freely as intact atoms do in ordinary gases.) As a result, nuclear fusion takes place at the surface of the small core of helium at the Sun's center—the helium that was part of the Sun originally plus an additional quantity formed by hydrogen fusion over the last 4.55 billion years.

If, then, we wanted to have a fusion reaction here on Earth, how could we manage it? How could we get a temperature and pressure high enough?

Once the fission bomb had been developed, it could be seen that this was one way of developing the necessary temperatures and pressures. If the fission bomb included a quantity of hydrogen in some form, the first few instants of the fission reaction might raise the hydrogen to the temperatures and pressures required to ignite a fusion process.

In 1952, both the United States and the Soviet Union developed a successful nuclear fusion bomb, which is more popularly known as a hydrogen bomb or an H-bomb. It is sometimes called a thermonuclear bomb, where thermo- is from the Greek word for "heat," in that the fusion bomb is ignited by extreme heat rather than by neutron bombardment.

The vast energies of the fusion bomb cause it to explode with enormous fury. The explosion is so powerful that in any war in which it is freely used, civilization will surely be destroyed almost at once; perhaps humanity in general will be, and, in the extreme, most or all of life.

The Sun itself is, in effect, a vast fusion bomb, but it does not blow apart. The Sun, which is 333,000 times as massive as the Earth, has an enormous gravitational field that holds it together against all of the fury of fusion energy in its interior. And so we sit and bask in the welcome light and warmth of this cosmic bomb—although it is well that we are at a safe distance of 93,000,000 miles from it.

Can nuclear fusion be ignited and kept going in a *controlled* fashion? Can it be made to develop energy slowly, energy that can be used and that will not be destructive? If this can be done in practical fashion, then we will have a form of nuclear energy in which the fuel is easy to get and easy to handle. Instead of having to obtain uranium and thorium from rocks in which it is thinly spread, we would get hydrogen-2 out of the ocean. (Hydrogen-2 is much rarer than the ordinary hydrogen-1, but hydrogen-2 will undergo fusion more easily, and there is enough of it, comparatively rare though it is, to last us for billions of years.)

Then, too, fission energy requires a rather sizeable minimum supply of fissionable material, creating the possibility of runaway reactions and meltdowns if not properly controlled. Fusion energy can be carried through with small quantities of fuel, which would make the possibility of a major accident a thing of the past. Finally, fusion energy would not produce radioactive materials with the profusion that fission energy does.

In order to produce controlled nuclear fusion, however, we must subject hydrogen-2 to high temperatures and pressures. We cannot yet manage the pressures, however, and we must raise the temperature now possible still higher, while at the same time keeping the hydrogen confined within a magnetic field.

For more than thirty years scientists have labored at

the task of producing controlled fusion, and have gotten ever closer—but they have not yet reached this goal.

Breakdown Particles

It was mentioned earlier that helium-4 is a particularly stable nucleus. Thus, soon after the Universe was formed, the four simplest nuclei were formed. Hydrogen-1 existed first because its nucleus was a mere proton. By adding a neutron, it became hydrogen-2 (1 proton, 1 neutron); which, by adding a proton, became helium-3 (2 protons, 1 neutron); which, by adding a neutron, became helium-4 (2 protons, 2 neutrons). Hydrogen-2 and helium-3, while stable nuclei, have fairly low packing fractions, which is why they had a considerable tendency under the conditions of the early Universe to change over to the stable helium-4. The result was that 90 percent of the atoms in the Universe today are hydrogen-1 and 9 percent are helium-4. Everything else makes up the remaining 1 percent.

Moreover, the early Universe never got past the helium-4 level. Helium-4 is so stable that it has virtually no tendency to add either a proton, a neutron, or another helium-4. The nuclei that would form—lithium-5 (3 protons, 2 neutrons), helium-5 (2 protons, 3 neutrons), or beryllium-8 (4 protons, 4 neutrons)—are all so unstable that they have half-lives of anywhere from a hundredth of a trillionth of a second to less than a billionth of a trillionth of a second. As a result, all nuclei with mass numbers beyond 4 have been formed (as I mentioned earlier) at the centers of stars, where conditions make such formation not only possible, but likely.

Among the higher mass numbers, nuclei that might be viewed as made up of helium-4 units are particularly stable.

Carbon-12 (6 protons, 6 neutrons—3 helium-4 units) hangs together tightly. So does oxygen-16 (8 protons, 8 neutrons—4 helium-4 units). Both have lower packing fractions than do their neighbors.

As nuclei grow more massive, this helium-4 effect diminishes. Nevertheless, nuclei such as neon-20 (5 helium-4 units), magnesium-24 (6), silicon-28 (7), sulfur-32 (8), and calcium-40 (10) are particularly stable. All of these nuclei, from helium-4 to calcium-40, are the most common isotopes of their elements.

Beyond calcium-40, however, the helium-4 unit seems to lose its stabilizing effect. Apparently, as the number of protons in the nucleus grows, it is no longer sufficient to have an equal number of neutrons in order to make the nucleus stable. There must be an excess of neutrons.

Thus, in iron-56, the most common iron isotope, there are 26 protons and 30 neutrons, which makes the neutron/proton ratio 1.15. In tin-118, the most common tin isotope, there are 50 protons and 68 neutrons, a neutron/proton ratio of 1.36. In gold-197, the only stable gold isotope, there are 79 protons and 118 neutrons for a neutron/proton ratio of 1.49. The most massive stable nucleus is bismuth-209, with 83 protons and 126 neutrons for a neutron/proton ratio of 1.52.

Beyond bismuth-209, no excess of neutrons will suffice to keep a nucleus stable. Thus, uranium-238 has 92 protons and 146 neutrons for a neutron/proton ratio of 1.59, but even that large excess does not suffice to keep the nucleus entirely stable.

The German-American physicist Maria Goeppert-Mayer (1906–1972) tackled the problem of why some stable nuclei are more stable than others. She suggested that there are nuclear shells and subshells, as there are electron shells and subshells. She worked out the numbers of protons and neutrons it took to fill these shells, and pointed out that filled shells produce nuclei more stable than their neighbors.

The number required to fill a nuclear shell is a shell number, which is sometimes called a *magic number*. (This last term is inappropriate in that there is no "magic" in science—but scientists are as prone to be dramatic as other human beings.) The German physicist Johannes Hans Daniel Jensen (b. 1907) worked out the notion of shell numbers independently of Goeppert-Mayer, and the two shared a Nobel prize for this work in 1963.

When a massive nucleus is so massive that it is unstable, there is a natural tendency for it to lose particles in order to become a less massive, and therefore more stable, nucleus. An efficient way for it to do this is to get rid of an alpha particle (a helium-4 nucleus) as a breakdown particle. Such a nucleus clings together so tightly that it is easy to expel as a unit, and the mass number decreases by four at a stroke. Therefore, uranium-238, uranium-235, thorium-232, radium-226, and many other nuclei more massive than bismuth-209, emit alpha particles.

Nuclei less massive than bismuth-209 do not usually emit alpha particles. Neodymium-144 is about the lightest nucleus that does, but it emits very few because its half-life is about 2,000 trillion years.

Of course, radioactive atoms often emit beta particles in the course of breakdown, which raised a problem. In the 1920s, the existence of beta particles was considered excellent evidence that the nucleus contained electrons. After all, if a dime falls out of your change purse, it can only be because a dime was in your change purse to begin with.

However, such homey analogies don't always work, which is why "common sense" is so often a dangerous guide as applied to science. By 1932, scientists were convinced that the nucleus contained only protons and neutrons—no electrons. Where, then, did the beta particles come from? If there are no electrons in the nucleus, we can only as-

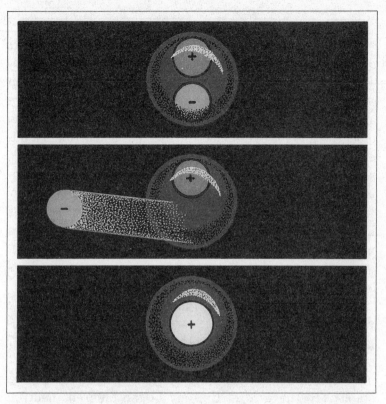

A neutron in its stable position inside a nucleus can be thought of as a neutral combination of a proton and an electron. Separated from the nucleus, the neutron breaks down into a proton and an electron, with a half-life of twelve minutes.

sume that one is created there and is immediately ejected. But how?

Suppose we consider the neutron an uncharged particle, not because it has no electric charge, but because it has both a positive and a negative charge that neutralize each other. If the negative charge is ejected in the form of an electron, then the positive charge is left, whereby the neutron has become a proton. (The situation is actually more complicated than that, as we shall see, but for now this viewpoint will do.)

Why should such a change take place, however? It doesn't take place in the various nuclei we find to be stable. These could remain unchanged for indefinite periods, even eternally, for all anyone could tell in the 1930s. Yet there are nuclei that eject electrons—some slowly, some quickly—and for every electron ejected, a neutron in a nucleus has changed to a proton. To answer the question, let's suppose that a nucleus of a given atomic number must contain a fixed number of protons equal to that atomic number. In addition, it must contain a certain number of neutrons if the nucleus is to be stable. Sometimes only one certain number of neutrons will do. Thus, in the case of fluorine, every atomic nucleus must have exactly 9 protons and exactly 10 neutrons if it is to be stable.

However, sometimes there is a certain flexibility where neutron numbers are concerned. Thus, every nitrogen nucleus must have 7 protons, but it can have either 7 or 8 neutrons and be stable. Every oxygen nucleus must have 8 protons, but can have 8, 9, or 10 neutrons and be stable. (In the case of tin, any of ten different neutron numbers will keep the nucleus stable.)

But what if a nucleus has too many neutrons to make stability possible? For instance, hydrogen-1 is stable, with its nucleus consisting of 1 proton alone. So is hydrogen-2, with a nucleus consisting of 1 proton and 1 neutron. Hydrogen-2 is less stable than hydrogen-1, but, nevertheless, stable—and if left to itself will exist unchanged indefinitely.

(How can one nucleus be less stable than another, yet stable? Imagine a coin resting in the center of a table. Left to itself it will remain there indefinitely. Another coin might be resting near the edge of the table. It, too, will remain there indefinitely—yet, while stable, it is less stable than the centered coin for it will take a smaller disturbance to knock it off the table. In the same way, hydrogen-2 can more easily be made to undergo fusion than hydrogen-1, which is why there is so much less hydrogen-2 in the Universe than hydrogen-1, even though both are stable.)

Hydrogen-3, with its nucleus consisting of 1 proton and 2 neutrons, is not a stable nucleus; there are too many neutrons for that. One might expect the hydrogen-3 nucleus to eject a neutron, but that would involve a high energy of activation—energy not available to the hydrogen-3 nucleus under ordinary conditions. A second alternative is for one of the neutrons of the hydrogen-3 nucleus to be converted to a proton, and for a beta particle to be ejected. This requires a rather small energy of activation; the half-life of hydrogen-3 is only about 12¼ years. After the ejection of the beta particle, the nucleus contains two protons and one neutron, having become the stable isotope helium-3.

Similarly, there are only two stable carbon isotopes: carbon-12 (6 protons, 6 neutrons) and carbon-13 (6 protons, 7 neutrons). In 1940, the Canadian-American biochemist Martin David Kamen (b. 1913) identified carbon-14 (6 protons, 8 neutrons), which has one neutron too many. The emission of a beta particle, converting one of the neutrons to a proton, however, produces stable nitrogen-14 (7 protons, 7 neutrons). There are many other examples of this type.

A neutron is more massive than a proton and an electron put together. Therefore, if a neutron is converted into a proton via an ejected electron, there is an overall loss of mass and a dissipation of energy. Will a free neutron undergo a spontaneous conversion to a proton, then, giving off electrons as it does so?

It was at first difficult to test this hypothesis because even if a stream of neutrons was produced, the neutrons generally collided with, and were absorbed by, other nuclei before they had a chance to break down. It was not until 1948 that the difficulty was overcome by a strong beam of neutrons passed through a large, evacuated cylindrical tank. There was an electric field around the tank so that any electrons that might be produced would curve off in one direction, while protons would curve off in the other. The proposed breakdown was then indeed observed, with the neutron breaking down into a proton and electron with a half-life in the neighborhood of 12 minutes. (This is not the full description of what happened, but it will do for now.)

If this is the case, why don't neutrons break down in every nucleus, until there is nothing left but protons? Apparently, neutrons within the nucleus are in close association with protons, and under such conditions, provided the number of neutrons is neither too many nor too few, the neutrons are stable. (There will be more to say about this later.)

Spontaneous breakdowns of isolated particles, when they take place, always seem to result in a reduction of mass—meaning that a neutron can break down to the less massive proton, but a proton cannot break down to the more massive neutron.

But in that case, why doesn't the proton break down to the less massive electron, releasing all but $1/1836$ of its mass as energy? The answer to this question is that there are conservation laws that seem always to be obeyed. There is, for instance, the conservation of electric charge, which states—if uncounted numbers of laboratory observations are to be trusted—that a positive electric charge, left to itself, can neither be created nor destroyed. The same is true for a negative electric charge.

In the 1930s, the only two charged particles known were the proton, which was positively charged, and the electron, which was negatively charged. (This is no longer the situation, but let us suppose it is and answer the question on that basis. We can always qualify the answer later.) If the only way the proton can lose mass is to be converted to an electron, then the positive charge of the proton must be destroyed and the negative charge of the electron must be created. Neither is possible. For this reason, the proton cannot break down into any other particle, for there is no less massive particle that contains a positive charge (so it was thought).

In the same way, an electron cannot break down into smaller particles, in that the only smaller particles known in the 1930s were the photon and the graviton, both of which have zero mass and zero charge. The electron, in breaking down into either of these, must have its negative charge destroyed—and this is impossible. For this reason, an electron cannot break down.

Notice that when a neutron breaks down into a proton (whether we are talking of a free neutron or of one that is part of a nucleus), an electron is formed at the same time. In this way, the uncharged neutron (0) forms a positive charge (+1) and a negative charge (−1). The two, taken together, are still zero: $0 = (+1) + (−1)$. (The law of conservation of charge will allow the production or destruction of pairs of opposite charge, but will not allow the production or destruction of one without the other.)

We might ask, then, why a photon can't turn into a graviton, or vice versa. In the case of these two particles, there is no electric charge to worry about. There is, however, spin, or angular momentum. The law of conservation of angular momentum tells us that spin can be neither created nor destroyed. A photon with a spin of 1 can't break down into a graviton with a spin of 2, or vice versa. There might be other factors that would prevent the change, but the matter of spin is, in itself, sufficient.

Thus, in the early 1930s there were just five known particles of which the Universe might be composed. Four of these, the proton, electron, photon, and graviton, are stable. The fifth, the neutron, is unstable. This view was not to endure for long.

8

ANTIMATTER

Antiparticles

The list of particles making up the universe—proton, neutron, electron, photon, and graviton—seems peculiar in one way. Why should the positive electric charge be housed in protons and the negative electric charge in electrons when protons are 1,836 times as massive as electrons?

The two particles, so different in mass, have electric charges that are precisely equal in size, even if they are opposite in nature. We can tell this because the hydrogen atom, made up of 1 proton in its nucleus and 1 electron outside the nucleus, is exactly neutral. No excess electric charge, either negative or positive, has ever been detected to even the tiniest degree.

Nor have scientists found any basic difference in the two types of charge that make it necessary for the electron to be associated with very little mass, while the proton is associated with much greater mass. In short, the proton and electron make an unlikely and puzzling pair.

The puzzle began to come together in the late 1920s, when Dirac attempted to study the properties of the electron by working out the mathematics of its wave properties. It seemed to him that it should be possible for the electron to exist in one of two energy states, one the opposite of the other. Naturally, Dirac's first thought was that the electron itself represented one of the energy states, and the proton the other. This would be delightful, if true, for it would simplify the Universe still further, in that the electron and proton would then be simply two different states of one fundamental particle.

That was too good to be true, however, for Dirac quickly saw that the equations would not be truly satisfied unless the two states were exactly similar in every way but for some crucial orientation. They would have to be mirror images of each other, so to speak, like your two hands, which are in overall form exactly alike in every way, except that one is thumb-rightward and the other is thumb-leftward.

If electric charge were subject to mirror image variation in the electron—positive in one state and negative in the other—that was all that could vary. Everything else had to be identical. Not only would the size of the charge in the two states have to be identical, but so, too, would its mass. In 1930, then, Dirac suggested that there must exist a particle exactly like the electron in every way, except that it carried a positive electric charge of precisely the same size as the electron's negative electric charge.

The same sort of argument would lead to the conclusion that there must exist a particle exactly like the proton in

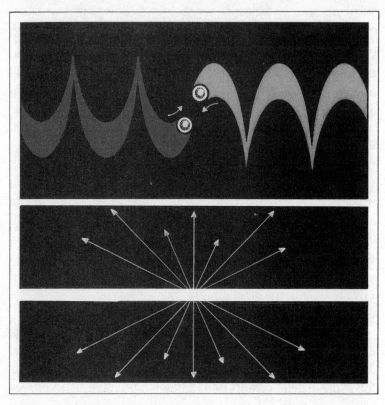

The meeting of particle and antiparticle is something like having two opposite waves cancel each other out. The total charge of the two particles is +1 + −1, or 0. In the course of mutual annihilation their mass is completely converted into energy in the form of gamma rays.

every way, except that it carried a negative electric charge of precisely the same size as the proton's positive charge.

In general, a particle that is exactly like another except for being opposite in one key respect has come to be called an antiparticle, where the prefix anti- is from a Greek word for "opposite." The positively charged electron would be an antielectron, and the negatively charged proton would be an antiproton.

If a particle and an antiparticle meet, it is something

like having two opposite waves (one going up, where the other goes down, and vice versa) meeting. Just as two waves can cancel each other out into a straight line going neither up nor down, so that there is no wave at all, the particle and antiparticle cancel each other out, leaving no particle at all. This is called mutual annihilation.

Interestingly, this phenomenon does not violate the law of conservation of electric charge, for when a particle and antiparticle meet, the total charge of the two particles is $(+1) + (-1)$, or 0. Once they undergo mutual annihilation, the electric charge that remains is still 0; therefore, the conservation law has not been violated. It is only the positive charge *alone* or the negative charge *alone* that can be neither created nor destroyed. Positive charge *and* negative charge *together* can be created or destroyed in any quantity.

It is only the electric charge that disappears in mutual annihilation, of course, for that is the only part of the particle and antiparticle that is of opposite character. Both the particle and antiparticle have identical mass, and this double mass cannot disappear. Mass, however, is a form of energy, and it can change its form. In the course of mutual annihilation, then, gamma rays appear with an energy exactly equivalent to the mass that existed before the annihilation.

The opposite change can also take place. If enough energy can be concentrated into a tiny spot, it can be converted into matter. When this takes place, however, a particle alone or an antiparticle alone cannot be formed, for in either case, electric charge would have been created out of nothing. One can only produce a particle *and* an antiparticle *together*, so that the total electric charge remains zero. This is called pair production.

Dirac's theory was an extremely interesting one, mathematically, yet mathematics, however interesting, does not carry much weight if it can't be matched with reality. For instance, scientists are certain, for a variety of mathematical and theoretical reasons, that gravitons (or gravitational waves, depending on whether you look at them as particles or as waves) must exist. To be sure, theory also tells us that gravitational waves are so lacking in energy that detecting them might be next door to impossible. (It would be rather like trying to pick up a single dust particle with a monkey wrench. You can't do the job until you devise a sufficiently delicate pair of tweezers.)

That is exactly what scientists are trying to do. Despite their certainty as to the existence of gravitational waves, there are those who bend their efforts toward the construction of devices that will actually detect these waves. For years they have failed, but scientists are confident that someday they will succeed. When success comes, it will crown theory with observation, and this will mean great jubilation and a Nobel prize for the detectors.

So it was with Dirac's suggestions concerning antiparticles. In the world about us there were electrons in countless numbers, but no antielectrons had ever been observed at the time Dirac's theories were announced. Until the antielectron was actually detected, the work could not be taken entirely seriously. However, the antielectron was soon detected, under circumstances that make it appear as though it would have been detected even without Dirac's published results. To see how this was done, we must again backtrack.

Cosmic Rays

A charged electroscope with its gold leaves spread wide apart tends to lose its charge slowly, even when there are no radioactive materials in the near vicinity. This was not a great surprise to early investigators because it was quite likely that small quantities of radioactive material were widespread in common soil. Even if they were present in quantities too small to detect in ordinary ways, the occasional speeding particle would drain off a bit of the electroscope charge, and eventually the electroscope would be entirely discharged.

But then investigators found that there seemed no way of stopping the discharge. If the electroscope were taken over stretches of water well away from land, the charge still disappeared slowly. If the electroscope were encased in a thickness of lead sufficient to block the passage of radiation, the charge still disappeared, although more slowly than before.

The Austrian physicist Victor Franz Hess (1883–1964) thought he would investigate the subject by sending an electroscope high into the atmosphere in a hot-air balloon. In that the radiation sources were assumed to be almost entirely in soil, moving far up above the soil ought to stop the loss of charge more efficiently than anything else tried hitherto.

In 1911, Hess made the first of ten balloon ascensions, taking the electroscopes up to six miles above Earth's surface. To his astonishment, Hess found that the electroscopes lost charge *faster* the higher he went. Hess could see no way of explaining this except to suppose that a very penetrating radiation must be coming from outer space. For this discovery, Hess received a share of a Nobel prize in 1936.

In 1925, the American physicist Robert Andrew Millikan (1868–1953) interested himself in this penetrating radiation from the sky, to which he gave the name cosmic rays because they seemed to originate from somewhere in the cosmos. It seemed quite certain to Millikan that cosmic rays were a form of electromagnetic radiation even shorter

When charged, the gold leaves of an electroscope spread wide apart in mutual repulsion. The closer they are to a radioactive source, the quicker they lose their charge and collapse. In 1911, Victor Franz Hess took one up in a hot-air balloon. He was attempting to remove

in wavelength, and therefore higher in energy and more penetrating, than gamma rays.

On the other hand, Compton rather suspected that cosmic rays were streams of electrically charged subatomic particles that, if moving rapidly enough, could also be more penetrating than gamma rays. (This resembled the argument, a generation before, over whether cathode rays were waves or speeding particles.)

it from the natural radioactivity of common soil. He actually found that it lost its charge faster the higher it went. In the illustration, the vertical lines represent cosmic rays.

How could the dispute be settled? Well, if cosmic rays were a form of electromagnetic radiation, as Millikan thought, they would be electrically uncharged and would be unaffected by Earth's magnetic field. If they came from all parts of the sky in equal quantity, they would reach all parts of Earth's surface in equal quantity. If, however, cosmic rays were charged particles, they would be deflected by the Earth's magnetic field and would reach Earth in greater quantities the farther you went from the equator and the closer you came to the magnetic poles. In other words, the higher the latitude, the greater the concentration of cosmic rays. This was called the latitude effect.

Compton traveled over the world measuring cosmic ray incidence here and there to see if the latitude effect really existed. By the early 1930s, he was able to show that it *did* exist and that the cosmic rays *were* electrically charged particles.

In 1930, the Italian physicist Bruno Benedetto Rossi (b. 1905) pointed out that if cosmic rays were positively charged, more of them would come from the west than from the east. If they were negatively charged, the reverse would be the case. In 1935, the American physicist Thomas Hope Johnson (b. 1899) was able to show that more came from the west, and that cosmic-ray particles carried a positive electric charge.

We now know that cosmic ray particles are speeding atomic nuclei issuing from stars. In that stars are composed mostly of hydrogen, cosmic-ray particles are mostly hydrogen nuclei; that is, protons. They also contain some helium nuclei and a thin scattering of more massive nuclei.

Our Sun emits a constant stream of speedy protons and other charged particles, as Rossi demonstrated in the 1950s. This is now called the solar wind. Particularly violent disturbances on the solar surface, such as solar flares, produce a shower of such particles with greater energy than usual. The greater the energy, the greater the velocity, and when these velocities approach the speed of light they are classified as cosmic-ray particles. The Sun occasionally emits particles that just barely fall into this classification.

Hotter and more violent stars than our Sun are more copious emitters of cosmic-ray particles; supernova explosions are particularly good sources. Once cosmic-ray particles are speeding through space, they can be accelerated and made even more energetic by the magnetic fields of the stars they pass, as well as by the overall magnetic field of the Galaxy.

As a result, cosmic ray particles are more energetic than the radiations obtained from radioactive materials. This offered nuclear physicists a new and more powerful tool, for cosmic ray particles can bring about nuclear reactions that radioactive radiations are not energetic enough to start.

To balance this, however, cosmic ray particles are not as easily handled as radioactive materials, which can be concentrated and worked with in the laboratory; radioactive radiations can be called on at will and carefully aimed. Cosmic ray particles, however, come at their own time and can only be dealt with in more concentrated form by climbing mountains or going up in a balloon.

The American physicist Carl David Anderson (b. 1905), a student of Millikan, also studied cosmic rays. He allowed them to pass through a cloud chamber and hoped that by following the curvature of the lines of fog droplets they produced he could learn something about them. The cosmic rays were so energetic, however, that they passed through the cloud chamber too quickly to have time to curve appreciably in response to the magnetic field. Anderson therefore devised a cloud chamber with a lead barrier running across the center. Cosmic rays striking the lead would be energetic enough to smash through, but they would lose enough energy in the process to curve markedly thereafter in response to the magnetic field.

In 1932, Anderson noted, emerging from the lead barrier, a curved pathway that looked precisely as though it had been caused by a speeding electron. However, it curved in the wrong direction! Anderson realized that he was observing the path of a particle just like an electron, but that carried a positive charge. It was the antielectron that Dirac

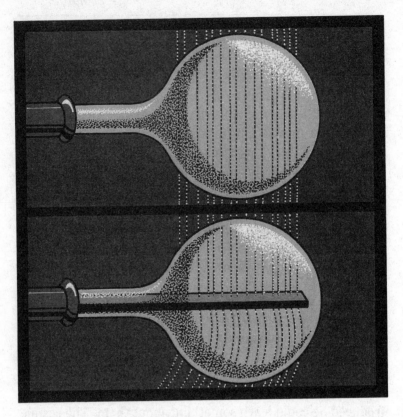

Cosmic rays pass through a cloud chamber so quickly that their path is not significantly changed by a magnetic field. Carl David Anderson placed a lead barrier inside the chamber, slowing them down enough to be studied.

had suggested, in theory, two years before. As a result, Anderson shared in a Nobel prize with Hess in 1936.

The particle Anderson found was referred to as a positive electron, or positron. This, in my opinion, is a faulty word formation, as well as a poor choice. For one thing, the common ending for subatomic particles is -on. We have electron, neutron, proton, photon, and graviton as examples. The *r* in electron and neutron belongs to the root of the word, as in *electricity* and *neutral*. For this reason, if the positive electron were to be given a subatomic name, it should be called a positon, without the *r*, because *positive* contains no *r*. Moreover, whether positon or positron, such a name obscures the relationship to the electron. The particle should be called an antielectron, for all other antiparticles, without exception, add the prefix anti- to the name of the particle that is their opposite. Nevertheless, the name positron has been used so commonly and so often that there is no longer any hope of changing it.

(It often happens that a poor name is given to an object or a phenomenon to begin with, either out of ignorance or out of bad judgment. Sometimes, it can be changed in time, but often the ill-chosen name is used so commonly by so many that it becomes inconvenient or even impossible to change it.)

The positron behaves exactly as Dirac's theory suggested. It quickly undergoes mutual annihilation when it encounters one of the numerous electrons in its immediate environment, producing gamma rays of energy exactly equal to that of the combined mass of the electron and proton. It was quickly found, too, that if alpha particles were allowed to strike lead, some of the particle energy could be converted into an electron-positron pair that would emerge from the lead, with tracks curving in opposite directions. This takes us back to the question we had raised earlier. What happens if we have a radioactive isotope that contains too few neutrons for stability?

The easiest way of producing an additional neutron inside a nucleus is to convert one of its protons into a neutron. This yields a nucleus with one additional neutron and one fewer proton, which is what might be called for for stability.

For example, phosphorus-30 has 15 protons and 15 neutrons in its nucleus. The only stable phosphorus isotope is phosphorus-31, with 15 protons and 16 neutrons in its nucleus. In other words, phosphorus-30 has too few neutrons for stability. Suppose, however, that one of the protons in the phosphorus-30 nucleus is converted into a neutron. The positive charge on the proton cannot be destroyed (in line with the law of conservation of electric charge), so it must show up elsewhere. But what if the nucleus emits a positron, a type of positive beta particle? This takes care of getting rid of the positive charge. If phosphorus-30 emits a positron, then instead of having 15 protons and 15 neutrons in its nucleus, it has 14 protons and 16 neutrons, which is the stable silicon-30.

Thus, in 1934, when the Joliot-Curies discovered artificial radioactivity in the form of phosphorus-30, they were also producing a new type of radiation that turned out to be a stream of speeding positrons. Here was a way of producing positrons without having to bombard material with either cosmic ray particles or alpha particles. They had formed a neutron-deficient nucleus and had allowed it to undergo radioactive decay.

Positron emission by a nucleus accomplishes results that are just the opposite of electron emission. Whereas electron emission causes the atomic number to go up by one as an additional proton is formed from a neutron, positron emission causes the atomic number to go down by one as a proton is lost through conversion into a neutron.

There might seem to be a puzzle here. Because the neutron is slightly more massive than the proton I have emphasized that the neutron spontaneously decays into a proton, but that a proton does not decay, "uphill," into a neutron.

This, however, is true only if we are speaking of free particles. In a nucleus, where protons and neutrons exist in association, what counts is the mass of the entire nucleus. In a neutron-deficient nucleus, the total mass of the nucleus can go down if a proton changes into a neutron because the packing fraction increases. Therefore, the change can occur.

That is what makes particular isotopes unstable. If the total mass of a nucleus is going to decrease if a proton changes into a neutron, or if a neutron changes into a proton, the appropriate change will take place. If an isotope has a mass that will be increased if either a neutron changes into a proton or a proton into a neutron, then neither change will take place and the isotope will remain stable.

As it happens, when there are either 43 or 61 protons in a nucleus, no matter how many neutrons there are present, a neutron-proton change, one way or the other, will always lower the total mass. This is why there are no stable isotopes of technetium (43) or promethium (61).

Another way in which a proton within a nucleus can be changed into a neutron is for the nucleus to pick up one of the electrons outside the nucleus, thus neutralizing the electric charge of one of its protons and converting it into a neutron. Almost always, the electron is picked up from the K shell, which is the electron shell nearest the nucleus. The process is therefore known as K capture, which was first observed in 1938 by the American physicist Luis Walter Alvarez (1911–1988). This is much less likely to happen than positron emission, however.

There is no theoretical reason why the reverse of proton-to-neutron conversion wouldn't be possible. To convert a neutron into a proton, a nucleus, instead of emitting an electron, might pick up a nearby positron. The only trouble here is that in the ordinary matter about us there are no positrons to speak of, so the chance of positron capture is nil.

Particle Accelerators

Once the antielectron, or positron, was produced and observed, scientists could feel entirely confident that an antiproton must also exist. Confidence, however, is not enough. They wanted to observe one in existence.

However, antiprotons don't seem to exist about us any more than do antielectrons. They must be formed in some type of nuclear reaction, and then observed, but this is easier to say than to do. A proton is 1,836 times as massive as an electron; we can, therefore, be reasonably certain that an antiproton must be 1,836 times as massive as an antielectron.

Scientists could form electron-antielectron pairs by smashing alpha particles into lead, but in order to form a proton-antiproton pair, one must find projectiles with 1,836 times the energy of the alpha particles that suffice for the smaller task. Such energetic alpha particles, unfortunately, do not exist.

There undoubtedly exist cosmic ray particles sufficiently energetic to do the job, but they are a great deal fewer in number than those energetic enough to bring about the production of a positron. It would be a long wait for one of the rare particles to happen to come along and form an antiproton exactly where it could be detected.

In the late 1920s, however, physicists began to work at the task of creating energetic projectiles of their own. For the purpose, you would want a massive particle to begin with, in that the energy of a speeding particle increases with its mass. That means something at least as massive as a proton, which is a natural choice because all one need do to obtain a supply of protons is to remove the outer electron from hydrogen atoms, something that can be done without much trouble. Alpha particles would be more massive still, of course, but they be from the nuclei of helium, which is a substance much rarer than hydrogen and much more difficult to strip down to its bare nucleus.

Given a supply of protons, one would then pass them through a magnetic field to accelerate them and make them move faster. The stronger the magnetic field, the more sharply a proton is accelerated. If the accelerated proton is allowed to smash into an atomic nucleus, a nuclear reaction is possible. A device that accomplished this was called an atom smasher by the newspapers in the early days of this work, but the term is overly dramatic. The proper name is the soberly descriptive particle accelerator.

The first useful particle accelerator was devised in 1929 by the British physicist John Douglas Cockcroft (1897–1967) and his co-worker, the Irish physicist Ernest Thomas Sinton Walton (b. 1903). Using their particle accelerator, they bombarded nuclei of lithium-7 (3 protons plus 4 neutrons) with energetic protons. In the process, a proton would smash into a nucleus and remain there, forming beryllium-8 (4 protons plus 4 neutrons). Beryllium-8 is, however, extremely unstable, and in about a billionth of a trillionth of a second splits up into two helium-4 nuclei (2 protons plus 2 neutrons). This was the first nuclear reaction brought about by an accelerated particle, for which Cockcroft and Walton shared a Nobel prize in 1931.

In the years immediately following this feat, other types of particle accelerators were developed. The one that yielded the most fruitful results was devised in 1930 by the American physicist Ernest Orlando Lawrence (1901–1958). He tackled the problem of an ordinary magnetic field causing a proton to accelerate forward in a straight line, quickly passing beyond the field and accelerating no farther. The field would have to be extended a great distance in order for the proton to continue to accelerate.

Lawrence devised a way of flipping a field back and forth so that it forced a proton to follow a curved path one way, and then to follow another curved path the opposite way, completing one "cycle" and staying within the magnetic field throughout. By repeating this over and over, the particle would travel in slowly expanding cycles. Although the particles would have to cover a greater and greater distance as the cycle expanded, they would go just enough faster to complete the circuit in the same length of time, staying in step with the field as it flipped back and forth. The particles would remain within the field for a considerable length of time, then, even though the device itself wasn't very large. In this way, a small device could produce particles of unexpectedly high energies. Lawrence called his device a cyclotron, receiving a Nobel prize in 1939 for its development.

Larger and more powerful cyclotrons were quickly built. New designs were used in which the magnetic field grew more powerful as the particles speeded up. This kept them moving in tight cycles so that they didn't expand out of the field until scientists were ready to have them do so. These proton synchrotrons created even more energetic particles. It was then possible to have two cycling sets of particles moving in opposite directions and eventually colliding head on. This doubled the former energy production possible through a single stream of particles hitting a stationary object.

In 1987, the United States began to consider spending some six billion dollars for the construction of a superconducting super collider, a particle accelerator in which particles will be sent around a track 52 miles long, producing ten times as much energy as any other particle accelerator in existence. (Later in the book there will be occasion to mention what it is hoped will come of such an unimaginably powerful device. Of course, there are very occasional cosmic ray particles with several million times the energy that even this accelerator will produce, but waiting for such particles to turn up would be a very long and thankless job.)

As long ago as the early 1950s, particle accelerators had been built that could produce particles energetic enough to form a proton-antiproton pair. Naturally, if they struck an appropriate target, such energetic particles would produce all types of nuclear reactions and result in the formation of all types of particles. This mass of various particles could then be allowed to stream through a magnetic field. All of the positively charged particles would curve in one direction, and all of the negatively charged ones in another. The most massive negatively charged particles were expected to be antiprotons; they would curve the least. At a considerable distance from the target, all of the particles would curve away and out of the field, with only the antiprotons (if any had formed) remaining.

In 1955, Segrè, the discoverer of technetium and now an American citizen, along with the American physicist Owen Chamberlain (b. 1920), did locate antiprotons in this manner. They shared a Nobel prize in 1959 for this work.

Baryons

Despite the fact that antielectrons (positrons), once formed, disappear almost at once through mutual annihilation with the first electron encountered, it is fair to consider an antielectron a stable particle. After all, an antielectron, *left to itself*, will remain an antielectron forever, as far as we know. It will never, of its own accord, change into anything else.

The reason for this is that any spontaneous change in a subatomic particle involves a loss of mass and the conversion of that mass into energy. The electron, however, remains, nearly a century after its discovery, the least massive object known that can carry a negative electric charge, while the antielectron is the least massive object known that can carry a positive electric charge. The only objects less massive than an electron or an antielectron are particles that, as far as we know today, have no mass at all—and none of them carries an electric charge. Therefore, the law of conservation of electric charge prevents electrons or antielectrons from undergoing any spontaneous change, although one can annihilate the other by canceling out its respective charges.

But now the problem arises as to why the proton is stable. As long as the antielectron was not known, one might have maintained that the proton was the least massive object capable of carrying a positive electric charge so that it had to be stable by the law of conservation of electric charge. After 1932, however, this was no longer a valid argument. Why couldn't the proton decay into an antielectron? The positive electric charge would still exist, and almost all of the proton's mass would be converted to energy. Similarly, why couldn't the antiproton decay into an electron? The answer is that they just don't (and a lucky thing, too, or the Universe as at present constituted couldn't possibly exist, and neither could we).

From the fact that these particles don't decay in this manner, and from data gathered concerning all of the nuclear reactions that scientists have studied in all of the decades since subatomic particles began to be studied, it seems fair to conclude that what keeps protons stable is a conservation law. If it is not the conservation of electric charge, then it is the conservation of something else.

Electrons and positrons are examples of leptons (from a Greek word meaning "small"); they have small masses. Protons, neutrons, and antiprotons are baryons (from a Greek word meaning "heavy"); they are much more massive than leptons. The law of conservation that keeps a proton stable is the law of conservation of baryon number, which works as follows. A proton and a neutron are each given a baryon number of $+1$, and an antiproton a baryon number of -1. In that the electron and the antielectron are not baryons, they are each given a baryon number of 0. A proton and an antiproton, taken together, have a total baryon number of $(+1) + (-1) = 0$. Hence, the two can undergo mutual annihilation, leaving behind no proton and no antiproton for a total baryon number of 0, which is what they started with. The law of conservation of baryon number is not broken. In the same way, a large quantity of energy can create a proton-antiproton pair—baryon number 0 before and baryon number 0 after. Again, the law of conservation of baryon number is not violated. (In both cases, the law of conservation of electric charge is not violated either. Obeying one conservation law doesn't mean you don't have to obey the others. *All* conservation laws have to be obeyed where they are applicable. We shall see cases, however, in which a conservation law might *not* be applicable.)

A neutron (baryon number $+1$) can break down into the slightly less massive proton (baryon number $+1$) and an electron (baryon number 0), without breaking the law of conservation of baryon number or the law of conservation of electric charge.

A single proton (baryon number $+1$), however, cannot break down into a positron (baryon number 0) without breaking the law of conservation of baryon number; such a breakdown appears not to be possible. The law of conservation of electric charge is not broken, but that isn't enough. For the same reason, it would seem, a single antiproton (baryon number -1) does not break down into an electron (baryon number 0).

You might, of course, ask *why* there is such a conservation law. At the moment, scientists cannot answer this question. All they can say is that observation of nuclear reactions demonstrates that such a conservation law exists. (In recent years, however, there has been some question as to whether this conservation law is absolute or whether, under certain conditions, it can be broken.)

The proton is stable because it is the baryon with the

smallest possible mass, incapable of breaking down without losing its baryon status. The antiproton is stable because it is the antibaryon with the smallest possible mass. Of course, whenever you have a conservation law, you are bound to keep your eyes open for any apparent violation that might reveal some new fact or necessary modification of the law rendering a fairer (meaning "more just" and "more beautiful") view of the Universe.

For instance, a possible violation of the law of conservation of baryon number was uncovered in 1956 when a group of physicists discovered that when a proton and antiproton skimmed by each other closely without actually colliding, it was possible for the electric charges to cancel each other, while leaving the rest of the particles apparently untouched.

Without their electric charges, both the proton and the antiproton become neutral; thus it might be supposed that each has been converted into a neutron. But this can't be so. The proton and antiproton in combination have a baryon number of 0, while two neutrons have a total baryon number of $+2$. How can this be?

The answer is to suppose that when the electric charges cancel, the proton becomes a neutron (baryon number $+1$), whereas the antiproton becomes an antineutron (baryon number -1). The proton-antiproton pair (baryon number 0) becomes a neutron-antineutron pair (baryon number 0), with the law of conservation of baryon number upheld.

But how can there be an antineutron? The antielectron has an electric charge opposite to that of the electron, and the antiproton has an electric charge opposite to that of the proton. But neither neutron nor antineutron has an electric charge. What is it, then, that distinguishes the two?

Both neutron and antineutron, although they have no overall electric charge, have small local charges here and there that balance out over the entire particle (as we shall see). Both neutron and antineutron spin, and the effect of the spin on the small local charges is to create a small magnetic field. The neutron has its north magnetic pole pointing in one direction, and the antineutron, spinning in the same direction as the neutron, has its north magnetic pole pointing in the other direction. It is the direction of the magnetic fields that is opposite, not the spin, and it is this that differentiates the neutron and the antineutron.

The proton, neutron, and electron form the atoms, the planets, the stars—all of the matter we know. If the antiproton, antineutron, and antielectron existed in quantity in a place of their own, they would undoubtedly fulfill all of the functions of protons, neutrons, and electrons. The antiparticles could form antiatoms, antiplanets, antistars and, in general, antimatter.

This is not entirely theory. There are some (admittedly very simple) observations that support this view. In 1965, an antiproton and an antineutron were combined to form an antihydrogen-2 nucleus. Later on, two antiprotons and an antineutron were combined to form an antihelium-3 nucleus. (Each was actually simply an antinucleus.)

If we consider the various conservation laws, we know that however the matter of the Universe came into being, an equal quantity of antimatter must have been formed at the same time. If so, where is it?

We certainly know that Earth consists entirely of matter. In fact, we can be certain that the entire solar system, and even our galaxy, is formed entirely of matter, with antimatter existing, if at all, only in insignificant traces. If this were not so, there would be interaction between matter and antimatter now and then, and the constant formation of gamma rays. We don't, in fact, detect gamma rays reaching us from outer space in anything like the quantities we would expect if the galaxy contained substantial amounts of antimatter.

Some scientists have suggested that "original" matter and antimatter were separated, somehow, after formation, whereby there are even now clusters of galaxies made of matter and clusters of antigalaxies made of antimatter. They remain separate, therefore producing no significant level of gamma rays. Yet even this view seems unlikely because, if it were so, there should be a significant quantity of antiprotons, and antinuclei generally, in cosmic ray particles, some of which are sure to come from other galaxies—and there is not.

The reality might even be that two Universes were originally formed, one of matter and one (an "antiuniverse") of antimatter, with no communication between them possible. Undoubtedly, in such a case, the intelligent inhabitants of the antiuniverse, if there were any, would consider their own Universe to consist of matter and ours to be of antimatter. They would have as much right to maintain this as we have to maintain the reverse.

In recent years, however, as we shall see, new views on this subject have come into prominence. Scientists are willing to consider the possibility that matter and antimatter might *not* have formed in equal quantities to begin with.

9

NEUTRINOS

Saving the Laws of Conservation

All of the laws of conservation are important signposts in the understanding of nuclear reactions and of the behavior of subatomic particles. Anything that defies a law of conservation ought not to happen, which leaves only a limited number of possibilities of action. In other words, the laws of conservation prevent total anarchy and tell scientists, in effect, what they should look for.

Anything that *seems* to violate a law of conservation is therefore very unsettling. This is especially true if the law is the one considered most basic, most important, and, therefore, most inviolate. This is the law of conservation of energy, which, in the 1920s, seemed to be shaken.

In general, subatomic particles behave strictly in accordance with this law. If an electron and a positron annihilate each other, the energy produced, in the form of gamma rays, is exactly equal to the energy that had been present in the form of the mass of these two particles plus their kinetic energy as they approached each other. The same is true when a proton and antiproton annihilate each other.

Again, when a nucleus undergoes radioactive breakdown and emits an alpha particle, the new nucleus plus the alpha particle have a total mass slightly lower than the original nucleus. (This is why the nucleus originally breaks down, spontaneously, in this fashion.) The decrease in mass makes its appearance as the kinetic energy of the emitted alpha particle.

This means that all of the nuclei of a particular isotope that break down by alpha-particle emission give off alpha particles all traveling at the same velocity, and all equally energetic and penetrating. Measurements have shown that alpha-particle production always conserves energy.

Occasionally, a nucleus undergoing radioactive change produces two or more batches of alpha particles traveling at different speeds and possessing different energies. This implies that in such a situation the nucleus could exist at one of two or more energy levels. The one at the higher levels would produce alpha particles going at greater velocities than the ones at the lower energy level.

The situation is quite different where beta particles are concerned. When a nucleus gives off a beta particle (that is, a speeding electron), the new nucleus plus the beta particle is slightly less massive than the old nucleus. The difference in mass should be accounted for by the kinetic energy of the beta particle.

Sometimes the beta particle does move so quickly that it possesses a kinetic energy that just about balances the loss of mass. A beta particle never moves faster than this; it never produces kinetic energy greater than the loss in mass. If it did so, it would be creating energy out of nothing and would be violating the law of conservation of energy.

However, the beta particle usually moves more slowly, even much more slowly, than it should, and has less kinetic energy than is needed to balance the loss in mass. This violates the law of conservation of energy; too little is as bad as too much. When nuclei (all of the same sort) break down by emitting beta particles, they do so within a range of resultant particle speeds and kinetic energy. On average, the kinetic energy of the beta particles is only about one-third the quantity represented by the loss in mass. Energy, it seemed to those first studying the phenomenon, simply wasn't conserved.

For twenty years this range of velocity in the case of beta rays was observed and studied, remaining an absolute puzzle. Niels Bohr was so anxious about it he suggested that the law of conservation of energy might as well be abandoned, at least where beta particles were concerned. Few physicists, however, were ready to do that. (A general rule that works under all conditions but one should not be thrown out until every effort has been made to explain the exception.)

Pauli came up with, in 1930, an explanatory theory in regard to the beta emission conservation of energy problem. He suggested that whenever a nucleus gave off a beta particle, it gave off a second particle that carried off whatever quantity of energy the electron didn't. The kinetic energy of *both* particles, taken together, would then exactly account for the loss of mass in beta-particle production. The only problem with Pauli's notion, however, was this: If a second particle is produced, why is it never detected? The answer was that the electron carried off all of the charge required in the conversion of a neutron into a proton; the second particle, therefore, would have to be electrically neutral, which is a type of particle much more difficult to detect than an electrically charged one.

The neutron was detected—a neutral, massive particle capable of knocking protons out of nuclei. It was this capability that helped researchers locate the particle. In the case of this newly suggested particle in beta production, the small amount of energy it carries off is just enough to account for its speed; so that it could not have more than the merest trace of mass. In fact, some beta particles are given off at just about sufficient speed to account for *all* of the loss in energy to the nucleus, so that the second particle might have no mass at all.

A particle with no electric charge and no mass might seem to be difficult to detect, but this is not borne out by the example of the photon, which has neither electric charge nor mass and is very easy to detect. But then, the photon is a type of fuzzy wave packet that easily interacts with any bit of matter it encounters. What if the beta particle's companion was a minute particle that did not interact with matter?

In 1934, Fermi took up the subject of this particle, giving it the name neutrino (from the Italian word for "little neutron"). Fermi worked out in considerable detail what the particle's properties ought to be. He believed the particle to indeed be something with no mass, no electric charge, and virtually no tendency to interact with matter. It was a "nothing particle," or, as it was sometimes referred to, a ghost particle. It might as well not have been there, except that it served to balance the law of conservation of energy—a view that was not, in itself, very impressive. It could be argued (and some did argue in this fashion) that the position was invented just to save appearances. If the only way to save the law of conservation of energy was to invent a ghost particle, the law wasn't worth saving. However, the neutrino saved other conservation laws.

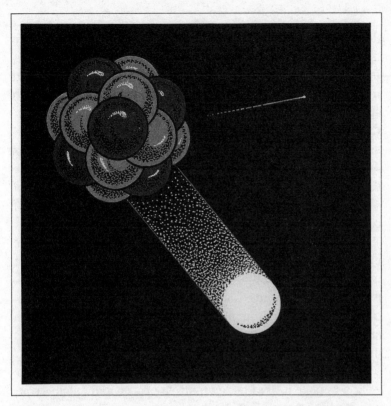

When a nucleus gives off a beta particle, the new nucleus plus the beta particle is slightly less massive than the old nucleus. The difference in mass, not completely accounted for by the kinetic energy of the beta particle, is assigned to the almost undetectable "ghost particle," the neutrino.

Consider, for example, a motionless neutron. Its velocity is zero; therefore, its momentum (which is equal to its mass times its velocity) is zero. There is a law of conservation of momentum that scientists understood even before the law of conservation of energy was grasped. In other words, whatever happens to a motionless neutron, the total momentum of the particles it gives rise to must remain zero—provided the rest of the universe doesn't interfere in any way.

After a certain time, the motionless neutron will break

down into a proton and an electron. The electron will go flying off in some direction at great speed and will, therefore, have a sizeable momentum. The neutron, now changed into a proton, will recoil in the opposite direction at a much slower speed, but will have a much greater mass. Ideally, the momentum of the electron (small mass × high velocity) should be equal to the momentum of the proton (large mass × low velocity). If the two dart off in exactly opposite directions, one has a momentum of $+x$ and the other of $-x$. These two momenta add up to 0; therefore, the law of conservation of momentum should be conserved.

But this is *not* what happens. The momentum of the electron is usually too low, and it and the proton do not move in precisely opposite directions. There is a bit of momentum not accounted for. If, however, we allow the existence of a neutrino, it might be moving in such a direction as to not only account for the missing energy, but for the missing momentum.

Then again, the neutron has a spin of either $+\frac{1}{2}$ or $-\frac{1}{2}$. Suppose it breaks down into a proton and an electron, and nothing more. The proton has a spin of $+\frac{1}{2}$ or $-\frac{1}{2}$, as does the electron. The total spin of the proton and the electron is always either $+1$, -1, or 0, depending on how you pick the signs. The proton and electron together can *never* have a total spin of either $+\frac{1}{2}$ or $-\frac{1}{2}$, as the original neutron had. This means that the law of conservation of angular momentum (another very familiar and rigidly observed law of conservation) is broken.

If, however, the neutron breaks down into a proton, an electron, and a neutrino, all three of which have spins of either $+\frac{1}{2}$ or $-\frac{1}{2}$, and the sum of the three ($+\frac{1}{2}$, $+\frac{1}{2}$, $-\frac{1}{2}$, for instance) adds up to $+\frac{1}{2}$, which would be that of the original neutron, angular momentum is conserved.

There is a fourth law of conservation, discovered much later than the others: the law of conservation of lepton number. A neutron and a proton each have a lepton number of 0, while an electron has a lepton number of +1 and a positron a lepton number of −1.

A neutron, then, starts off with a baryon number of +1 and a lepton number of 0. If it breaks down into a proton (baryon number equal to +1 and a lepton number equal to 0) and an electron (baryon number equal to 0 and a lepton number equal to +1), the baryon number is conserved, but lepton number is *not*.

But suppose a neutrino is also formed with a lepton number of −1. In this case, the neutron (baryon number +1 and lepton number 0) breaks down into a proton (baryon number +1 and lepton number 0), an electron (baryon number 0 and lepton number +1), and a neutrino (baryon number 0 and lepton number −1). Here, you start with a neutron (baryon number +1 and lepton number 0) and end with three particles with a total baryon number of +1 and a total lepton number of 0. Lepton number is now conserved, as well as baryon number.

Of course, for the neutrino to have a lepton number of −1, it should by rights be the mirror image (antineutrino) of the particle, but this is all right. An antineutrino conserves the laws of energy, electric charge, momentum, and angular momentum just as well as a neutrino. The antineutrino also conserves lepton number.

An undetectable particle designed simply to save a single law of conservation is not very convincing. Four different undetectable particles designed to save each of four laws of conservation are even less convincing. However, a single undetectable particle that happens, in itself, to save each of four conservation laws—energy, momentum, angular momentum, and lepton number—becomes very convincing. As the years went by, physicists, more and more, took the attitude that neutrino and antineutrino, whether detected or not, must exist.

Detecting the Antineutrino

Physicists would not feel completely comfortable about the existence of the neutrino and antineutrino until they had been detected. (Usually, the term neutrino is used, for simplicity's sake, to include the antineutrino.)

To be located, a neutrino must interact with another particle, and the interaction must be detectable, as well as distinct from other interactions. In other words, you must recognize an interaction as being caused by a neutrino and by nothing else.

This is not easy because the neutrino hardly ever interacts with anything. The average neutrino can pass through 3,500 light-years of solid lead, it is calculated, before being absorbed.

This is the *average* neutrino. Individual neutrinos might continue avoiding direct hits by sheer chance and travel twice as far as that, or a million times as far, before being absorbed. Others might just happen to make a direct hit and be absorbed after having traveled only half the average distance, or one-millionth. This means that if you deal with a beam containing trillions upon trillions of neutrinos and have them all pass through a quantity of matter in the laboratory, a very few might just happen to strike some particle in that matter and interact.

To have any chance of detecting a neutrino, then, you must have a rich source of them. That rich source became available once nuclear reactors, based on fissioning uranium nuclei, were devised.

Uranium nuclei, being very complex, need a great many neutrons even to be nearly stable; 143 neutrons to 92 protons, in the case of uranium-235. When the uranium nucleus breaks into two smaller fragments, each requires fewer neutrons to be stable, and so some neutrons are set free. With time, many of these break down into protons, liberating antineutrinos as well. A typical fission reactor might easily give off a billion billion antineutrinos every second.

The next problem would be to decide what to expect the antineutrino to do. We know that a neutron breaks down into a proton by emitting an electron and an antineutrino. Can we reverse this process by having a proton absorb an electron and an antineutrino at the same time to become a neutron again?

This is asking a great deal. An antineutrino hitting a proton is very unlikely in itself. To expect it to hit a proton just as an electron is also hitting the same proton is asking entirely too much. It would happen so rarely as to be quite an impractical process. However, having an electron hit a proton is the equivalent of having a proton emit a positron. (This is like saying that having someone give you a dollar is equivalent to having someone pay off a dollar debt that you might have. Either way, your assets go up by one dollar.)

This means that it is possible for an antineutrino to hit a proton, which will then give up a positron and become a neutron. This conserves the baryon number because a proton becomes a neutron, both with a baryon number of $+1$. It also conserves the lepton number, for an antineutrino disappears and a positron appears, each with a lepton number of -1. It conserves electric charge because a proton disappears and a positron appears, each with a change of $+1$. The laws of conservation of energy, momentum, and angular momentum are also upheld.

Suppose, then, that an antineutrino does hit a proton and produces a neutron and a positron. How can you tell that it has happened? It only happens at long intervals and, meanwhile, all sorts of other interactions are taking place, drowning out the antineutrino action. But if a neutron and a positron are produced, the positron is bound to combine with any electron it meets within a millionth of a second, undergoing mutual annihilation. In the process, two gamma rays are formed of equal strength, traveling in opposite directions, with a total energy equivalent to the mass of the two particles. As for the neutron, it could be quickly absorbed by the nucleus of a cadmium atom (if there are any in the neighborhood). The nucleus will gain enough energy in the process to emit three or four photons with a fixed total energy.

No other known interaction will produce exactly this result. If, then, you locate photons emitted all at the same time, in the proper directions, and with the proper energies, you have detected the interaction of a neutrino with a proton, and nothing else.

A team of American physicists led by Frederick Reines (b. 1918) and Clyde Lorrain Cowan (b. 1919) began to tackle the problem along these lines in 1953. They had the use of a fission reactor and saw to it that as many of the antineutrinos as possible struck large tanks of water, which contained trillions of trillions of protons in the hydrogen and oxygen nuclei making up the water molecules. The chemical cadmium chloride had been dissolved in the water. The cadmium nuclei would act to pick up any of the neutrons given off. Finally, the setup included devices that would detect the gamma ray photons and determine their directions and energies, so that all they had to do was to wait for the right combination of photons.

Naturally, in order to detect the right combination as easily as possible, they had to cut out as many wrong combinations as possible, so they shielded the entire apparatus more and more efficiently as time went on. In the end, virtually nothing got in *but* antineutrinos. Eventually, they managed to cut down enough of the "background noise" to be sure that they were detecting the occasional interaction of an antineutrino with a proton. In 1956, twenty-six years after Pauli's suggestion, Reines and Cowan announced that they had detected the antineutrino.

Other scientists instantly tried to repeat the experiment, or to try modifications, and there was no question about it: given the proper equipment, anyone could detect antineutrinos. It was no longer a ghost particle, but exactly the particle that Pauli and Fermi had deduced as being necessary to explain the details of beta-ray radiation and the breakdown of the neutron. (What a testimonial this was to the value of the use of logic in science. It also showed how important it is to stick to a good theory—such as the various laws of conservation—as long as is reasonably possible. Of course, the time might come when an idea that has seemed as firm as steel must be given up—even laws of conservation—and we shall come across such cases. It is the glory of science that it occasionally corrects itself, however reluctantly. No other variety of human intellectual endeavor seems to have quite the same built-in machinery with which to do so.)

Detecting the Neutrino

Just as fission produces a flood of antineutrinos because of the numerous conversions of neutrons to protons it makes necessary, so fusion produces a flood of neutrinos because of equally massive conversions of the same two particles. In the fusion of hydrogen to helium, for instance, four hydrogen nuclei, made up of four protons altogether, are converted into one helium nucleus made up of two protons and two neutrons. In the process, two positrons are formed, and with the positrons, two neutrinos.

Whereas we have working fission reactors to provide

us with floods of antineutrinos, we do not, as yet, have working fusion reactors to provide us with floods of neutrinos. The uncontrolled fusion of a hydrogen bomb produces floods of neutrinos for a while, but working close enough to such an explosion in order to take advantage of the burst is not a very practical notion. However, we do have an enormous and continuously "exploding" hydrogen bomb about 93 million miles away: the Sun. It produces an incredible number of neutrinos every second, and has been doing so for about four and a half billion years.

Produced at the center of the Sun, where fusion is taking place, are photons. The photons react very readily with matter, so that they are absorbed, reemitted, absorbed again, and so on indefinitely. It takes a tremendous length of time for photons to make their way from the Sun's center to its surface, where they are launched into space, some to reach the Earth. So much has happened to these photons in their travels within the Sun, however, that nothing about them is likely to tell us very much about what happens at the Sun's center.

Neutrinos are another case. They interact so slightly with matter that, after being formed, they move from the Sun's center to its surface in a little over two seconds. (Because neutrinos are massless, they travel at the speed of light, as do photons and gravitons.) Once the neutrinos reach the Sun's surface, they continue going and, if they happen to be moving in the direction of the Earth, reach us in eight minutes.

Because the neutrinos reach us directly from the Sun's center, there is at least a chance that, from their properties, we might be able to garner information about that center that is unavailable in other ways. To detect solar neutrinos, then, is much more than to simply prove their existence. It is to investigate the Sun.

In order to detect neutrinos, we have to make use of

236

a particle interaction that is the reverse of the one used to detect antineutrinos. To detect antineutrinos, we have them strike protons to produce neutrons and positrons. To detect neutrinos, we must have them strike neutrons to produce protons and electrons. In detecting antineutrinos, then, we need a target rich in protons, such as water. In detecting neutrinos, we need a target rich in neutrons, and for that we have to use a neutron-rich nucleus.

A particularly neutron-rich nucleus suggested by the Italian physicist Bruno M. Pontecorvo (b. 1913) is chlorine-37, which has a nucleus made up of 17 protons and 20 neutrons. If chlorine-37 absorbs a neutrino, one of its neutrons will be turned into a proton, with the emission of an electron.

But why this neutron-rich nucleus rather than any other? Because when the chlorine-37 nucleus loses a neutron and gains a proton, it becomes argon-37 (18 protons and 19 neutrons), which is a gas that can be easily removed from the material containing the chlorine-37 nuclei. The recovery of such a gas indicates a neutrino absorption and nothing else.

It might seem that the best way of getting a chlorine-37 target is to use chlorine itself, but chlorine is a gas, and it would be difficult to separate very small quantities of another gas from it. One might liquefy the chlorine (and argon would still be a gas at the temperature of liquid chlorine), but that would require refrigeration. It would be much better to use a compound that is liquid at room temperature and that has molecules containing many chlorine atoms.

One such compound is perchloroethylene, in which each molecule is made up of two carbon atoms and four chlorine atoms. It is a chemical used as a common dry-cleaning compound, and is not very expensive. If even a few atoms of argon-37 are formed, they can be flushed out of the liquid by a stream of helium, and can then be detected, for argon-37 is radioactive and can be identified even in minute traces by its characteristic form of breakdown.

It was by using this interaction that the American physicist Raymond E. Davis showed that the neutrino actually exists.

Beginning in 1965, Reines, one of the discoverers of the antineutrino, began working on the detection of solar neutrinos. He made use of large vats of perchloroethylene buried deep in a mine, with a mile or so of rock between it and the surface. The rock would absorb all radiation, even cosmic-ray particles, *except* for neutrinos, which could easily pass through the entire Earth. (There might also be some particles originating from radioactive material in the rocks immediately surrounding the experiment.)

It is odd to think that the Sun must be studied from a vantage point a mile underground, but that is what Reines proceeded to do. However, whatever way he tried to improve his techniques and to refine his instruments, Reines never obtained more than about ⅓ of the neutrinos he expected to detect.

Why? It might have been that the nature of the observations were in some ways inadequate; or that we don't know all there is to know about neutrinos; or that our theories about what is going on at the center of the Sun are mistaken. No solution has yet been reached on this "mystery of the missing neutrinos," but, when it comes, it is sure to be exciting.

If the Sun produces neutrinos, so, we may be sure, do other stars. However, even the nearest star, Alpha Centauri (a system containing two Sun-like stars, and one dim dwarf star), is some 270,000 times as far from us as our Sun. The number of neutrinos reaching us from the Alpha Centauri stars can be, at best, about one fifty-billionth as many as those supplied us by the Sun. We can just barely detect the neutrino emission of the Sun, so we haven't the slightest chance of detecting any from other, normal stars.

But not all stars are normal. Every once in a while, a star explodes as a gigantic supernova, in which radiation of all types suddenly increases a hundred-billion-fold.

In February 1947, such a supernova appeared in the Large Magellanic Cloud, 150,000 light-years away. It was 33,000 times as far away as Alpha Centauri, but the floods of neutrinos it produced more than made up for that. It was the closest supernova to us in nearly 400 years, and the first for which we had "neutrino telescopes," such as that Reines had been working with.

One such neutrino telescope was placed under the Alps. A team of Italian and Soviet astronomers there detected a sudden burst of seven neutrinos the night before the supernova was detected by eye. It turns out, then, that as astronomers improve their ability to detect and study neutrinos, they will not only learn more about what goes on in the center of our Sun, but also about what goes on in colossal star explosions and, perhaps, about other facets of astronomical lore.

Other Leptons

We have thus far described four leptons: the electron, positron, neutrino, and antineutrino. The electric charges are, respectively, -1, $+1$, 0, and 0. The masses are (if we set the mass of the electron at 1), respectively, 1, 1, 0, and 0. Their spins are either $+\frac{1}{2}$ or $-\frac{1}{2}$; it is this half-spin that makes them all fermions. (Photons and gravitons have 0 mass and 0 electric charge, but they have spins of 1 and 2, respectively; the whole-number spin makes them bosons.)

This was the situation as late as 1936, when the neutrino, the antineutrino, and the graviton had not yet been actually detected, but seemed certain of existence in each case. At the time, there were also four baryons known: the proton, neutron, antiproton, and antineutron.

Add to these the photon and graviton and you have ten particles altogether, which seemed to account for every piece of matter in the Universe, as well as all of the interactions that scientists had observed. It would have been nice to end there, for a ten-particle Universe is reasonably simple.

In 1936, however, Anderson, who had discovered the positron four years earlier, was still studying cosmic rays in the mountains, and noticed particle tracks that curved in an odd manner. The curve was less sharp than that of an electron, so it had to be more massive than an electron (assuming that the new particle had the same electric charge). It was sharper than that of a proton, however, indicating that it had to be less massive than a proton. Moreover, there were curves of this sort, otherwise exactly alike, in both directions, indicating that some were particles and some antiparticles.

The conclusion was that there existed particles and antiparticles of intermediate mass, between those of the known leptons and the known baryons. Measurements showed that the new particles were 207 times as massive as electrons and, therefore, about ⅑ the mass of a proton or neutron.

Anderson at first called the new particle a mesotron, the prefix meso- coming from a Greek word meaning "middle," or "intermediate." Notice again that improper -tron ending. This time, the ending did not stick, fortunately, and the term *meson* came to be used, instead, as a general term for all particles of intermediate mass. Because Anderson's particle turned out, eventually, to be only one of a number of such intermediate-mass particles, each had to be distinguished from the others. Anderson's particle came to be called a mu meson, where mu is a letter of the Greek alphabet, equivalent in sound to the English *m*. Then it turned out, as I shall explain later, that the mu meson was different from other intermediate-mass particles in a very fundamental way. The term meson was therefore restricted to the others and *did not include the mu meson*. Anderson's particle is, therefore, now called a muon.

The muon was the first particle discovered that did not have an obvious use either in forming part of the structure of atoms, in preserving the laws of conservation, or in mediating subatomic interactions. The Austrian-American physicist Isidor Isaac Rabi (1898–1988) is supposed to have asked, on hearing of the muon, "Who ordered that?"

The muon has an electric charge of − 1, precisely that of the electron, whereas the antimuon has an electric charge of + 1, precisely that of the positron. In fact, the negative muon, except for its mass and one other property, is in all respects identical to the electron, while the antimuon is similarly identical to the positron. This is true of such things as electric charge, spin, and magnetic field. A negative muon can even replace an electron in an atom, producing a short-lived muonic atom.

In order to conserve angular momentum, a muon must have the same angular momentum as the electron it replaces. In that the muon has much more mass than an electron, which would add to the muon's angular momentum, it must decrease it by moving in closer to the nucleus than the electron ever is. We can also see that this must be so because the muon, with far greater mass than the electron, has a much shorter associated wave, which can squeeze into a much tighter orbit.

Because muonic atoms are, for these reasons, much smaller than electronic atoms, two muonic atoms can get much closer to each other than electrons can. The nuclei of

241

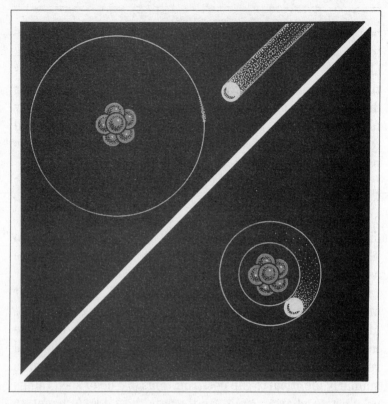

A negative muon, except for its mass and one other property, is identical to an electron. It can even replace an electron in an atom, producing a short-lived muonic atom.

muonic atoms, therefore, have a far greater tendency to fuse than do ordinary electronic atoms. Muonic atoms might thus seem a possible route for practical fusion—except for one enormous catch that we will come to later.

Muons and antimuons can undergo mutual annihilation, producing 207 times as much energy as do electrons and positrons. Similarly, if 207 times as much energy is concentrated into a tiny area as would suffice to form an electron-positron pair, a muon-antimuon pair can be formed.

But what happens if a muon is produced without its

antimuon, as an electron is produced without a positron by the breakdown of a neutron? There are negatively charged particles more massive than a muon (particles I will get to later), which break down to form a muon *without* also forming an antimuon. Similarly, there are positively charged particles that will form an antimuon without a muon.

This does not violate the law of conservation of electric charge, but, as in the case of neutron breakdown, there is a violation of the laws of conservation of energy and momentum. In addition, the heavier particles that break down into muons are themselves neither baryons nor leptons, but the muon *is* a lepton; therefore, while the breakdown does not violate the law of conservation of baryon number, it does violate the law of conservation of lepton number by seeming to form a lepton out of nothing. Here, too, as in the case of neutron breakdown, the simultaneous formation of neutrinos and antineutrinos preserves the laws of conservation, but with an added complication I will get to later.

We can think of a muon as merely a massive electron, and of the antimuon as a massive positron, but why should they exist, and why should they have masses equal to 207 times that of an electron or positron, rather than any other mass?

Let's make an analogy. Suppose an electron is a golf ball lying at the bottom of an energy valley. There is no way for it to descend any lower, so it just stays there. If, however, energy is added to it (as would happen if a golf ball in that position were struck by a golf club), the added energy would cause the golf ball to roll up the hillside. It would reach some maximum height, then roll down to the valley bottom again, giving up the added energy as it did so.

The harder the golf ball is hit, the higher up the hillside it rolls before descending again. If it is hit hard enough, it might just roll high enough to reach a ledge on the hillside where it might remain. It would still be a golf ball (i.e., electron), but it would have gained sufficient energy to be far above its normal position at the bottom of the valley, or, in subatomic terms, it would have gained far more mass than it had at the bottom of the valley.

The ledge happens to be at an energy height equivalent to 207 times the mass of the electron. Why is it just at that height? We don't know, but our inability to advance an exact reason is not dreadfully disturbing. (Science has not succeeded in explaining everything about any significant branch of knowledge and, perhaps, never will. Scientists have discovered innumerable answers, or apparent answers, to innumerable problems, but each answer supplies us with new and more subtle, and, perhaps, more intractable problems.)

Unstable Particles

The electron, as we know, is a stable particle. By stable, I don't mean that nothing can happen to it. If an electron meets a positron, both undergo annihilation and are converted into photons. If an electron collides with particles other than a positron, it can undergo other types of changes.

However, if an electron is isolated in space and does not encounter any other particles, it will (as far as we know) remain in existence, retaining all of its properties unchanged, forever.

The same is true of a positron, as well as of a neutrino and an antineutrino. All four leptons known prior to the discovery of the muon are stable particles. (The same is true of the two bosons known in the 1930s, the photon and the graviton.)

Of the particles that are *not* leptons, and that were

known prior to the discovery of the muon—the proton, antiproton, neutron, and antineutron—the proton and antiproton seem to be stable (although there are now doubts about this as we shall see).

The neutron and the antineutron are *not* stable. If a neutron is isolated from all other particles, it will nevertheless break down into a proton, an electron, and an antineutrino, while an antineutron will break down into an antiproton, a positron, and a neutrino. However, this is a relatively slow process, taking several minutes on average. In addition, when a neutron forms part of a nonradioactive nucleus, it is stable and can remain there, unchanged, indefinitely. Muons, however, break down into electrons almost at once. The average muon, when left strictly to itself, breaks down in only $\frac{1}{500,000}$ of a second.

Why should the muon endure for so short a time? Consider the analogy I have already used, in which an electron is driven up the hillside of mass and comes to rest on a ledge representing 207 times the mass at the bottom of the valley. We can picture the ledge as a narrow one, and the muon resting on that ledge as vibrating, or trembling. As a result of this tremble, sooner or later, the muon falls off the ledge, slides down to the valley below, becoming an electron again. The "sooner or later" turns out, from the narrowness of the ledge and the magnitude of the tremble, to be $\frac{1}{500,000}$ of a second.

All objects, including you and me, exhibit a type of tremble, dictated by the fact that quantum mechanics shows all objects to be associated with a wave aspect. For ordinary objects, the tremble is so excessively minute as to be of no importance, but the smaller the mass, the more marked the tremble relative to the size of the object. Subatomic particles have so little mass that the tremble gains considerable importance and must be taken into account in any study of their properties.

The electron also has a tremble—one even more marked than that of the muon—but the electron is at the bottom of the valley. It has no way of falling any farther, and is therefore stable.

In 1975, the American physicist Martin Perl detected an electron-like particle, even more massive than the muon, in the debris produced by collisions within accelerators. He called this particle a tau lepton, where tau is a letter in the Greek alphabet equivalent to the English *t*. It is also called a tauon.

The tauon shares all of its properties but two with the electron and the muon. Of the two properties that are distinctive, one is mass. The tauon is a super-massive electron, with a mass about 3,500 times that of an electron, and nearly 17 times that of a muon. It is nearly twice as massive as a proton or a neutron, and yet from the way it behaves it is clear that it is a lepton, even though that name is used chiefly for the far better known particles of little or no mass. (It might sound confusing and contrary to common sense that a name that implies smallness should be applied to a very massive particle, but consider this as an analogy. Reptiles, including alligators, anacondas, and the extinct dinosaurs, are *much* larger than insects, if one considers each group as a whole. There are, however, Goliath beetles as large as your fist, while there are lizards small enough to fit on your fingertip; yet the Goliath beetle is an insect and the lizard is a reptile.)

The second way in which a tauon is distinct is in its instability. It is far more unstable than the muon, for it breaks down in only a five-trillionth of a second. In doing so, it changes into a muon, which, in turn, changes into an electron.

We can imagine the tauon resulting from a gain in energy that drives it far higher up the hillside of mass. It reaches a much higher and much narrower ledge than the muon attained. The tauon remains on that ledge for the barest instant before falling off. Can we expect to find additional leptons, then, more massive and more unstable than the tauon? Are there an infinite number of ledges on our allegorical hillside, each higher and narrower than the one before?

Apparently not. Physicists have reason to believe, on the basis of some recent, rather involved observations, that three is the limit and that we now have located all of the leptons there are.

Tauonic atoms, if they existed, would be even smaller than muonic atoms, and therefore still easier to fuse. By now, though, you probably see the catch. These heavy leptons are far too unstable to serve as practical routes to fusion. They would be gone before we could do much with them.

Neutrino Varieties

Let's go into the breakdown of the muon in some detail, because there's a problem there involving the laws of conservation. Suppose the muon breaks down into an electron and an antineutrino. This conserves electric charge and momentum. However, angular momentum is *not* conserved. The muon has a spin of $+\frac{1}{2}$ or $-\frac{1}{2}$, which is also true of the electron and the antineutrino. An electron and an antineutrino, taken together, have a spin of either $+1$, 0, or -1, depending on the signs of their spins. The two cannot possibly have a total spin of either $+\frac{1}{2}$ or $-\frac{1}{2}$.

Why does the muon breakdown seem to break the law of conservation of angular momentum, even with an antineutrino taken into account, when the neutron breakdown does not? This is because the neutron breaks down into

247

three particles—a proton, electron, and antineutrino—and three half-values can add up to a total half-value. The muon breakdown as we have described it so far produces only two particles—an electron and an antineutrino—and two half-values can add up to only an integral value, never a half-value. In other words, we have to suppose that the muon, upon breaking down, also produces three particles— an electron and *two* antineutrinos perhaps.

Unfortunately, this does not necessarily mean that conservation is preserved. The muon has a lepton number of $+1$. The electron and the antineutrino each have a lepton number of $+1$, so that we begin with $+1$ and end with $+2$, violating lepton number conservation. If we add a second antineutrino, we start with $+1$ and end with $+3$, which is still worse. However, suppose the muon breaks down into an electron, an antineutrino, and a neutrino. The neutrino has a lepton number of -1, so the three particles that are produced have lepton numbers of $+1$, $+1$, and -1, for a total of $+1$, which is the lepton number of the original muon. If, then, we suppose that a muon, in breaking down into an electron also produces both an antineutrino and a neutrino, all of the laws of conservation are preserved.

Happy ending? Yes, except for one small point. All of the neutrino-producing interactions physicists had, until the discovery of the muon, observed produced *either* a neutrino *or* an antineutrino. Muon breakdown is odd in that it produces *both* a neutrino *and* an antineutrino.

Can it be that there are two types of neutrino? Can it be that one is produced only by electrons and the other only by muons, so that one can speak of an electron neutrino (and antineutrino) and a muon neutrino (and antineutrino)? Can it be that when the muon breaks down into an electron the muon and electron each produce a variety of neutrino and that this is why the muon breakdown involves two neutrinos? This is called the two-neutrino problem.

If an electron neutrino and a muon neutrino are different in nature, this must be because of some difference in properties—but physicists have not been able to find any. It is even more difficult to study the neutrinos produced by muons than those produced by electrons, but as nearly as physicists can tell, the two types of neutrino are identical. Both have 0 charge, 0 mass, either $+\frac{1}{2}$ or $-\frac{1}{2}$ spins, and so on.

Does this settle the question? Of course not. It might well be that there is a difference in some respect that no scientist has ever thought of, and therefore that no scientist has ever tried to measure, assuming we have the devices with which to measure it.

But if we can't spot any difference directly, perhaps we can spot one indirectly by having the particles make the judgment themselves. Suppose, for instance, that an antineutrino produced by an electron meets a neutrino produced by a muon. If they are identical in all respects, except for being mirror images, they ought to undergo mutual annihilation and produce a minute pulse of energy. If they differ in *any* respect other than being mirror images, they should *not* undergo mutual annihilation. If there is no annihilation, the particles recognize a difference between themselves, and that is good enough. We will take their word for it, even if we don't know what the difference consists of.

However, neutrinos are such infinitely minute, nonreactive particles, that the chance of two of them encountering each other is essentially zero. Even if they did, the energy produced might well be too small to be observed. There is, however, another way in which neutrinos can be made to reveal their nature. If an electron produces only electron neutrinos and a muon produces only muon neutrinos, if the interactions are reversed, the electron neutrino should bring about the production only of electrons and the muon neutrinos only of muons. If the two neutrinos are truly identical, however, they should produce electrons and muons in equal numbers.

Such an experiment was carried out in 1961 by a team headed by the American physicist Leon Max Lederman (b. 1922). They began by hurling high-energy protons into a target of the metal beryllium. This produced a vast number of particles, among which were high-energy muons that decayed to produce high-energy muon-neutrinos. This vast melange of particles was then hurled at a 12-meter-thick slab of steel, which absorbed everything but the neutrinos (which can pass through anything). On the other side of the steel, the stream of energetic muon neutrinos entered a device that could detect neutrino collisions. There wouldn't be many, of course, but over a period of eight months, 56 such collisions were noted, and every one of them produced a muon.

This made it clear that a muon neutrino could *not* produce an electron and was therefore different in some distinctive way (whether we knew the way or not) from an electron neutrino. Lederman got a Nobel prize for this in 1988.

Lederman's work meant that the conservation of lepton number is a little more complicated than had been thought. There is a conservation of electron number and a conservation of muon number separately. Thus, an electron has an electron number of $+1$, and a positron -1. An electron neutrino has an electron number of $+1$, and an electron-antineutrino -1. All four have a muon number of 0. In the same way, a muon has a muon number of $+1$, and an antimuon -1. A muon neutrino has a muon number of $+1$, and a muon antineutrino -1. All four have an electron number of 0.

When a muon, with a muon number of $+1$ and an electron number of 0, breaks down, it forms an electron

(electron number +1, muon number 0), an electron anti-neutrino (electron number −1, muon number 0) and a muon neutrino (electron number 0, muon number +1). All three breakdown particles added together have a muon number of +1 and an electron number of 0, which is what is true of the original muon. The breakdown, therefore, conserves both electron number and muon number.

The tauon also produces a neutrino, which has been little studied so far, that physicists suspect has all of the obvious properties of the other two neutrinos but that is somehow distinct from them. It seems unavoidable to suppose that there is such a thing as conservation of tauon number.

Physicists now speak of three "flavors" of leptons. These are (1) the electron and the electron neutrino, (2) the muon and the muon neutrino, and (3) the tauon and the tauon neutrino. There are also three flavors of antileptons: (1) the antielectron (positron) and the electron antineutrino, (2) the antimuon and the muon antineutrino, and (3) the tauon and the tauon antineutrino. (The term *flavor* is, in some ways, an unfortunate one. It is used in common English to distinguish objects by taste, as in different flavors of ice cream. It isn't quite right to give nonscientists the notion that the differences between subatomic particles are "shades" of differences rather than absolute, measurable ones. However, scientists are human and sometimes reach for a dramatic or even humorous term. For example, some atomic nuclei are easier to hit with subatomic projectiles than others. Those nuclei that are particularly easy to hit were said by some whimsical scientists to be as easy to hit "as the side of a barn." As a result, the nuclear cross-section that gives the measurement of the ease with which a particular nucleus can be targeted is given in a unit called a barn.)

There are, then, twelve leptons and antileptons all

together. They are fundamental particles (or at least are currently considered such) because they will not spontaneously break down into particles that are simpler than leptons. The tauon and the muon break down into electrons, whereas the antitauon and the antimuon break down into positrons. The electron, the positron, the three neutrinos, and the three antineutrinos don't seem to break down at all.

Why are there twelve leptons, when the Universe seems to contain only electrons and electron neutrinos in appreciable numbers? Electron antineutrinos are produced only in radioactive transformations, which are few in number in the Universe as a whole. Positrons are produced in some radioactive transformations, but less often even than electron antineutrinos. The heavier leptons and their neutrinos are produced, as far as we know, only in the laboratory by such things as cosmic-ray bombardment.

Why, then, doesn't the Universe get by on just electrons and electron neutrinos? Why needlessly complicate things? Because, my instinct tells me, the complications are not needless. The Universe is built in such a way that every interaction must play its role. We might not see what possible use the tauon has, for instance, but I have the strong feeling that whatever it is that makes our Universe work as it does requires the tauon's existence; that without the tauon the Universe would not be the Universe we live in and might not even have the capacity to exist.

10

INTERACTIONS

The Strong Interaction

Putting leptons to one side, what about the baryons? What about the particles that make up the atomic nucleus? These represented a serious problem once the neutron was discovered and the proton-neutron structure of the nucleus was advanced. The problem is summed up in the following question: "What holds the nucleus together?"

Until 1935, there were only two interactions known that could hold objects together: the gravitational and the electromagnetic. Of these, the gravitational interaction was so weak that it could be entirely disregarded in the world of subatomic physics. It only makes itself seriously felt when a huge mass is accumulated. It is important on the

253

level of satellites, planets, stars, and galaxies, but certainly not where atoms and subatomic particles are concerned.

That left the electromagnetic interaction. The electromagnetic attraction between positive and negative electric charges is perfectly adequate to explain how molecules are held together in crystals, how atoms are held together in molecules, and how electrons and nuclei are held to each other in atoms, but when scientists got all the way down to the atomic nucleus, they encountered a problem.

As long as they thought that nuclei were made up of protons and electrons, there seemed no problem. The protons and electrons attracted each other strongly; the more strongly, in fact, the closer together they were. In the nucleus, they were virtually in contact. The protons were also virtually in contact with each other, as were the electrons. Between objects of the same electric charge there was a repulsion just as strong as the attraction between objects of opposite charge.

Within the nucleus, then, it might be expected that the protons would repel each other and the electrons would also repel each other, but, presumably, they were intermingled and placed in such a way that the attractions were more effective than the repulsions. This was true in crystals, which were often built up of an intermixture of positively charged ions and negatively charged ions, with the opposite charges distributed so that attractions overwhelmed repulsions and the crystals hung together. In short, the electrons within the nucleus acted as a type of cement for the protons, and vice versa. Between the action of the two cements, the nucleus held together.

But the nature of the nuclear spin and the necessity of conserving angular momentum cast serious doubt on the validity of the proton-electron theory of nuclear structure. With the discovery of the neutron, it quickly became obvious that it was necessary to assume a proton-neutron structure that would solve all of the difficulties raised by the proton-electron structure—except one. The cement had disappeared.

If we consider the electromagnetic interaction only, then, the only force that could make itself felt inside a proton-neutron nucleus is the extremely strong repulsion of each proton for all others. The neutrons, being electrically uncharged, neither attract nor repel the protons and are merely innocent bystanders, so to speak. The strong proton-proton repulsion should suffice to explode the nucleus instantly into individual protons.

Yet this does not happen. The atomic nucleus remains quietly in place, quite stable, with no sign of mutual destructive repulsion among protons. Even in the case of those nuclei that are radioactive, explosions take place in a strictly limited fashion, turning a proton into a neutron or vice versa, eliminating a two-proton, two-neutron alpha particle, or, in extreme cases, breaking into two halves. All of this happens relatively slowly, sometimes exceedingly slowly. *Never* does any nucleus explode, instantaneously, into individual protons.

The natural conclusion we must come to is that there is some interaction involved that is neither gravitational nor electromagnetic—one that human beings have never thought of, let alone studied—and that it is this interaction that holds the nucleus together. It might be called a nuclear interaction.

The nuclear interaction, whatever it is, must introduce a strong attractive force—one that is far stronger than the repulsive force produced by the positive charges on the various protons. In fact, as it eventually turned out, the nuclear interaction produces an attractive force over 100 times as intense as that of the electromagnetic interaction. This is, in fact, the strongest force that is known to exist between subatomic particles (and, it is thought, the strongest that can exist). It is therefore usually called the strong interaction. But what is the strong interaction? How does it work?

The first to consider the strong interaction was Heisenberg, who, in 1932, had first suggested the proton-neutron structure of the nucleus. This is not surprising. When a scientist advances a startling notion that solves a great many problems, but which has a gaping hole in it, he is bound to do his best to mend that hole. After all, it's his baby.

Heisenberg worked up the notion of exchange forces. This is something that classical physics, the type of physics that existed before quantum mechanics was devised, could not deal with or understand. If quantum mechanics is used, however, exchange forces are seen to be possible, and effective.

To explain it without mathematics, we can imagine protons and neutrons constantly exchanging something. Let us say that what they exchange (as Heisenberg first suggested) is electric charge. This means that the positive electric charges inside a nucleus are constantly being transferred from a particle possessing a positive charge to a particle that does not. This means that every baryon would be a proton and a neutron in exceedingly rapid alternation. No proton can feel repulsion because before it has time to react to the repulsion it is a neutron. (It's like bouncing a hot potato rapidly from hand to hand to keep from being burned.)

Such an exchange force would set up a powerful attraction and keep the nucleus together; but on closer examination Heisenberg's suggestion proved inadequate, unfortunately. Then the Japanese physicist Hideki Yukawa (1907–1981) tackled the job. It seemed to him that if exchange forces worked inside the nucleus for the strong interaction, they would have to work for all interactions. He

applied quantum mechanics to the electromagnetic interaction and, it appeared to him, that what was exchanged
was a particular particle, the photon. It was the continual,
extremely rapid exchange of photons between any two particles with electric charge that produced the electromagnetic interaction. Between particles with the same charge,
the exchange produced a repulsion, and between those with
opposite charges, it produced an attraction.

Between any two particles with mass, there was also
a rapid exchange of gravitons. (These particles have never
been detected because they are so weakly energetic that
nothing we have yet developed is sensitive enough to demonstrate their existence unequivocally; however, no physicist doubts their existence.) Since there seems to be only
one type of mass, the gravitational interaction produces
only an attraction.

Within the nucleus, then, there would have to be another such exchange particle that dashed endlessly between
the protons and neutrons within the nucleus. Here, however, there was a difference. Both the electromagnetic and
gravitational interactions were long-range effects whose intensity declined only slowly with distance. The effect of
electromagnetism, possessing both attraction and repulsion, is muted, but we can see clearly what this means in
connection with gravitation where masses can be huge and
where there is *only* attraction. The Earth holds the Moon
firmly, even at a distance of nearly 400,000 kilometers
(237,000 miles). The Sun holds the Earth firmly, at a distance of 150,000,000 kilometers (93,000,000 miles). Stars
hold together in galaxies and galaxies in clusters over thousands and even millions of light-years.

The strong interaction, however, decreases in intensity
with distance *much* more rapidly than does either gravitation or magnetism. Double the distance, and the latter
two interactions decrease to one-fourth their intensity;

however, double the distance, and the strong interaction decreases to less than 1 percent of its intensity. This means that the strong interaction is very short-range indeed, and cannot be felt measurably except in the immediate vicinity of the particle that gives rise to it.

In fact, the effective range of the strong interaction is only about a ten-trillionth of a centimeter, or about $\frac{1}{100,000}$ the width of an atom. The only way, then, that protons and neutrons can feel the attractive effect of the strong interaction is to remain in close contact. This is why atomic nuclei are so small. They are just large enough to come under the umbrella of the strong force. Indeed, the largest nuclei known are so wide that the strong interaction has difficulty stretching the required distance, therefore there is a tendency for these nuclei to undergo occasional fission.

It is because of this difference in range that only gravitational and electromagnetic interactions are part of the common experience of human beings. The former has been known to exist from the very dawn of human intelligence, and the latter from the days of ancient Greece. The strong interaction, however, expressing itself only over nuclear distances, couldn't possibly be experienced until the nucleus was discovered and its structure understood—that is, not until the 1930s.

But *why* this difference in range? It seemed to Yukawa that, from quantum mechanical considerations, a long-range interaction required a massless exchange particle. In that the photon and the graviton, both massless, had electromagnetic and gravitational forces that were long-range, the strong interaction had to have an exchange particle with mass because it acted over a very short range. In fact, Yukawa calculated the mass of the exchange particle to be about 200 times that of the electron.

At that time, no particles were known with masses in this range, which gave Yukawa the depressing feeling that his theory was wrong, but he published it anyway in 1935. Almost at once, however, Anderson discovered the muon and it was precisely in the range of mass that Yukawa had predicted for his exchange particle of the strong interaction. Naturally, everyone thought that the exchange particle had been discovered. Interest in Yukawa's theory flared up.

However, the interest died down quickly. The muon showed no tendency to interact with protons and neutrons, and thus could *not* be the exchange particle. In fact, it was simply not subject to the strong interaction at all, and that was the chief reason for classifying it as a lepton. None of the leptons are subject to the strong interaction. In fact, once the muon was recognized as merely a massive electron it was realized that it could no more be the exchange particle than was the electron.

Disappointment did not last forever. The British physicist Cecil Frank Powell (1903–1969) was studying the effect of cosmic rays on the atmospheric atoms and molecules they struck. He, too, like Anderson, invaded the mountain heights in order to do so. In the Bolivian Andes, for instance, he was at a high enough altitude so that the cosmic ray intensity (coming from outerspace and not absorbed by the lowermost thicknesses of Earth's atmosphere) was ten times higher than at sea level. Powell made use of special instruments of his own devising, more sensitive than Anderson's, to detect, in 1947, the curved tracks of particles of intermediate mass.

The new particle was judged, from its curvature, to have a mass equal to about 273 times that of an electron (close to Yukawa's prediction), being about a third more massive than the muon. The new particle was also just about as unstable as the muon, breaking down, on average, in about $1/400,000$ of a second.

These similarities were purely coincidental, however, for there was a deep and fundamental difference between the particles. Powell's particle was *not* a lepton. It was subject to the strong interaction and it interacted with protons and neutrons readily. In fact, it was the exchange particle that Yukawa had predicted.

The new particle was called the pi meson (pi being a Greek letter equivalent to the English letter *p*, standing for Powell, I presume). It was the first of a new class of particles, all subject to the strong force, called mesons (a name originally given to the muon, but withdrawn because the muon turned out to be a lepton). Despite the fact that the pi meson has the right to be called a meson, its name is frequently shortened for the sake of convenience to pion.

There is, of course, a positively charged pion, with a charge exactly equal to that of a proton or a positron, and a negatively charged antipion, with a charge exactly equal to that of an antiproton or an electron. The pion breaks down into a muon and a muon antineutrino, and the antipion breaks down into an antimuon and a muon neutrino, which conserves muon number because the muon and the muon antineutrino have muon numbers of $+1$ and -1, respectively, whereas the antimuon and the muon neutrino have numbers of -1 and $+1$, respectively. In that the pions have muon numbers of 0, the muon number is 0 both before and after the breakdown.

There is also a neutral pion, which has no electric charge and only about $29/30$ of the mass of the charged pions. It is less stable than the charged pions, has an average lifetime of only about a millionth of a billionth of a second, and breaks down into two gamma rays. The neutral pion is one of the few particles that, like the photon and the graviton, have no antiparticles; or, to look at it another way, it serves as its own antiparticle.

Mesons, by the way, have spins of 0, and are therefore not fermions; there is no law of conservation of meson number. Mesons can appear and disappear at will.

As a result of the discovery of the strong interaction, Yukawa received a Nobel prize in 1949, and Powell received one in 1950.

The Weak Interaction

While the strong interaction was an extremely dramatic discovery, it was not the first new interaction to be discovered in the 1930s. In 1933, Fermi, who was later to bombard uranium with neutrons (with consequences of enormous importance), grew interested in the work that Dirac had done on the electromagnetic interaction. In trying to describe the manner in which photons were emitted in electromagnetic interactions, Dirac had come upon the notion of antimatter.

It occurred to Fermi that the manner in which neutrons gave off electrons and neutrinos might be treated in the same way, mathematically, as was the manner in which particles gave off photons. He came to the conclusion that the mathematics worked, but that it indicated an interaction that was far different from the electromagnetic interaction that governed the release of photons. This new interaction, which was at first called the Fermi interaction, was much weaker than the electromagnetic interaction. It was, in fact, only about a hundred-billionth as intense as the electromagnetic interaction. (It was less than ten-trillionth as intense as the later-discovered strong interaction.)

The Fermi interaction was very short range, being felt only at distances equal to about a thousandth of the width of the atomic nucleus. It therefore played no role to speak of in the nucleus, but was of importance in the case of single particles. It was a second nuclear interaction (in the sense that it was a second short-range one involving only subatomic particles). After Yukawa's theory was accepted, people spoke of the strong nuclear interaction and the weak nuclear interaction, the latter replacing the earlier Fermi interaction.

But then, in the interest of word economy, I suppose, the "nuclear" was dropped and scientists began to speak of the strong interaction and the weak interaction.

(This last is, to my way of thinking, not entirely appropriate, for though the weak interaction is far weaker than the strong and the electromagnetic reactions, it is, nevertheless, ten thousand trillion trillion times as intense as the gravitational interaction. It is the gravitational interaction that has a right to the name of weak.

(And yet I might be wrong in this. In that we are only truly acquainted with the gravitational interaction in connection with the huge mass of the Earth and of other astronomical bodies, there is no way in which we can consider gravitation weak in a practical, rather than a subatomic, sense. We have only to fall down to put weakness out of our mind in connection with gravitation. And, indeed, if enough mass is accumulated and squeezed into a small enough volume, the total gravitational intensity becomes stronger than anything conceivable—so intense that even the strong interaction could not fight it. If we stop to think of it that way, the weak interaction is the weakest, at least in the manner in which we ordinarily encounter it—so perhaps the name is not a bad one, after all.)

Some individual particles undergo changes, such as breakdowns, or interactions with each other, that are mediated by the strong interaction, and some undergo changes that are mediated by the weak interaction. Naturally, those events that are mediated by the strong interaction take place much more rapidly than those mediated by the weak interaction, just as a baseball travels more rapidly if thrown by a major-league pitcher than if thrown by a five-year-old child.

In general, events mediated by the weak interaction are likely to take place in a millionth of a second or so, while those by the strong interaction take less than a trillionth of a second—sometimes taking place in just a few trillionths of a trillionth of a second.

Baryons and mesons can both respond to either the strong interaction or the weak interaction, but leptons respond to the weak interaction only. (Baryons, mesons, and leptons can all respond to the electromagnetic interaction if, and only if, they are electrically charged. Neutrons, neutral pions, and neutrinos do *not* respond to the electromagnetic interaction.) This is why lepton events, such as the decay of the pion into a muon, that of the muon into an electron, or that of the radioactive productions of beta particles, all tend to happen in what seems slow motion on the subatomic scale. The neutral pion, which does not break down into muons, is affected by the strong interaction and therefore breaks down much more rapidly than do the charged pions.

The weak interaction differs from the other three in that it alone is not involved in some very obvious force of attraction. The gravitational interaction holds astronomical bodies together and allows the solar system to exist. The electromagnetic interaction holds atoms and molecules together and allows the Earth to exist. The strong interaction holds baryons together and allows atoms to exist.

The weak interaction does not hold anything together. It merely mediates the conversion of certain particles into other particles. This is not to be taken lightly, however. For one thing, it mediates the processes whereby protons join each other to form helium nuclei. This is the nuclear fusion process that keeps the Sun shining and makes life on Earth possible.

The weak interaction raised a problem, however. If the other three interactions all exert their effects by means of exchange particles, then the weak interaction must also have an exchange particle. In that the weak interaction is short-range, its exchange particle should have mass. In fact, in that the weak interaction is considerably shorter in range than the strong interaction, it ought to have an exchange particle considerably more massive than the pion.

A theory advanced first in 1967 (which I'll get to later) indicated that the weak interaction ought to have three exchange particles—one positively charged, one negatively charged, and one neutral—and that these must be perhaps 700 times as massive as the pion and 100 times as massive as the proton.

The exchange particles were referred to as W particles, the W standing for weak. The electrically charged particles were symbolized as W^+ and W^-, and the neutral particle Z^0.

It was important to find these exchange particles, not just to add them to the collection of particles scientists knew about, but because their existence would verify the theory that predicted them. This would be undeniably true if their masses were really as unbelievably enormous as the theory predicted. The theory was, as we shall see, an important one, and the detection of the exchange particles was vital to its importance.

The catch lay in the huge masses of the particles. Correspondingly huge energies would have to be disposed of to create particles that could be detected. It was not until 1955 that a large enough energy production capability became available to produce and detect the antiproton. To do the same for W particles would require the concentration of at least a hundred times as much energy.

It was not until the 1980s that particle accelerators were devised that would supply the necessary energies. A group of American scientists at Fermilab in Batavia, Illinois, was striving for it, and a group of European scientists at the European Center for Nuclear Research (CERN) near Geneva, Switzerland, was doing the same at this time.

The task for these two groups was not just a matter of energy. If the particles appeared, they would endure too brief a time to be detected directly. They would have to be identified by their breakdown products (muons and neutrinos), which would have to be detected among large numbers of other particles formed at the same time.

The race between the two laboratories was, therefore, an intricate one. As it happened, Fermilab ran out of money and had less adequate equipment. The CERN was, on the other hand, under the demonically energetic leadership of the Italian physicist Carlo Rubbia (b. 1934), and it won out.

Rubbia modified existing instruments to do the job and, in 1982, obtained 140,000 particle events that might conceivably have resulted in the production of W particles. Making use of computers, these were reduced to but five events that could be explained only as W^- particles in four cases, and W^+ in one. Furthermore, they managed to measure the energy of these particles, and from this worked out their mass, which turned out to be right on the nose, exactly as the theory had predicted.

This was announced on January 25, 1983. Rubbia continued to look for the Z^0 particle, which was 15 percent more massive than the W particles, and therefore harder to detect. In May 1983 it was detected, and the announcement made in June. In 1984, Rubbia received a Nobel prize for this work.

There is one other particle that might exist in connection with the theory. This is the Higgs particle, after the British physicist Peter Higgs, who first proposed its existence. The theory is not clear on what its mass and other properties are. The thought is that it is considerably more massive than the W particles; therefore, no one is really sure when it can be detected. It remains an achievement for the future.

The Electroweak Interaction

We have examined four interactions: strong, electromagnetic, weak, and gravitational, in order of declining intensity. Are there any more? Scientists, generally, are of the strong opinion that there are no more.

But it might be difficult to be sure of this. After all, as recently as 1930, scientists knew of only two interactions, electromagnetic and gravitational, and then the two nuclear interactions turned up. However, the new interactions were not unexpected. The very existence of radioactivity was unsettling in this respect, in that it was clear that neither gravitation nor electromagnetism was useful in explaining it. Once the structure of the atomic nucleus was worked out, there was a shouting need for something new.

The situation today is quite different. In the half-century since the two nuclear interactions were discovered, there has been intense research into every facet of sub-atomic physics with instruments of unprecedented power and subtlety. In addition, scientists have studied the larger world about us and have probed the Universe with instruments and devices undreamed of in the 1930s.

Discoveries have been made in vast numbers that no one could have predicted, and it is clear that more has been done in the way of scientific investigations and findings in the last fifty years than in all of the thousands of years before.

And yet all of the scientific investigations of the last half century have not isolated a single phenomenon anywhere in the entire range from the Universe to the neutrino that can't be explained by one of the four interactions. The need of a fifth interaction has never shown up, and it is this that leads scientists to believe that four interactions are all there are.

In the late 1980s, to be sure, there was some talk of a fifth interaction that was even weaker than gravitation, that had a range intermediate between that of the nuclear reactions and the other two, and that varied with the chemical compositions of the material involved. For a while, some interest was stimulated by the announcement, but the properties of the interaction were such a tissue of complexity that to me, at least, it seemed very unlikely from the start. As it happened, it faded out rapidly.

Of course, it remains possible that some aspect of the Universe might yet be discovered that lies well outside our knowledge, and that will come as a complete surprise (such as the discovery of radioactivity in 1896). Such a discovery might make necessary the development of additional interactions useful under conditions that until now we have never had occasion to study, but the chances of this seem small.

On the whole, then, the question of the number of interactions is not "Why aren't there more interactions?" but "Why are there as many as four?" Scientists have the definite feeling that there is a principle of economy, so to speak, in the structure of the Universe; that its workings are as simple as possible; that two tasks are not accomplished by two utterly different pathways if they can be done by a single pathway suitably modified to fit both cases.

Thus, as late as 1870, there seemed to be four different phenomena that could make themselve felt across a vacuum: light, electricity, magnetism, and gravitation. All four seemed quite distinct. Nevertheless, Maxwell, as mentioned earlier in the book, in one of the great scientific insights of all time, prepared a set of equations that governed both electricity and magnetism and showed them to be inextricably related. Moreover, if the electric field and the magnetic field were combined into an electromagnetic field, it turned out that light was a radiation intimately and inextricably related to that field. Maxwell could predict an entire family of light-like radiations, from radio waves to gamma rays, with his new insight—radiations not actually discovered until a quarter century later.

It seemed only natural to try to expand Maxwell's treatment further to include the gravitational interaction. Einstein spent the final third of his life at this task, but failed, as did everyone else. Then, in the 1930s, the situation was complicated by the discovery of the two nuclear interactions, the strong and the weak, with science finding itself with four fields again. But this didn't mean scientists gave up trying to find ways of describing all of the fields with one set of equations (a unified field theory). The lure of showing the Universe to be as simple as possible is too strong to be ignored.

In 1967, the American physicist Steven Weinberg (b. 1933) worked out a set of equations that would cover both electromagnetic and weak interactions. The two seemed so different in nature, and yet Fermi had worked out the theory of weak interactions by using the type of mathematics Dirac had used for electromagnetic interactions, so there must be *some* similarity.

Weinberg came up with a treatment that placed the two interactions under a single umbrella, which showed that for what might be called the electroweak interaction there had to be four exchange particles. One was massless, and was undoubtedly the photon. The other three had mass, a great deal of mass, and were what came to be called the W^+, W^-, and Z^0 particles. (There was also the Higgs particle, but that was less certain.)

At about the same time, the Pakistani-British physicist Abdus Salam (b. 1926) produced an almost identical theory, quite independently. (This is not very surprising. It often happens in science that when scientific information reaches a certain level in some field, some startling advance is crying to be made—the time is ready for it, so to speak—and more than one person responds. The most startling case of this sort of thing took place in 1859, when Charles Robert Darwin and Alfred Russel Wallace, independently and simultaneously, made ready to publish the theory of biological evolution by natural selection.)

The electroweak interaction did not achieve immediate recognition and acclaim. The mathematics was incomplete in certain respects, and it was only a few years later that the Dutch physicist Gerard 't Hooft refined the mathematics appropriately.

If the electroweak interaction existed, there should be neutral currents. In other words, there should be particle interactions involving an exchange particle of the weak interaction that did not involve a shift of electric charge from one particle to another. It was for this neutral current that the Z^0 particle was necessary. In 1973, such neutral currents were actually detected, and suddenly the electroweak theory began to look very good. In 1979, Weinberg and Salam received Nobel prizes for it. The actual detection of the weak interaction exchange particles in 1983 put the cap on the theory.

You might wonder, if there is a single electroweak interaction, why the electromagnetic interaction and weak interaction aspects of this single phenomenon are so widely different. Apparently, this is the result of our living at low temperatures. If the temperature were high enough (far, far higher than exists in our environs today), there would really be only one interaction. As the temperature drops, however, the two aspects separate. They are still a single interaction, but are manifested in two widely different forms.

We can make use of an analogy. Water exists in three

forms: liquid water, ice, and vapor. To people unfamiliar with our world, these would seem like three entirely different substances, unrelated to one another.

Now, suppose the temperature were high enough so that all the water would be in the form of vapor. Water would clearly be a single substance with a single set of properties. But let the temperature drop, and some of the vapor would condense into liquid, and the liquid and vapor would remain in equilibrium. Now there would be, apparently, two different substances, with two widely different sets of properties.

If the temperature dropped still lower, some of the water would freeze and you would have ice, water, and vapor all in equilibrium; all three quite different in appearance and properties; and yet all three still the same substance, fundamentally.

There is the thought, therefore, that when the Universe first came into being, it was at an extraordinarily high temperature, something like ten million trillion trillion trillion degrees, and at that time and under those conditions there was only one interaction. As the temperature dropped (very rapidly, as we measure time today), gravitation split off as an apparently separate interaction, which grew weaker as the temperature continued to drop. Then the strong interaction split off, and finally the weak and the electromagnetic interactions split apart.

This makes it seem, naturally, as though the process might be reversed mathematically and that a single treatment might draw all four interactions under one umbrella. Various plans for unifying the electroweak and strong interactions have been advanced, and many scientists are confident of success in developing such a "grand unified theory." So far, however, all attempts to also include gravitation have failed. The phenomenon remains an intractable problem (of which I will have more to say later).

11

QUARKS

The Hadron Zoo

Let's consider the various subatomic particles as we've now described them. First, there are the leptons, which are subject to the weak interaction, and, if electrically charged, to the electromagnetic interaction, but *not* to the strong interaction. These seem to be fundamental particles that have never been shown to have any internal structure. They include three flavors: the electron and its neutrino, the muon and its neutrino, and the tauon and its neutrino. There are also antiparticles for each of these, which brings us to twelve leptons altogether. Scientists don't expect to find any more.

Second, there are the exchange particles, which mediate the four interactions: the graviton for the gravitational interaction, the photon for the electromagnetic interaction, the W particles for the weak interaction, and the pion as Yukawa's exchange particle for the strong interaction. The graviton and the photon are single particles, but the W particles and the pion exist in positively charged, negatively charged, and neutral varieties. This means that there are eight exchange particles altogether. Scientists don't expect to find any more.

This leaves the particles that are subject to the strong interaction. The longest known of these are the baryons; that is, the proton and the neutron, whose existence, cheek by jowl, in the atomic nucleus was the occasion for the development of the strong-interaction theory. In addition, the pions, which are mesons, are subject to the strong interaction.

The particles subject to the strong interaction, the baryons and the mesons, are lumped together as hadrons, from the Greek word for "thick," or "strong." Hadrons are thus a good opposite for leptons, which, as I explained earlier, is from the Greek word for "weak."

If the proton and neutron and their antiparticles, plus the three pions, were all the hadrons there were, there would be seven, a reasonable number. In that the three pions could be counted among the exchange particles, this would mean that leptons, exchange particles, and hadrons, in both their normal form and their antiform, would come to only twenty-four particles altogether, which scientists could live with under a view of the Universe as simple.

However, as particle accelerators grew larger and more efficient, capable of disposing of more and more energy, physicists found that the energy available coalesced, so to speak, into numerous particles that simply didn't exist except under high-energy conditions. These particles were all extremely unstable, enduring at most a millionth of a second, and for the most part for much shorter periods of time.

The new discoveries included the tauon and its neutrino among the leptons, and the W particles among the exchange particles, with all of the rest of the numerous discoveries among the hadrons.

In 1944, for instance, a new particle was discovered that was identified as a meson. It was called the K meson, or, often, the kaon. It had a mass three and a half times that of a pion, and roughly half that of a proton.

In 1947, the first of a group of particles more massive than a proton or neutron was discovered. These were called hyperons, from a Greek word meaning "beyond," in that their masses were beyond those of the proton and neutron, which, until this time, had been thought to be the most massive particles.

This sort of thing continued, and eventually over a hundred hadrons of different sorts were discovered, which implied the existence of a hundred different antihadrons. Some of them endured for only a few trillionths of a trillionth of a second before breaking down, but they were particles just the same.

Scientists were troubled. Every sign had pointed to a satisfactory simplicity of the Universe, and now the "hadron zoo" had reduced things to an apparently meaningless complexity again. Naturally, attempts were made to find order among all of these hadrons; to find ways of grouping them in a meaningful way. If this could be done, one could deal not with many individuals, but with a few groups.

For instance, as early as 1932, Heisenberg pointed out that if one ignored electric charge, the proton and the neutron could be viewed as a single particle in two different states. It was impossible to describe the difference between the states in ordinary terms, but it sufficed to call one state positive and the other negative.

In 1937, the Hungarian-American physicist Eugene Paul Wigner (b. 1902) proposed that the proton and neutron were analogous to isotopes in the periodic table of the elements, and that the two states might be pictured as spins of some sort, in that two spins would account for the difference in states. He called Heisenberg's states isotopic spin, which is now usually shortened to isospin. In 1938, the Russian physicist N. Kemmer pointed out that the three pions—positive, negative, and neutral—could be treated as the same particle in three different isospin states.

The isospin was important, first, because it did group some of the particles and helped ameliorate the hadron complexity, and second because it was conserved among hadrons. This helps make some sense out of the hadron zoo, because all of these particles do not undergo changes and interactions at random, but must conserve various properties. This limits the number of permissible changes. The greater the number of conserved properties that can be worked out, the greater the limitations and the easier it is to understand what is happening.

For instance, kaons and hyperons last a surprisingly long time. It takes a millionth of a second for kaons to break down, and nearly a billionth of a second for hyperons to break down. The mechanics of their production indicates clearly that they are formed through the mediation of the strong interaction, and therefore ought to break down the same way—in a minute fraction of a trillionth of a second.

But they don't; they last for thousands, even millions of times as long as they ought to, and therefore must break down by way of the weak interaction, which seemed strange. In fact, they came to be called strange particles.

In 1953, the American physicist Murray Gell-Mann (b. 1929) suggested there must be a characteristic possessed by strange particles not possessed by other hadrons. He called the characteristic, naturally enough, "strangeness."

The proton, neutron, and various pions each had a strangeness number of 0, but kaons and hyperons did not. The strangeness number was conserved in the strong interaction. Kaons and hyperons could not break down by way of the strong interaction because they formed pions and protons with strangeness numbers of zero, which meant the disappearance of strangeness and a resultant violation of the law of conservation. Kaons and hyperons must therefore break down by means of the weak interaction, in which strangeness need not be conserved. This is why strange particles endure so long.

Studies of the hadrons did not always succeed in establishing or preserving laws of conservation. In one case, the revision of a law of conservation was forced on scientists.

As early as 1927, Wigner had advanced the law of conservation of parity. Parity cannot be explained literally, but we can deal with an analogy here, in terms of odd and even number. Two even numbers always add up to an even number, and two odd numbers always add up to an even number. However, an even number and an odd number always add up to an odd number. If, then, we call some particles even and some odd, the permissible changes must adhere to the same rules: even + even = odd + odd = even; while even + odd = odd + even = odd.

But then, in the early 1950s, it was found that a particular variety of kaon had a peculiar way of decaying. Sometimes it decayed into two pions and sometimes into three. The two pions added up to even parity, but the three added up to odd parity. The question was, then: How could the kaon be both odd and even?

The easiest way out was to suppose that there were actually two very similar particles, one of which was odd parity and one even parity. These were named tau meson and theta meson after two letters of the Greek alphabet. This would have settled the issue except that there seemed no way to distinguish between the tau meson and the theta meson.

This is not a deadly situation, however. The muon neutrino cannot be distinguished from the electron neutrino by any measurable property, but only by their behavior in various interactions. Perhaps this was true of the tau meson and theta meson, too.

In the case of the two neutrinos, however, there had seemed no alternative but to accept an indistinguishable difference. In the case of the two mesons, there was. What if parity was not always conserved?

The Chinese-American physicists Chen Ning Yang (b. 1922) and Tsung-Dao Lee (b. 1926) worked out the theoretical consequences of this in 1956, convinced that parity was *not* conserved, at least in those reactions mediated by the weak interaction. But how could this be demonstrated?

The answer lay in the fact that, in a way, the conservation of parity was equivalent to the notion of left-right symmetry. In other words, if parity is conserved and if a certain interaction produces a stream of particles, these particles will come off to the left and to the right in equal numbers. If, however, parity is not conserved, the particles will come off only to the left, or only to the right. (One of the reasons why scientists found it so difficult to believe that parity was not conserved was that they saw no reason why the Universe should distinguish between left and right.)

An experiment was, therefore, arranged at Columbia University, with another Chinese-American physicist, Chien Shung Wu, in charge. She worked with a sample of the radioactive isotope cobalt-60, which broke down to yield beta particles, mediated, of course, by the weak interaction. These beta particles came off in all directions, partly because the atoms themselves faced in all directions. Wu therefore placed the material in a strong magnetic field so that all of the atoms would line up in the same direction. That would give them a chance to fire off the beta rays in one direction only if parity was not conserved. Of course, at ordinary temperatures, the atoms would wriggle themselves into different directions despite the constraint of the magnetic field; therefore, Wu cooled the cobalt-60 very nearly to absolute zero.

If parity was not conserved, the beta particles should come off only on one side. By January 1957, there was no doubt: the beta particles were coming off in only one direction, and parity was not conserved in weak interactions. That year Yang and Lee received a Nobel prize.

Parity was conserved in other types of interaction and, even in the weak interaction, a more general law of conservation could be substituted. If a particular particle was "left-handed," in terms of parity (P), its antiparticle with an opposite charge (C) was right-handed. This meant that if a particle and its antiparticle were taken together, the property of CP (parity and charge both taken into consideration) would be conserved.

But then, in 1964, the American physicists Val Logsden Fitch (b. 1923) and James Watson Cronin (b. 1931) showed that even CP wasn't always conserved. The property of time (T) had to be added. If CP wasn't conserved in one direction of time, it wasn't conserved, in the opposite way, in the other direction. It is now believed that CPT symmetry is what is conserved in weak interactions. Cronin and Fitch shared the Nobel prize in 1980 for this work.

In 1981, Gell-Mann began using a number of conserved characteristics to group hadrons in symmetrically formed polygons containing eight, nine, or ten individuals. He thus set up families of particles and introduced something analogous to the periodic table of elements. At the same time, the Israeli physicist Yuval Ne'eman (b. 1925) was doing the same thing.

It was difficult for scientists to take Gell-Mann's arrangements seriously, just as it had been difficult for them, a century earlier, to take Mendeleev's periodic table seriously. Mendeleev, however, had won them over when he used the table to predict the properties of undiscovered elements—and had proved to be right.

Gell-Mann envisioned a triangle of ten particles, arranged so that the values of different conserved properties varied in a fixed and regular way from point to point. However, the point at the apex did not correspond to any particle known at the time.

From the arrangement, it could be seen that the missing particle had peculiar properties, including an unusually high mass and an unusually high strangeness number. It was called the omega minus particle, whose existence had to be taken with some skepticism.

From the nature of the properties of the omega minus particle, Gell-Mann believed it must be produced by the interaction of a negative kaon with a proton. These would have to be smashed together at energies high enough to produce a particle with the unusually large mass attributed to the omega minus.

Gell-Mann then had to persuade someone in control of a large particle accelerator to try the experiment. In December 1963, the team at the accelerator at Brookhaven, Long Island, began smashing K mesons into protons. On January 31, 1964, an event was detected that could only have involved an omega minus particle, for a particle showed precisely the properties predicted by Gell-Mann. In 1969, Gell-Mann received a Nobel prize for his work. Now Gell-Mann's groupings of hadrons had to be taken seriously. The hadron zoo was yielding to order.

Inside Hadrons

Merely dividing hadrons into groups and setting up a type of subatomic periodic table wasn't enough. Mendeleev's periodic table wasn't satisfactorily explained until the internal structure of atoms was worked out and the significance of the difference in electron arrangements within their shells was understood.

It seemed to Gell-Mann, then, that there had to be an internal structure in regard to hadrons that would account for their existence in groups. This was by no means an untenable idea. The leptons were fundamental particles that behaved as though they were simple points in space without internal structure, but this was not necessarily true of the hadrons.

What Gell-Mann tried to do, then, was to make up a group of particles that would perhaps be fundamental, with properties such that if they were put together properly they would form all of the various hadrons with *their* properties. One combination would yield the proton, another the neutron, still another the various pions, and so on.

Gell-Mann set about the task and quickly found that he could not manage to do this if he stuck to the principle that every particle must have electric charges either equal in size to the familiar charge on the electron or proton or a multiple thereof. He found, instead, that the constituent particles of the hadrons would have to have fractional charges.

At this, Gell-Mann quailed. In all of the time people had been working with electrically charged particles, dating back to the very beginning of Faraday's researches on electrochemistry a century and a third earlier, charges had seemed to come in even multiples, the smallest (and apparently indivisible) of which had for three quarters of a century been considered that on the electron.

In 1963, however, Gell-Mann decided to publish anyway. He suggested that there would be three fundamental particles making up the hadrons, and three antiparticles making up the antihadrons. Each hadron was made up of either two or three of these fundamental particles. The mesons were made of two, and the baryons of three.

Gell-Mann called these fundamental particles quarks. (It might have been intended as a bit of whimsy taken from James Joyce's *Finnegans Wake*, in which appears the phrase, "Three quarks for Muster Mark." This I have always taken to mean, in Joycean language, "Three quarts for Mister Mark," and supposed it to represent an order for beer. To Gell-Mann, it seemed there were "three quarks for Muster Hadron." The name should not have been kept, in my opinion. It is inelegant. However, the name stuck— possibly to Gell-Mann's own surprise—and is now ineradicable.)

Gell-Mann specified three types of quarks, which were whimsically called the up quark, the down quark, and the strange quark. (The adjectives are not to be taken literally, of course. One can speak of the u quark, the d quark, and the s quark, or symbolize them simply as u, d and s. The s is sometimes said to stand for "sideways" quark to have it harmonize with *up* and *down*, but *strange* is better because it is more significant.)

The u quark has an electric charge of $+\frac{2}{3}$, and the d quark of $-\frac{1}{3}$. (Naturally, the u antiquark has a charge of $-\frac{2}{3}$, and the d antiquark of $+\frac{1}{3}$.) Each type of quark has a series of numbers representing the various characteristics it conserves. The quarks have to be put together in such a way that the hadron they form has all of the various proper numbers for its characteristics.

Naturally, it is the fractional charges that one must be most careful of. Quarks also have to be put together in such a way that the total electric charge on the hadron comes out to be $+1$, -1, or 0. For instance, a proton is built of two u quarks and one d quark; therefore, its total charge is $+\frac{2}{3}$ and $+\frac{2}{3}$ and $-\frac{1}{3}$, or $+1$. An antiproton is built of two u antiquarks and one d antiquark ($-\frac{2}{3}$ and $-\frac{2}{3}$ and

+ ⅓), for a total charge of − 1. A neutron is built of one u quark and two d quarks (+ ⅔ and − ⅓ and − ⅓), for a total charge of 0, and an antineutron is built of one u antiquark and two d antiquarks (− ⅔ and + ⅓ and + ⅓), for a total charge of 0.

A positive pion is built of a u quark and a d antiquark (+ ⅔ and + ⅓), for a total charge of + 1, and a negative pion is built of a u antiquark and a d quark (− ⅔ and − ⅓), for a total charge of − 1.

The s quark goes into the making of the strange particles, which is where it got the s of its name. The s quark has an electric charge of − ⅓ and a strangeness number of − 1. The s antiquark has an electric charge of + ⅓ and a strangeness number of + 1.

The positive K meson contains a u quark and an s antiquark (+ ⅔ and + ⅓), for a total charge of + 1 and a strangeness number of + 1. The negative K meson consists of a u antiquark and an s quark (− ⅔ and − ⅓), for a total charge of − 1 and a strangeness number of − 1.

A lambda particle (a neutral hyperon) consists of a u quark, a d quark, and an s quark (+ ⅔ and − ⅓ and − ⅓), for a total electric charge of 0, whereas an omega minus is made up of three s quarks (− ⅓ and − ⅓ and − ⅓), for an electric charge of − 1. Both the lambda and the omega minus are strange particles.

In this sort of way, the various hadrons are built up, and no combination is possible that doesn't yield a total charge of either 0, + 1, or − 1.

But is all of this really true? Do the quarks really exist, or is this just bookkeeping? After all, a dollar bill is worth any of various combinations of coins—half-dollars, quarters, dimes, nickels, and pennies—but if the dollar bill is torn to bits it turns out there are no coins making up any part of its structure.

Suppose, then, you pull a hadron apart; will quarks

come tumbling out? Or is *this* just bookkeeping? Unfortunately, no one has yet succeeded in tearing a hadron apart or in unequivocally producing a free quark. If one were produced, it would be easy to identify because of its fractional charge. However, there are some scientists who think that it is impossible, even in theory, to pull a quark out of a hadron. And even if it were possible, we certainly don't yet dispose of energies high enough to turn the trick. However, there is indirect evidence for the actual existence of quarks.

In 1911, Rutherford described the experiments he had done in bombarding atoms with alpha particles. The alpha particles, for the most part, passed through the atoms as though they were nothing but empty space, but there was some scattering. Every once in a while they hit some small objects within the atoms and were deflected. From this Rutherford deduced that there was a minute massive point within the atom—the atomic nucleus.

Might it not be possible to bombard protons with very high-energy electrons and thereby make them scatter? From the results, might it not be possible to deduce that there were scattering points within the proton, and therefore that quarks really existed in there?

Such experiments were carried out at the Stanford (University) Linear Accelerator in the early 1970s by Jerome Friedman, Henry Kendall, and Richard Taylor, who received the Nobel prize in physics in 1990 as a consequence. The results were satisfactorily interpreted by the American physicist Richard Phillips Feynman (1918–1988), who had already received a Nobel prize in 1965 for something I'll mention later. By 1974, it was clear that quarks really *did* exist, even if they were never spotted in the free state.

Feynman referred to these particles inside the protons as partons. (This, to my way of thinking, is a much better name than quarks. Either Feynman thought, as I do, that quark was uneuphonious, or he thought that Gell-Mann's quark theory was not quite the way it ought to be.)

But now comes the possibility of trouble. We got down to atoms and there proved to be so many different types of them that simplicity was lost. We went down a notch to subatomic particles to restore the simplicity, and there proved to be so many different types of these that simplicity was lost a second time. Now that we are down to quarks, will it turn out that there are a great many different types of them?

Some people thought there ought to be at least one more quark. One among them was the American physicist Julian Seymour Schwinger (b. 1918), who had shared a Nobel prize with Feynman in 1965. It seemed to Schwinger that the quarks were fundamental particles like the leptons. He believed that they were point particles without internal structure (whose diameter is zero as nearly as we can determine) and that there ought to be symmetry between these two sets of fundamental particles.

Two flavors of leptons were known at the time—the electron and its neutrino, and the muon and its neutrino—and thus two flavors of antileptons. There should thus be two flavors of quarks. One flavor was the u quark and the d quark (and its antiquarks, of course). The second flavor was the s quark and—what? If a fourth quark existed, particles containing it had not been found, but that might be because the fourth quark and the particles containing it were so massive that considerable energy was needed to produce it.

In 1974, a team led by the American physicist Burton Richter (b. 1931) made use of the powerful Stanford (University) Positron-Electron Accelerating Ring, producing a particle that was massive indeed—three times as massive as a proton. A particle that size should break down in the merest evanescence of a second, but it didn't, it hung on. Therefore, it had to contain a new quark—one that, like the s quark (but much more massive than the s quark), prevented breakdown by way of the strong interaction.

The new particle was called a charmed particle because it lived so long and, presumably, contained a "charmed quark" or c quark, which was the fourth quark that Schwinger had been looking for. It was, indeed, more massive than the other three. The same work was done, and the same conclusions reached, at the same time by the American physicist Samuel Chao Chung Ting (b. 1936) at Brookhaven. Richter and Ting shared a Nobel prize in 1976.

By this time, however, a third flavor of leptons had been discovered, in the form of the tauon and the tauon neutrino (and its antiparticles). Did this mean there should be a third flavor of quarks?

In 1978, a fifth particle was indeed discovered, which was called the bottom quark, or b quark. There must be a sixth, which scientists call the top quark or t quark, but it hasn't been located yet, presumably because it is excessively massive. (Some scientists prefer to have the b and t stand for "beauty" and "truth.")

Quantum Chromodynamics

We now have three flavors of quarks, as we have three flavors of leptons. In each flavor there are two leptons, or quarks, and two antileptons and antiquarks. This means that there are 12 leptons and 12 quarks altogether. There are 24 particles that, with the exchange particles, make up the entire Universe (or so it now appears). This puts us back to a tolerable simplicity—at least for now. As I shall explain later, the situation might not last.

The similarities between the two types of particles are

interesting. In the case of leptons, the first flavor consists of an electron with a charge of -1 and an electron neutrino with a charge of 0. This pattern is repeated in the other two flavors: a muon with a charge of -1 and a muon-neutrino with a charge of 0, and a tauon with a charge of -1 and a tauon-neutrino with a charge of 0. Naturally, this is reversed for the antileptons, where all three flavors have charges of $+1$ and 0.

In the case of quarks, the first flavor includes the u quark ($+\frac{2}{3}$) and the d quark ($-\frac{1}{3}$). This pattern is repeated in the second and third flavors with the c quark ($+\frac{2}{3}$) and the s quark ($-\frac{1}{3}$) and with the t quark ($+\frac{2}{3}$) and the b quark ($-\frac{1}{3}$). Again, this pattern is reversed in the case of antiquarks.

Of course, the comparison isn't exact. The leptons include particles with integral charge and zero charge, while quarks have neither, including only particles with fractional charge.

Again, the masses of the particles goes up with the flavor in the case of the charged leptons (the uncharged neutrinos are massless). If we set the electron's mass at 1, the mass of the muon is 207 and that of the tauon is about 3,500. The mass goes up with flavor in the case of the quarks, also, but there are no massless quarks, perhaps because there are no uncharged quarks.

In the case of the first flavor of quarks (if we still consider the electron's mass 1), the u quark, which is the least massive of all of the quarks, has a mass of 5, and the d quark has a mass of 7. In the case of the second flavor, the s quark has a mass of about 150 and the c quark has one of about 1,500. The c quark is almost as massive as a proton, which is why it takes so much energy to produce charmed particles and why their discovery came so late.

The third flavor is more massive still. The b quark has a mass of about 5,000, or nearly three times the mass of a proton, which is why it was discovered even later than the c quark. The t quark, not having been located, has no reliable figure for its mass, but the estimate is that it must be up to at least 25 times the mass of a proton, which is why it hasn't been discovered.

It is not enough, of course, simply to list all of the quarks. One has to make sense out of the mechanism by which they work. In 1947, for instance, three physicists worked out, independently, three somewhat different ways of describing exactly what happens in the interaction of electrons and photons, thus explaining the mechanism of the electromagnetic interaction. All three ways were valid, and were essentially equivalent.

Two of the physicists were Schwinger and Feynman. The third was the Japanese Sin-itiro Tomonaga (1906–1979). (It might be that Tomonaga had it first, but World War II was raging and Japanese scientists were isolated. Tomonaga could not publicize his ideas until after the war.) All three shared a Nobel prize in 1965.

The theory is called quantum electrodynamics, which turned out to be one of the most successful theories ever devised. It predicted events involving the electromagnetic interaction with phenomenal accuracy, and has not been improved on since its formulation.

Naturally, scientists thought that the techniques used in quantum electrodynamics could be used to work out the details of the strong and weak interactions, but in this they were at first disappointed. Finally, Weinberg and Salam were able to unify the electromagnetic and weak interactions, but the strong interaction continued to present problems.

For instance, quarks have half spins and are therefore fermions, just as the leptons are. There is an exclusion principle first worked out by Pauli in 1925 that states that two fermions cannot be grouped into the same system if all of their quantum characteristics are identical. There always has to be some difference in the quantum numbers assigned them. If an attempt is made to squeeze together two fermions with identical quantum numbers, there is a repulsion between them far larger than electromagnetic repulsion. Nevertheless, it turns out that in some hadrons three identical quarks can be squeezed into one hadron, just as though the exclusion principle did not exist. The omega minus particle, for instance, is built of three s quarks.

There was, however, great reluctance to give up the exclusion principle, which worked everywhere else in subatomic physics, and scientists were anxious to save it in the case of quarks. It might be that there was some distinction among quarks that were otherwise apparently the same. If there were, for instance, three varieties of s quarks, one of each variety might be squeezed into a hadron without violating the exclusion principle.

Beginning in 1964, several physicists—among them Oscar Greenberg at the University of Maryland, Japanese-American physicist Yoichiro Nambu (b. 1921) at the University of Chicago, and Moo-Young Han at Syracuse University—worked on this matter of quark varieties.

They decided that the varieties were not something analogous to anything else in subatomic physics, and could not really be described. It could only be given a name, and the manner in which it worked detailed. The name given the distinction was *color*.

In a way, of course, this is a bad name, for quarks do not have color in the everyday sense. In another way, though, it is perfect. In color photography and color television, it is well known that the colors red, green, and blue will combine to give the impression of colorlessness; that is, whiteness. If every quark comes in red, green, and blue varieties, a combination of one of each leads to a disappearance of color, to whiteness. Every quark combination in hadrons must produce a white result. No hadron is known in which there is a color because the quark content is color imbalanced.

This explains why there are three quarks to every baryon, and two quarks (or, rather, a quark and an antiquark) in every meson. These are the only combinations that are colorless.

Once color was taken into account, several observations that had been anomalous without it could be seen to be right on the nose with it. For this reason, the notion of colored quarks was quickly adopted by scientists.

Of course, if there are six different quarks and six different antiquarks among the three flavors, and if each quark comes in three different colors, then there are thirty-six colored quarks altogether. This increases the complexity of the situation, but it gives scientists a handle with which to evolve a theory of quark behavior that approximates the value of quantum electrodynamics. The new theory is called quantum chromodynamics, where chromo- is from the Greek word for "color." Much of this new theory was worked out in the 1970s by Gell-Mann, who had first suggested the quark concept.

The strong interaction is essentially that between quarks. Hadrons, which are made up of quarks, experience the strong force secondarily *because* they are made up of quarks. Pions, which seem to be the exchange particle for this secondary hadron interaction, are exchange particles only because they, too, are made up of quarks. In other words, all of the emphasis on the *fundamental* strong interaction must be shifted to the quarks.

If this is so, there must be some exchange particle that exists on the quark level. It was Gell-Mann who came through with a name for this new exchange particle. He called it a gluon because it was the glue that held quarks together.

Gluons had properties that were quite unusual. Thus, in the case of the other exchange particles, the greater the distance between particles subject to the interaction, the

If you try to pull two quarks apart, the number of gluons between them increases. This is the attractive force between quarks, which actually increases with distance. Thus quarks can only move about freely within the hadron. Scientists suspect that they may never be able to study free quarks.

fewer the exchange particles bouncing between them and the weaker the interaction. The gravitational and electromagnetic interactions grow weaker at a value the square of the distance between the two objects subject to these interactions. The weak interactions, and the secondary strong interaction between hadrons, decline in intensity even more rapidly with distance.

In the case of quarks and gluons, however, it is quite the other way around. If you try to pull two quarks apart, the number of gluons bouncing between them *increases*. This is equivalent to saying that the attractive force between quarks increases with distance.

Within the hadron, the quarks move about freely and easily. They stiffen, however, as they move apart. This means that quarks undergo particle confinement; they can only exist comfortably inside the hadrons. It is for this reason that scientists suspect we are not ever going to be able to study free quarks. There's no way of making them leave the hadrons. Of course, the hadrons themselves can change from one to another, carrying their load of two or three quarks (which may themselves change from one color to another) within themselves.

Gluons are more complex than other exchange particles in another way. Gravitons are exchanged by particles with mass, but gravitons themselves have no mass. Photons are exchanged by particles with electric charge, but photons themselves have no electric charge. Gluons, which themselves have color, are exchanged by particles with color; therefore, gluons can stick to each other. This is another reason why the name gluon is a good one. (Some scientists refer to a pair of gluons that stick together as a glue ball.)

The gluon has the capacity of changing the color of a quark (but not the flavor). There is a gluon that changes a red quark to a green quark of the same flavor, another that changes red to blue, and so on. To account for all of the color changes possible, there must be eight different gluons. This is an added complexity. With one gravitational exchange particle, one electromagnetic, three weak, and now eight strong, we have thirteen exchange particles altogether.

Nevertheless, quantum chromodynamics based on quarks in three flavors and three colors, with eight colored gluons (forty-four particles altogether), is a successful theory, and scientists expect that it will continue to explain all of the ins and outs of hadrons and their behavior.

12

THE UNIVERSE

The Mystery of the Missing Mass

Naturally, the observations and experiments scientists have made in connection with subatomic physics are, for the most part, conducted right here on Earth. How do we know that the results we get are applicable to other worlds—to the stars or to the Universe in general?

To begin with, we have made studies of the surface of the Moon, of Mars, and of Venus directly, and we have studied the surface of other objects in our solar system by rocket probes employing sophisticated instruments for the purpose—even if they have not made actual physical contact. We even have bits of extraterrestrial matter that arrive on Earth in the form of meteorites. None of these investigations have offered us any subatomic surprises. Scientists are quite certain that all of the planetary objects in the solar system are made up of the same matter that Earth is, and therefore must follow the same rules.

But what about the Sun, which seems so different from all of the other members of the solar system? Well, charged particles, mostly protons, reach us from the Sun, as do neutrinos, and they are what we expect them to be.

What about the Universe beyond the solar system? We have received neutrinos from the supernova that exploded in 1987 in the Large Magellanic Cloud, and we receive cosmic rays (mostly protons and alpha particles) from the Universe generally. They indicate that the Universe follows the rules that have been worked out here on Earth.

The most important information that reaches us from the Universe generally comes in the form of photons. We actually *see* the Sun and the stars, and we even see galaxies that are billions of light-years away. We can also detect photons that are too energetic, or unenergetic, for our eyes to see—gamma rays, X rays, ultraviolet, infrared, and radio waves.

The photons we get give a clear indication of the chemical structure of the objects that emit them. Astronomers are quite satisfied that the stars and galaxies are made up of matter like that that makes up our own Sun. Our Sun is made up of matter like that on Earth (allowing for the Sun's much higher temperature).

But do we really see, or sense, a fair sampling of all of the photons there are? Is there anything that exists in the Universe that *doesn't* radiate photons? Not really! Every object in the Universe that is surrounded by space at the average temperature of the Universe (about three degrees above absolute zero)—and that means just about every object—radiates photons. Some of the radiation, however, is either insufficiently intense or insufficiently energetic for us to pick up.

There are many stars that are so dim that unless they are fairly close to us we cannot see them, even with the best instruments we now have available. There are certainly planets in other star systems that are as surface-cold as the planets of our own solar system, and whose feeble radiation of radio waves is lost in the blaze of the stars they circle.

Nevertheless, it seemed fair to assume that by far the largest percentage of the mass of the Universe took the form of stars, and that the amount of mass we couldn't sense, because it was too cold and faint, was not significant. In our own solar system, for instance, all of the planets, satellites, asteroids, comets, meteors, and dust that circle the Sun make up only 0.1 percent of the total mass. The other 99.9 percent of the mass is found in the Sun. There would seem every reason to suppose that, by and large, other stars also predominated in this way over the objects circling them.

Naturally, it is possible that there are places in the Universe in which conditions are so extreme that the laws of nature we have worked out don't necessarily hold. The most likely place where this might be true are in black holes, where matter has collapsed into conditions of nearly infinite density, creating a small neighborhood of nearly infinite gravitational intensity. We can't study black holes in detail and, as yet, we have not even completely and unmistakably identified any. However, assuming they exist, they might be governed by laws outside those we know.

Another realm of uncertainty exists in the first instants after the Universe came into existence, when conditions were so extreme that our structure of physical theory might not apply. (I will have a few words to say about this later.) Yet nothing ever seems to be surprise-free. All of the photons from the outside universe that we study are the product of the electromagnetic interaction, and the surprise arose from the effects of gravitation, the other long-range interaction.

We can't detect gravitons, but we *can* detect the effect of gravitation on the movement of stars and galaxies. We can measure the speed of rotating galaxies in different parts of their structures, and we assume that this rotation is driven by gravitational forces within the galaxy, just as the rotation of the planets of the solar system is driven by the gravitational influence of the Sun.

Because 99.9 percent of the mass of the solar system is concentrated in the Sun, the solar gravitational influence overwhelms everything else in the solar system. Except for very minor corrections, this influence alone needs to be taken into account. The farther a planet is from the Sun, the less intense is the Sun's gravitational influence on it, and the more slowly it moves. The variation in motion with distance was first worked out in 1609 by the German astronomer Johannes Kepler (1571–1630), and was explained by the law of universal gravitation, advanced in 1687 by Newton.

Like the solar system, galaxies have their mass concentrated at the center, although not to quite such an extreme. We can *see* that the stars are more and more numerous as one approaches the center of a galaxy, and it seems a fair conclusion that about 90 percent of the mass of all large galaxies is contained in a relatively small volume at their cores. Therefore, we would expect the stars to be circling the center of their galaxies more and more slowly as one moved outward from the core. But this does not happen. Apparently, the stars in a galaxy move at about the same speed as one moves outward from the core.

No scientist wants to abandon the law of gravity (which has been modified and extended, but not replaced, by Einstein's theory of general relativity), in that it would seem that no alternative law can explain what goes on in the Universe generally. Therefore, we must suppose that the mass of the galaxy is *not* concentrated at the core, but is spread out much more evenly throughout the galaxy. Yet how can this be when we *see* that the mass, in the form of stars, *is* concentrated?

The only conclusion we can come to is that there is matter outside the core that we *don't* see. It is "dark matter" that doesn't send us anything in the way of perceptible photons, but that exerts its gravitational influence. In fact, we are forced to assume, from a gravitational standpoint, that the mass of a galaxy might be many times as great as it would seem to be from the photons it radiates. Until studies of galactic rotations were made, we were apparently missing most of the mass of the galaxies.

Another point: galaxies exist in clusters. Within the clusters (each of which is made up of anywhere from dozens to thousands of galaxies), the individual galaxies move about restlessly, like a swarm of bees. The clusters are held together by the mutual gravitational attraction of the galaxies that make them up, but the masses of the galaxies—if we go by just what we can *see*, by the photons we can detect—are simply not great enough to supply the necessary gravitational pull to hold the clusters together. Yet the clusters apparently *do* hold together. Again, there must be mass that we're not aware of. The larger the cluster, the larger the quantity of mass we cannot quantify. There might be as much as 100 times the mass in the Universe as that we can see. This phenomenon is what is called the "mystery of the missing mass." What is it?

The easiest answer is to suppose that every galaxy contains vast crowds of small, very dim stars, planets, and dust clouds. The trouble is that it isn't reasonable, from what we know about the Universe, to suppose that such material is likely to exist in such quantities that its mass would be a hundred times as great as that of the stars we can see.

Let's get down to the subatomic world, then. About 90 percent of the mass of the Universe, as far as we know, is made up of protons. The only other subatomic particles to match or exceed the number of protons are the electrons, which are equal to the protons in number, and the photons and electron neutrinos, each of which might, in numbers, be a billion times as many as the protons. However, the electrons have only minute masses, and the photons and electron neutrinos have no intrinsic mass at all. The electrons, photons, and electron neutrinos are all speeding along and have energies of motion equivalent to their masses, but the masses that produce the energy are extremely small—so small as to be neglected. This leaves only the proton as the mass material of the Universe.

Is it conceivable that the missing mass is made up of additional protons that we are not aware of? The answer to this seems to be no! Astronomers have ways of estimating the density of protons in the Universe, and therefore of determining how many there can be, seen or unseen, in the regions taken up by galaxies or by clusters of galaxies. The amount of protons present is, at most, only 1 percent of the missing mass. Whatever the missing mass is, then, it can't be protons.

This leaves the electrons, photons, and electron neutrinos. We are quite certain that the electrons and photons can't possibly supply the missing mass, but we are a little less certain about the electron neutrino.

In 1963, a group of Japanese scientists suggested that the electron neutrino might have a minute mass; just a small fraction of that of electrons, for instance. If this is so, the muon neutrino might have a slightly larger mass, and the tauon neutrino a still larger one. All of the masses might be very small, but not quite zero.

If this were so, the neutrinos would travel at less than the speed of light—though perhaps not much less—and each would travel at a slightly different speed. Therefore, the three neutrino flavors would oscillate, shifting from one flavor to another rapidly.

This would mean that if there were a beam of electron neutrinos starting from the Sun, some eight minutes later, when it had completed its 150-million-kilometer race to the Earth, it would appear on Earth as a beam of equal quantities of electron neutrinos, muon neutrinos, and tauon neutrinos.

This would be interesting indeed, in that Reines, who has been detecting neutrinos from the Sun for decades, uses detecting devices that work only for electron neutrinos. If the neutrinos are oscillating, he would be receiving a beam constituted of only one-third electron neutrinos, instead of entirely electron neutrinos. He would detect only one-third, which would explain why the electron neutrino count he received was always so low.

In 1980, Reines reported that he had conducted experiments that gave him reason to believe that oscillation was taking place and that neutrinos *did* have a very small mass. If so, this would explain not only the missing neutrinos from the Sun, but the mystery of the missing mass. There are so many neutrinos floating around the Universe that even if each one had a mass of only $1/10,000$ that of an electron, this would be enough to make the total mass of neutrinos a hundred times that of the mass of all of the protons in the Universe. Moreover, such slightly massive neutrinos might be used to explain how the galaxies formed in the first place, a problem that is giving astronomers a great deal of headache material right now.

The possibility of a slightly massive neutrino would thus very nearly solve a number of problems, and it makes one ache to believe that such a situation is true. The only trouble is that no one has confirmed Reines's report. In general, it is thought he was wrong. No matter how beautiful and desirable a theory is, if it doesn't match the Universe, it must be given up.

But even if the missing mass is not protons and not neutrinos, it still seems to exist. What is it, then? Physicists have, in recent years, been trying to work out theories that unify the strong interaction and the electroweak interaction. Some of these theories require the invention of new and exotic particles. Perhaps it is such particles, never actually observed, and existing, so far, only in the minds of some imaginative scientists, that account for the missing mass. If so, we must wait for observations that will back up these far-out theories.

The End of the Universe

To the casual observer, the Universe, whether seen by eye or by various instruments, might seem unchanging. What changes do take place are likely to be cyclic. If some stars explode, others are formed. There would seem to be no reason to think that the Universe necessarily has an end or a beginning, except for one overwhelming effect, which might not be cyclic: The Universe is expanding.

This story began in 1912, when the American astronomer Vesto Melvin Slipher (1875–1969) began studying the spectra of certain nebulas. These were actually distant galaxies lying far outside our own Milky Way, but that was not understood at the time. From the spectra, Slipher could tell whether the spectral lines were shifted toward the violet end of the spectrum (in which case the nebula was approaching us), or toward the red end (in which case it was receding from us).

By 1917, Slipher found that of the fifteen nebulas he

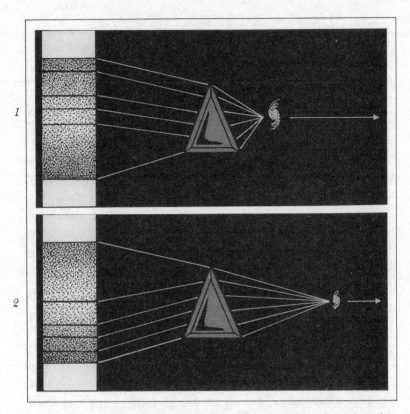

1. *Spectral lines in the light arriving from a nearby galaxy (moving away) will only be slightly shifted toward the red end of the color spectrum.*
2. *Spectral lines from the light of a distant galaxy (moving rapidly away) will be greatly shifted toward the red end of the color spectrum.*

had studied, all but two showed a red shift and were receding. Others took up the task, and when the nebulas were recognized as distant galaxies, it turned out that, barring the two approaches that Slipher had noted (which were two galaxies unusually close to us), all were receding. Moreover, the dimmer the galaxy, the more rapidly it was receding.

By the end of the 1920s, the American astronomer Edwin Powell Hubble (1889–1953) had collected enough data to be able to announce that the Universe was expanding, and that the clusters of galaxies that made up the Universe were receding from one another.

This made sense in the light of theory. In 1916, Einstein evolved his theory of general relativity, which described gravitation more accurately than Newton had done. The equations Einstein worked out to describe gravity were, in effect, the founding of the science of cosmology (the study of the Universe as a whole).

Einstein assumed at first that the Universe would have to be unchanging overall, and adjusted his equations to fit this assumption. In 1917, the Dutch astronomer Willem de Sitter (1872–1934) showed that the unadjusted equations, if properly solved, implied that the Universe was expanding. Hubble's observations proved this theory to be correct.

Now the question is: How long will the Universe continue to expand? Resisting expansion is the mutual gravitational attraction to one another of all parts of the Universe. The expansion, then, is taking place against the pull of gravity, just as an object hurled upward from the surface of the Earth moves against the pull of Earth's gravity.

It is our common experience that an object sent upward under ordinary circumstances is eventually defeated by the gravitational pull of Earth. Its speed of ascent decreases to zero, whereupon the object begins to be pulled back to Earth. The more forcefully the object is hurled upward—and therefore the greater its initial upward speed—the higher it climbs and the longer it takes to begin to fall back.

If an object is sent up from the Earth with sufficient force (sufficient initial speed), it *never* falls back. The gravitational pull of the Earth weakens as the object places more and more distance between itself and Earth's center. If the object moves upward rapidly enough (11 kilometers per second, or 7 miles per second), the declining gravitational intensity never suffices to bring it back. This means that 11 kilometers per second is the escape velocity from Earth's gravitational pull.

We might ask, then, whether or not the rate of the Universe's expansion outward against gravitation's inward pull had achieved escape velocity. If its expansion speed is over escape velocity, the Universe will expand outward forever. It would then be an open Universe. If, however, its expansion speed is below escape velocity, the expansion will gradually slow, and eventually come to a halt. After this, the Universe will begin to contract. It would then be a closed Universe.

It can't possibly affect us in our individual lifetimes, or even in the lifetime of our planetary system, whether the Universe is open or closed, whether it will end as an ever-expanding and thinning ball of matter, or whether it will end as a contracting and thickening ball of matter, but scientists are curious. In order to come to a decision, they try to estimate the rate of expansion. They also try to estimate the average density of matter in the Universe, which gives them an idea of the strength of the inward gravitational pull. Both determinations are difficult to carry through, and the results are only approximate. The conclusion is, however, that the density of the Universe is only about 1 percent or so of that required to end the expansion. The Universe would therefore seem to be open and to be expanding forever.

But wait! The determination of the density of the matter in the Universe is based on what we can detect—but what about dark matter? If it is true that the dark matter in the Universe, the matter whose nature we have not yet determined, is perhaps as much as a hundred times as massive as the matter we *can* detect, this might be enough to close the Universe. We end, therefore, by being uncertain as to whether the Universe is open or closed.

It is also possible that there is enough dark matter in the Universe to place it just on the boundary (or very near it) of open and closed, meaning that the Universe is "flat." This would be an extraordinary coincidence, and the feeling is that if the Universe is flat, there must be a reason for it.

You see, then, how important it is, from a cosmological standpoint, that we know whether or not dark matter is really there, and, if it is, just what it consists of. The answer, when it comes, is bound to arise out of the realm of subatomic particles. Thus, we see that the advance of knowledge is truly unitary. Knowledge of the greatest object we recognize, the Universe, depends on what we know of the smallest objects we recognize, the subatomic particles.

Another way in which subatomic particles might effect the end of the Universe arises out of the attempted unification of the strong and electroweak interactions. The first attempts in this direction began in 1973, when Salam, the cofounder of the electroweak theory, tackled the problem.

In that the electroweak interaction involves the leptons, and the strong interaction involves the quarks, a unified theory must imply that leptons and quarks have a basic, underlying similarity; that, under some conditions, one can be turned into the other. The natural assumption is that quarks can be turned into leptons, because this would be the direction of declining mass and energy.

Suppose, then, that a quark inside a proton is converted into a lepton. The proton would then no longer be a proton; it would have broken down into such less massive particles as kaons, pions, muons, and positrons (all positively charged, preserving the conservation of electric charge). The kaons, pions, and muons would eventually decay into positrons, meaning that, overall, protons would change into positrons.

This violates the law of conservation of baryon number.

However, all of the conservation laws are merely deductions from observations. We have never observed any change that alters the baryon number in an isolated system, so we naturally *assume* that such a change can never take place—and that gives us the conservation law. Nevertheless, however powerful and convenient the conservation laws are, they remain assumptions, and scientists must be ready now and then to accept the fact that a given conservation law might not work under all conceivable circumstances. They found this to be so in the case of the law of conservation of parity, as I explained earlier.

Still, scientists have been studying protons intensively for many decades, and no proton has ever been seen to decay. On the other hand, because scientists are quite convinced it cannot decay, they haven't put an emphasis on finding out for sure.

In addition, the unification of interactions that are now extant (there are several varieties) indicate that the half-life of the proton is extremely long. It would take 10^{31} years (ten million trillion trillion years) for half the protons in any given sample of matter to break down. In that the Universe is only about 15 billion years old, the half-life of the proton would be nearly 70 billion trillion times the age of the Universe. The number of protons that have broken down in the course of the entire lifetime of the Universe would thus be an insignificant fraction of the whole.

But it would not be zero! If you start with 10^{31} protons, which is what you would find in a tank holding some 20 tons of water, there would be an even chance of having one proton break down in the course of a single year. Detecting that one proton in the 20 tons of water and identifying its breakdown as due to the change of a quark into a lepton would not be an easy job, and scientists, who made some initial attempts to investigate the subject, have not yet succeeded in spotting such a breakdown.

Success or failure is important. Success will go a long

way toward establishing the validity of the interaction unification, the so-called grand unified theory; failure would
cast it into doubt.

Then, too, think of the light it would cast on the fate
of the Universe. If the Universe is open and expands forever, it will very slowly lose its proton content. It will
eventually become an unimaginably vast and thin cloud of
leptons—electrons and positrons (and, of course, photons
and neutrinos).

Of course, we also suspect that as the Universe ages
more and more of it will be concentrated into black holes—
and we haven't the faintest idea what the laws of nature
are like at the center of black holes. Will there be hadrons
of some sort at these centers? Will they decay, very, very
slowly, but very, very surely, and will the black holes eventually disappear? The puzzles continue—and will probably
continue forever.

The Beginning of the Universe

The Universe is at present expanding. Regardless of
whether it is open or closed, it is *at present* expanding. This
means it was smaller last year than it is now, and smaller
still the year before, and so on.

If we look into the future, it is at least conceivable that
there is unendingness about it, for the Universe might be
open, and might expand forever. If we look back at the
past, however, there is no chance of unendingness. The
Universe grows smaller and smaller, and at some moment
in the far past it can be viewed as having shrunk down to
some minimum size.

The first person to point this out in some detail was
the Belgian astronomer Abbé Georges Henri Lemaitre

(1894–1966). In 1927, he suggested that, in looking backward, there was a time when the matter and energy of the Universe were literally squashed together into one exceedingly dense mass. He called it the cosmic egg, thinking of it as unstable. It exploded in what we can only imagine to have been the most gigantic and catastrophic explosion the Universe is capable of affording. The effects of that explosion are still with us in the form of the expanding Universe. The Russian-American physicist George Gamow (1904–1968) called this the big bang, and the name stuck.

Naturally, there was some resistance to the notion of the big bang. Other scenarios were advanced that would account for the expanding Universe. The issue was not settled until 1964, when the German-American physicist Arno Allan Penzias (b. 1933) and the American physicist Robert Woodrow Wilson (b. 1936) studied radio wave radiation emanating from the sky.

In no matter which direction they looked, if they penetrated far enough they would detect radiation that had been traveling for so many billions of years that it must have originated in the big bang itself, if there was one. They found a faint radio wave background of identical intensity from every part of the sky, which was taken to represent the distant "echo" of the big bang. Physicists accepted this as establishing the big-bang theory, and Penzias and Wilson received a Nobel prize in 1978 for their work.

The big-bang theory has its problems, of course. For instance, when did it happen? One way to determine this is to measure the rate at which the Universe is now expanding and then work backward, allowing for the intensification of the gravitational pull as the Universe becomes smaller and denser.

It is a great deal easier to say this than to do it. There are several ways in which the rate of expansion can be

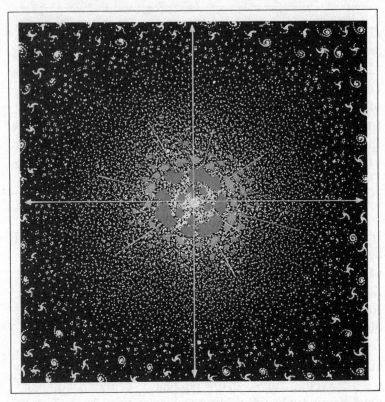

Today, radio astronomers can still listen to the distant "echo" of the big bang.

determined, in which the age of the oldest stars can be measured, and in which the distance of the farthest objects we can see can be determined (and, therefore, the amount of time it took radiation to reach us from these objects).

The results tend to conflict, and estimates of the time of the big bang vary from 10 to 20 billion years ago. Usually, people split the difference and assume that the Universe is, counting from the moment of the big bang, 15 billion years old, but I suspect that the truth is closer to 20 billion.

There are other more subtle difficulties. The radio wave background that Penzias and Wilson detected is extremely uniform in all parts of the sky, and represents an overall (average) temperature of the Universe of three degrees above absolute zero. This is puzzling, because to have a situation in which the temperature is the same everywhere, there usually has to be contact of some sort between the various parts so that heat can flow from here to there, evening out. This can't have happened in the Universe because different parts of it are separated by a greater distance than light can travel in the course of the entire lifetime of the Universe. Nothing else can travel faster than light, so what has evened out the temperature? What, in other words, makes the Universe so smooth?

Another problem is just the reverse. If the Universe *is* smooth, why didn't it stay smooth? Why isn't it just a featureless blob of subatomic particles, expanding endlessly? Why did the particles condense into huge pieces that became clusters of galaxies, with the galaxies condensing into individual stars. In other words, why is the Universe so smooth in some ways and so lumpy in others?

There are some other problems, too, but all of them—the age, the smoothness, the lumpiness, and so on—depend on just what happened at the very beginning of the Universe in the first instants after the great explosion. Naturally, no one was there to see, but scientists try to reason it out from what they know of the present state of the Universe, and from what they have learned concerning subatomic particles.

Thus, they assume that as they move backward in time, closer and closer to the big bang, the temperature grows higher and higher and the energy density is more and more enormous. Scientists feel that they cannot talk of times less than 10^{-45} second (one billion of a trillionth of a trillionth of a trillionth of a second) after the big bang. Less time than that brings about conditions so extreme that space and time themselves have no meaning.

However, the Universe cools rapidly in incredibly short fractions of a second. At first it was nothing but a sea of quarks, which existed freely because nothing else could exist, and because they had too much energy to even settle down enough to combine with each other.

By the time the Universe was a millionth of a second old, however, the quarks had split up into present-day quarks and leptons, and those quarks had cooled sufficiently to be able to combine so that the baryons and mesons were formed. Free quarks were never seen again. The interactions, which started as only one form, were splitting apart into the four we now recognize. When the Universe was one second old, it had rarefied to a point at which neutrinos had stopped interactions with other particles. They had begun to exist in free indifference of the rest of the Universe, and have done so ever since. Once the Universe was three minutes old, the simpler atomic nuclei began to be formed.

After a hundred thousand years, electrons began to circle nuclei. Atoms were formed. Thereafter, matter began to condense into galaxies and stars, and the Universe began to take on the shape we know.

Still, scientists could not prevent themselves from thinking of time zero, the actual instant of the big bang before the 10^{-45} limit. Where did the material of the cosmic egg come from?

If we consider the situation as it was before the cosmic egg was formed, we might visualize a vast illimitable sea of nothingness. Apparently, though, that is not an accurate description of what would exist. The nothingness contains energy. It is not quite a vacuum because, by definition, a vacuum contains nothing at all. The pre-Universe, however, had energy, and although all of its properties were otherwise those of a vacuum, it is called a false vacuum. Out of this false vacuum, a tiny point of matter appears where the energy, by the blind forces of random changes, just happens to have concentrated itself sufficiently for the purpose. In fact, we might imagine the illimitable false vacuum to be a frothing, bubbling mass, producing bits of matter here and there as the ocean waves produce foam.

Some of these bits of matter might disappear promptly, subsiding into the false vacuum from which they came. Some, on the other hand, might be large enough, or have been formed under such conditions generally, as to undergo a rapid expansion in a way that makes certain the Universe will form and survive, possibly, for many billions of years.

It might be, then, that we inhabit one of an infinite number of Universes in various stages of development and, for all we know, with different sets of laws of nature. However, there is absolutely no way of communicating with any other Universe, and we are forever confined to our own as a quark is confined to a hadron. This should not plague us unnecessarily. Our Universe is large enough all by itself, and varied enough, and puzzling enough for all purposes.

Viewing such a beginning of the Universe, the American physicist Alan Guth suggested in 1980 that in the very early stage of the Universe there was a rapid "inflationary" phase. This picture is described as the inflationary Universe.

It is difficult to grasp how brief the inflationary period is and how enormous is the inflation. The inflation starts about 10^{-35} second (ten trillionths of a trillionth of a trillionth of a second), with the Universe doubling in size for every 10^{-35} second thereafter. After a thousand doublings (only 10^{-32} second after the big bang), the inflation ceases. This difference in time (ten billionths of a trillionth of a trillionth of a second) was enough, however, to enable the Universe to grow 10^{50} times in volume. It ends the inflationary period with a hundred trillion trillion trillion trillion times the volume it had at the start. Moreover, in increasing its volume it incorporated more of the false vacuum and its energy content, thereby increasing its mass enormously. It can be shown that such a rapid initial inflation was what made the Universe smooth, and just about flat, possessing just the mass density that would place it on the boundary between being open and closed.

Guth's inflationary Universe didn't explain all properties of the present-day Universe. Scientists have been working to modify it so that it can give a better picture of what now surrounds us, especially as concerns the formation of the galaxies.

In order to do this, there must be a further unification. Not only must the strong and electroweak interactions be brought under a single umbrella, but gravitational interaction as well. Gravitation has so far resisted all attempts to incorporate it, but scientists are working with something called superstring theory, which they also call the "theory of everything."

It is not only that baryons and leptons are brought together as two different examples of something more fundamental, but fermions and bosons are unified and considered two different examples of something more fundamental. A new group of particles has been postulated in which there are new fermions analogous to our bosons, and new bosons analogous to our fermions.

Where this will go, I cannot say. There seems no point in trying to outline the current thinking, for it is bound to change and to be modified almost from day to day. Moreover, none of it has any observational backing at all, so that it remains merely speculation.

Still, there is the dream of a single set of equations that can cover all of the particles that exist in the Universe, as well as all of the interactions that involve them. This we need for a firm picture of a Universe that began with a single type of particle governed by a single type of interaction—a particle that as it gradually cooled, divided itself into the grand variety of effects we experience today.

And it all began with some ancients who questioned how far one could divide matter. It shows what asking the right questions can bring about.

INDEX

INSIGHT GUIDES

WASHINGTON D.C.

DISCOVERY CHANNEL

APA PUBLICATIONS

Part of the Langenscheidt Publishing Group

ABOUT THIS BOOK

Editorial

Edited by
Brian Bell
Updated by
Rosanne Scott

Distribution

United States
Langenscheidt Publishers, Inc.
46–35 54th Road, Maspeth, NY 11378
Fax: 1 (718) 784 0640

Canada
Thomas Allen & Son Ltd
390 Steelcase Road East
Markham, Ontario L3R 1G2
Fax: (1) 905 475 6747

UK & Ireland
GeoCenter International Ltd
The Viables Centre, Harrow Way
Basingstoke, Hants RG22 4BJ
Fax: (44) 1256 817988

Australia
Universal Press
1 Waterloo Road
Macquarie Park, NSW 2113
Fax: (61) 2 9888 9074

New Zealand
Hema Maps New Zealand Ltd (HNZ)
Unit D, 24 Ra ORA Drive
East Tamaki, Auckland
Fax: (64) 9 273 6479

Worldwide
**Apa Publications GmbH & Co.
Verlag KG (Singapore branch)**
38 Joo Koon Road, Singapore 628990
Tel: (65) 6865 1600. Fax: (65) 6861 6438

Printing

Insight Print Services (Pte) Ltd
38 Joo Koon Road, Singapore 628990
Tel: (65) 6865 1600. Fax: (65) 6861 6438

©2003 Apa Publications GmbH & Co.
Verlag KG (Singapore branch)
All Rights Reserved

First Edition 1992
Third Edition 2003

This guidebook combines the interests and enthusiasms of two of the world's best-known information providers: Insight Guides, whose titles have set the standard for visual travel guides since 1970, and Discovery Channel, the world's premier source of nonfiction television programming.

The editors of Insight Guides provide both practical advice and general understanding about a destination's history, culture, institutions and people. Discovery Channel and its popular website, www.discovery.com, help millions of viewers explore their world from the comfort of their own home and encourage them to explore it firsthand.

Insight Guide: Washington, D.C. is structured to convey an understanding of the city and its people as well as to guide readers through its sights and activities:

◆ The **Features** section, indicated by a yellow bar at the top of each page, contains essays covering the cultural history of the city.

◆ The main **Places** section, indicated by a blue bar, is a complete guide to all the sights and areas worth visiting. Major sites are coordinated by number with the maps.

◆ The **Travel Tips** listings section, with an orange bar, contains information on travel, hotels, shops, restaurants and more.

The contributors

Martha Ellen Zenfell, an Insight managing editor who has produced many of the series' US titles, created the original authoritative edition of this book, and it has now been

ABOVE: the Viking exhibit at the National Museum of Natural History.

thoroughly overhauled and updated by **Rosanne Scott** to paint a vivid portrait of Washington today. Scott arrived in Washington with a political science degree "not long after Jimmy Carter moved into the White House – that would be the mid-1970s; we tend to think in terms of administrations rather than calendar dates – and, in a city whose population turns over with every shift in the political wind, have been here long enough now to claim status as a native." She has worked in the federal government and in the graphic arts, and has written for Time-Life Books and for the *Washington Times*.

Much material has been retained from earlier editions. The chapters detailing the city's tumultuous history were written by **John Gattuso**, who has edited more than 20 books for Insight Guides and is based in New Jersey. Many of the features were written by **Martin Walker**, a former DC bureau chief for a British national newspaper, and his wife **Julia Watson**, a journalist and novelist. Educated at Oxford and Harvard universities, Walker first dipped his toes into political waters by serving on the staff of Senator Ed Muskie in 1970–71. The "How the Federal Government Works" essay was written by **B. Claiborne Edmunds**.

Two writers produced the bulk of the Places chapters. **Maria Mudd** has also written for *National Geographic Traveler* magazine, *Islands* and the *Washington Post*. **Alison Kahn** is a former *National Geographic* magazine staffer whose research skills were honed at the Smithsonian. The short piece on the C&O Canal was written by **Elaine Koerner**.

Among the most important aspects of an Insight Guide are the pictures, and two photographers whose work has graced many Insight Guides and many issues of National Geographic feature prominently here. **Catherine Karnow**, the daughter of local Pulitzer Prize-winning historian Stanley Karnow, alternated between the family home in Potomac and an apartment in lively Adams Morgan while doing this work. **Richard T. Nowitz** is based in Rockville, Maryland.

Hilary Genin tracked down some obscure photographs, **Zoë Goodwin** was cartographic editor, and the book's index was compiled by **Elizabeth Cook**.

Map Legend

—··—	International Boundary
——	State Boundary
—·—·—	National Park/Reserve
————	Ferry Route
Ⓜ	Metro
✈ ✈	Airport: International/Regional
🚌	Bus Station
❶	Tourist Information
⊠	Post Office
✝ ✝	Church/Ruins
✝	Monastery
☾	Mosque
✡	Synagogue
⌂ ⌂	Castle/Ruins
⌂	Mansion/Stately home
∴	Archaeological Site
∩	Cave
✦	Statue/Monument
★	Place of Interest

The main places of interest in the Places section are coordinated by number with a full-color map (e.g. ❶), and a symbol at the top of every right-hand page tells you where to find the map.

INSIGHT GUIDE
WaSHINGTON D.C.

CONTENTS

Spring cherry blossoms frame the Jefferson Memorial

Travel Tips

J.E.FRASER · 1949

A CAPITAL IDEA

With its emphasis on politics and its abundance of

monuments, Washington is unique among big American cities

"**W**ashington has only politics; after that, the second biggest thing is white marble," said an East Coast notable some years ago. It's certainly true that politics is much in evidence in Washington, and that the white marble monuments are striking signature images. But to sum up the capital of the United States by its most obvious facets is to ignore the undercurrents of an intricate city.

For instance, Washington is neither Southern nor Northern. It retains characteristics of both but remains aloof from each. The city has more black residents than white (although tourists can rarely tell), and more lawyers than it does doctors. Compared to other established East Coast cities there's little "old money" around, just new administrations every four years. It's a town of meetings and agendas, of ethnic groups and power struggles, both in the boardrooms and on the streets. Its boulevards are tree-lined and shady. In Washington, even some of the ghettos are attractive, their low-rise buildings of architectural significance. Yet, despite its beauty, Washington has one of the highest murder rates in America.

The "public face" of DC can slip in a matter of minutes. Taxis, for instance, are allowed to pick up more than one passenger at a time. On a sweltering summer's day, squashed in a cab with an Ethiopian driver, four stony-faced, perspiring strangers and a broken-down air-conditioning system, the cool corridors of Washington can feel very like a Third-World *barrio*.

The capital is not known for its restaurants, yet the variety of food is both excellent and expanding. Nor is it known for its cultural activities. A number of small museums, however, offer first-rate concerts, lectures and films, not to mention the events held in the grand museums on the Mall.

The city has a hot, sticky climate, but cool bathing beaches are only an hour's drive away. You can go hiking in the mountains, sailing in the bay, swimming in the ocean, or boating on the canal. This, in a town more noted for its indoor accomplishments than its outdoor pursuits.

Politics does propel Washington; it provides the impetus for achievement and its power residents with a job and a *raison d'être*. On moonlit nights, when strolling along the Mall with the Lincoln Memorial to the left, the Capitol to the right and the Washington Monument soaring high above, it does seem as if the city were composed of white marble.

But to dismiss the nation's capital in such a narrow way is like calling the president of the United States a public official. It's true – but it's not the whole story. ❑

PRECEDING PAGES: DC's breathtaking "reciprocity of sight"; the Jefferson Memorial, completed in 1943; the reading room of the Library of Congress.
LEFT: Washington is a city of classical statues.

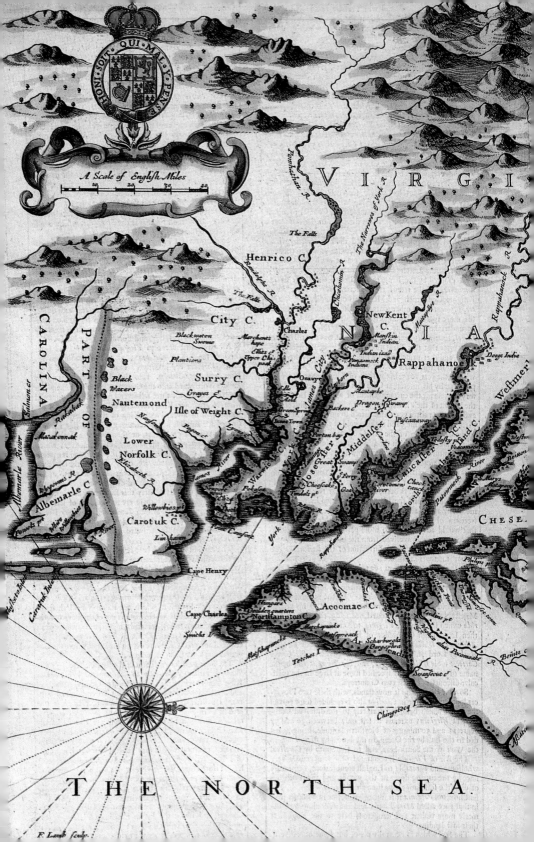

HONI · SOIT · QUI · MAL · PENSE

A Scale of English Miles

V I R G I

Pamunkey R.

The Falls

The Narrows of York R.

Henrico C.

Randolph R.

Chickahomini R.

The Falls

Matapony R.

City C.

NewKent

Charles

N

Rappahanock R.

Blackwater Swampe

Marchants hope

Cliffts ct Upper Chickahack

Westmer

Plantions

Mansfin Indian

Indian island

Pingaseock Indians

Doggs India

Surry C.

Onauye

City

Clayborn

Rappahanock

CAROLINA

Graves

Ile Fall

Matapke

PART OF

Black Waters

Green Spring

James Town

Dragon Swamp

Backers C.

Piscattaway

Nantemond

Isle of Weight C.

Naseway r.

Poquosin r.

Gloster C.

Middlesex

Lancaster C.

Jamberland C.

Mulberry Isle

Morton bay

Great Swamp

Checsake

Corotoman River

River

Albemarle River

Marchconnuk

Warwic C.

Cheesake

ye Ferry

Chech tanck

S. Marys

Lower Norfolk C.

Elizabeth r.

Poketon

Tindals pt.

God

Cotomen conibeley

Patomeck

Britton

Morattico

Roanoke R.

James River

Palmers pt

Albemarle C.

Willowbies pt

Crawford

Rappahanock

CHESE.

Carotuk C.

Linhaven

York

Phelps pt

Cape Henry

THE NORTH SEA.

Cape Charles

Wampara

Hidden quarters

Northampton C

Accomac C.

Matchapunto

Smicks I.

Matchapreck

Scarburgh's Quarophard

Ar

Tetches I.

Raskhagmock

Swanfecut ct

Chingoteeg I.

F. Lamb sculp.

Decisive Dates

THE EARLY YEARS

1608 Captain John Smith sails up the Potomac.
1662 The first land patent in the area is granted.
1749 Nearby Alexandria, Virginia, is established.
1751 Georgetown is established.
1775 The American Revolution begins.
1776 Members of the Continental Congress sign the Declaration of Independence.
1783 The Revolution ends with the surrender of General Charles Cornwallis at Yorktown, Virginia.
1789 The US Constitution is ratified.

1790 With land ceded by Maryland, the site that will be the nation's capital is chosen.
1791 Alexandria, Virginia, is added to the territory that makes up the Federal City; Major Pierre L'Enfant lays out a plan for the city.
1792 The cornerstone for the White House, the city's first public building, is laid.
1793 The cornerstone for the Capitol is laid.
1800 President John Adams is the first president to occupy the White House; the population of Washington is 14,000.
1802 The city of Washington is chartered.
1802 Eastern Market, the city's oldest surviving marketplace, is opened.
1812 The War of 1812 begins after several years of serious maritime trade disputes with the British.
1814 The British set fire to the White House, the Capitol and other public buildings.

WASHINGTON COMES INTO ITS OWN

1815 President James Madison signs the Treaty of Ghent ending the War of 1812. The City Canal is completed, along what is now Constitution Avenue.
1820 District's population slightly exceeds 33,000.
1822 Pennsylvania Avenue in front of the White House is cut through, establishing Lafayette Square.
1829 Wealthy British scientist James Smithson dies and leaves his fortune to the US for the establishment of an institution in his name.
1832 A severe cholera epidemic strikes the city.
1835 Baltimore and Ohio Railroad reaches DC.
1846 The Smithsonian Institution is established.
1848 Work begins on the Washington Monument.
1849 The first gas works is built and the White House now lit with gas. The District's population surpasses 51,000.
1850 The Chesapeake and Ohio canal is completed. Congress abolishes the slave trade in the District, although owning slaves is still legal.
1855 The Castle, the Smithsonian's first building, is opened on the Mall.
1861 The Civil War begins and a series of forts is erected around the District to protect it.
1863 President Lincoln signs Emancipation Proclamation, abolishing slavery. Capitol is completed.
1865 The Civil War ends. Lincoln is assassinated.
1867 Howard University, the prestigious black institution, is set up.

THE NEW ERA

1871 Congress votes $20 million for civic improvements such as paving streets and installing sewers. Georgetown is incorporated into Washington.
1874 The Corcoran Gallery, the city's first art museum, opens.
1874 After a brief experiment at self-government, control of the city reverts to Congress.
1877 The *Washington Post* begins publication. Black leader Frederick Douglass is appointed Marshal of the District and Recorder of Deeds.
1879 There are 400 telephones in operation in the city, and one operator.
1880 After the Civil War, citizens flock to the capital of the preserved nation and the population swells to more than 175,000.
1884 Washington Monument is finally completed.
1889 National Zoo is founded.
1897 The Library of Congress building opens.

1899 Congress limits the height of buildings in downtown, precluding the building of skyscrapers.
1890 Rock Creek Park, a 4-mile (6-km) stretch of woodland, is established.

MODERN TIMES

1902 Much needed restoration work begins on the White House; the West Wing, which includes the Oval Office, is added.
1908 The railroad station on the Mall is moved to Capitol Hill, becoming Union Station.
1912 Griffith Stadium opens and President William Howard Taft throws out the first baseball of the season, thus beginning an annual tradition.
1913 Last farm animal in the White House, a cow named Pauline, leaves with outgoing President Taft.
1914 The Lincoln Memorial is begun.
1918 War War I ends. The District's population exceeds 400,000.
1922 Lincoln Memorial is dedicated.
1924 Baseball's Washington Senators beat the New York Giants to win their first and only World Series at Griffith Stadium in the city.
1935 The Supreme Court takes up residence in its new building on Capitol Hill.
1939 Denied permission to perform at DAR Constitution Hall because she is black, famed operatic singer Marian Anderson performs for a crowd of 75,000 before the Lincoln Memorial.
1941 The US enters World War II. The first plane lands at National Airport. The National Gallery of Art opens on the Mall.
1943 The Pentagon is completed.
1945 World War II ends.
1949 The Whitehurst Freeway is completed along the Georgetown waterfront.
1952 President Truman moves back into the White House after extensive restoration work.
1961 John F. Kennedy is inaugurated as the 35th President. Congress gives District residents the right to vote in presidential elections.
1963 Martin Luther King, Jr. addresses a crowd of 200,000 in front of the Lincoln Memorial and gives his famous "I Have a Dream" speech.
1967 Congress establishes a new political structure for the District and appoints a mayor.

PRECEDING PAGES: map of the DC area, *circa* 1675.
LEFT: Abraham Lincoln, photographed in 1861.
RIGHT: this bronze of President Franklin D. Roosevelt in a wheelchair was added to the FDR Memorial in 2001 after campaigners demanded that his disablement from polio should be acknowledged.

1968 Rioting erupts following the assassination of Dr. King, costing 12 lives.
1971 The Kennedy Center opens.
1972 Burglars break into Democratic Party HQ at the Watergate, beginning the saga that leads to President Richard Nixon's resignation in 1974.
1975 The first elected mayor, Walter Washington, and a city council take office.
1976 Metrorail opens its first subway route.
1982 The Vietnam War Memorial is dedicated.
1983 The renovated Willard Hotel and the Old Post Office building open.
1990 The Washington National Cathedral, begun in 1907, is completed.

1993 The Vietnam Women's War Memorial is dedicated. The US Holocaust Museum is opened.
1995 The Korean War Memorial is dedicated; Washington hosts the Million Man March.
1997 Memorial to Franklin D. Roosevelt opens.
1999 President Bill Clinton impeached by House of Representatives, but acquitted by Senate.
2001 On September 11, Islamic terrorists crash a hijacked plane into the Pentagon, killing 189.
2002 Memorial to founding father George Mason, responsible for the Bill of Rights, is dedicated. Two snipers terrorise the city and the surrounding area for three weeks, killing 10 in random attacks.
2003 The White House and the Pentagon are nerve centers for the war against Iraq. ❑

BEGINNINGS

A regional compromise determined the site of the nation's capital on the banks of the Potomac, a marshland until Pierre L'Enfant envisioned a grand metropolis

Washington, DC was hacked out of the wilderness with one purpose in mind: to serve as the nation's capital. Appropriately, the whole thing started with a political deal. The architects of this deal were Thomas Jefferson and Alexander Hamilton, the nation's first Secretary of State and Secretary of Treasury and two of the young republic's savviest political operators. The year was 1790, about one year after George Washington's presidental inauguration; the place was Philadelphia, temporary headquarters of the fledgling government.

Despite months of bitter debate, Congress had still not agreed on a permanent location for the nation's capital. Northern delegates wanted to keep the capital in the North. Southern delegates wanted to move it to the South. And neither side seemed willing to compromise.

Northerners and Southerners were deadlocked over a second issue, too: namely, Alexander Hamilton's plan to consolidate federal finances by assuming the states' Revolutionary War debts. Again, the argument fell along regional lines: Northerners were in favor of the plan. Southerners were against it. Hamilton's plan "produced the most bitter and angry contest ever known in Congress," Jefferson wrote, noting with particular concern that some delegates were talking of "secession and dissolution." Convinced that the survival of the Union was at stake, Jefferson organized a meeting with Hamilton, Washington and several key members of Congress in the hope of striking a compromise.

The deal they came up with was simple: In exchange for the necessary Southern votes in favor of Hamilton's financial plan, the Northerners agreed to vote for a federal capital farther south than they previously wished for – that is, on the banks of the Potomac River. It was a classic case of one hand washing the other.

LEFT: the city was to strive for the highest ideals, embodied in the Supreme Court's architecture.
RIGHT: the Declaration of Independence having been signed in 1776, the hunt began for a new capital.

Within a year, President Washington was authorized by Congress to select a site, and the 10-mile-square District of Columbia was ceded to the federal government by the states of Maryland and Virginia. The president, a no-nonsense businessman, went to Suter's Tavern in Georgetown to negotiate personally with the land-

owners. He offered them $66.66 an acre, a modest amount even then, but Washington was nothing if not persuasive and they accepted. It was their land, after all, on which the "seat of empire" would be built.

Tidewater colony

More than 180 years earlier, in 1608, the first European to chart the inland waterways of the tidewater region described the "Patawomeck" flowing "downe a low pleasant valley overshadowed in manie places with high rocky mountains; from whence distill innumerable sweet and pleasant springs." The man was Captain John Smith, a founding member of

Jamestown, Virginia, the first permanent English colony in North America, located about 150 miles south of present-day Washington. Smith hoped to find gold along the Potomac or, failing that, a shortcut to the South Seas. What he found instead were several Indian villages, many willing to trade for much-needed provisions.

As Smith already knew, most of the villages in this region were members of a loose Indian confederacy led by Chief Wahunsonacoock, known to the English as King Powhatan. Several months earlier, Smith had had the unexpected pleasure of meeting Powhatan after he and a few of his men were captured by an

Indian hunting party. According to legend – and most of it is probably true – Powhatan's daughter, Pocahontas, begged the old chief to spare Smith's life. It seems Powhatan took a liking to the brash young captain and even offered to exchange corn and meat for English goods.

Despite the Indians' help, life in the early years of the colony was a constant struggle for survival. During the dreadful winter of 1609–10, known to the English as the "starving time," only 60 of the 500 colonists survived. "So great was our famine," a colonist reported, "that a Salvage [Savage] we slew and buried, the poorer sort tooke him up again and eat him." The same grisly report tells of an Eng-

lish settler who killed his wife, "powdered [salted] her, and had eaten part of her before it was knowne." To which the writer adds this gastronomical note: "now whether shee was better roasted, boyled or carbonado'd, I know not; but of such a dish as powdered wife I never heard."

The arrival of several hundred new colonists in 1610 put the wretched little settlement back on its feet and, thanks to the tyrannical leadership of the new governor, Sir Thomas Dale, who employed whipping, burning at the stake and exile to motivate slackers, Jamestown finally sunk roots into the tidewater's marshy soil. Relations with the Indians ran hot and cold during these years. "Blessed Pocahontas" was kidnapped by the English and married to a prominent colonist named John Rolfe. He later brought her back to England where she became an immediate sensation in London society, only to die of smallpox several months later.

Tobacco's effects

Rolfe was also responsible for introducing a strain of West Indian tobacco that was particularly well-suited to the tidewater's climate and soil. The "precious herb" was much in demand in England at the time, and it rapidly became Virginia's most profitable crop. But if tobacco was a blessing to the English, it was a curse to the Indians. As the number of colonists multiplied and the demand for land increased, the tidewater tribes were forced off their traditional homeland in greater and greater numbers.

Hostilities broke out in 1675 when a group of vigilantes launched a bloody campaign against the Nanticoke and Susquehannock Indians, violating the explicit orders of colonial Governor William Berkeley who wanted to avoid more trouble. The leader of the vigilantes was Nathaniel Bacon, Berkeley's younger cousin, who felt their struggle was as much against the Indians as it was against the privileged and wealthy "Parasites whose tottering Fortunes have bin repaired and supported at the Publique Chardg."

Bacon's makeshift army marched on Jamestown and set it on fire. Before the rebellion could escalate into all-out war, however, the young rebel died suddenly of the "bloody phlux" (probably tuberculosis), putting rather an anticlimatic finish on what some historians claim to be the first fitful steps toward American independence.

L'Enfant's city

The American Revolution came and went with little military impact on the Potomac, although the British blockade of Chesapeake Bay virtually shut down the tobacco trade out of Georgetown and Alexandria. After the British defeat at Yorktown, General George Washington had hoped to retire to his Mount Vernon estate near the Potomac, but destiny had different plans. In 1789, Washington accepted his election to the new office of president, and several weeks later took the oath of office.

Even before the Revolutionary War, George Washington had high expectations for the development of the Potomac River. He saw the Potomac as a gateway to the resources and markets of the Ohio and Mississippi valleys, which he felt were in imminent danger of falling to the Spanish or British, or, almost as bad, to the merchants of New York City. Immediately after the war, he undertook a 600-mile (960-km) survey trip in order to assess the Potomac's commercial potential, and then helped create the Potomac Company to make the upper river navigable. Thomas Jefferson, a fellow Virginian, shared Washington's enthusiasm for the region, and when the search began for a permanent federal capital it seemed only natural to them that the Potomac be placed at the top of the list. To design the new Federal City (the self-effacing Washington didn't like to refer to it as the City of Washington), the president selected Major Pierre Charles L'Enfant.

Several years earlier, at the age of 22, L'Enfant had left the French court of Versailles and arrived in America to serve as a private in the Continental Army. As a member of the Corps of Engineers, he endured the brutal winter of 1777–78 with General George Washington at Valley Forge, then distinguished himself at the battles of Savannah and Charleston. After the Revolutionary War, L'Enfant supervised an extensive renovation of Federal Hall in New York City, site of Washington's first inauguration and temporary seat of the infant US government.

When talk of a new capital reached L'Enfant, he wrote President Washington requesting to be involved in the city's founding. "The late

determination of Congress to lay the foundation of a Federal City which is to become the Capital of this vast Empire, offers so great an occasion for acquiring reputation… that Your Excellency will not be surprised that my ambition and the desire I have of becoming a useful citizen should lead me to wish to share in the undertaking."

L'Enfant was hired to lay plans for the capital in 1791. With memories of Versailles lingering in his mind and Thomas Jefferson (a masterful architect in his own right) inspiring him toward grandeur, L'Enfant conceived a city of broad boulevards, sweeping vistas, and

stately public buildings on the monumental scale of classical Greece or Rome. The plan also incorporated civic open spaces, markets, fountains, monuments, and squares. "No nation had ever before the opportunity offered them of deliberately deciding on the spot where their Capital City should be fixed," L'Enfant wrote to Washington. "The plan should be drawn on such a scale as to leave room for the aggrandizement and embellishment which the increase of the wealth of the nation will permit it to pursue at any period, however remote."

On Jenkins Hill, the city's highest prominence, L'Enfant placed the Capitol, seat of the Congress, the very embodiment of the young

---

LEFT: an English adventurer, Captain John Smith, sailed up the Potomac in 1608, at the age of 29.
RIGHT: architect Pierre L'Enfant (1754–1825).

nation's democratic principles. On the banks of the Potomac, commanding a view of Alexandria, L'Enfant placed the Executive Mansion, known later as the White House. Between these two hubs stretched Pennsylvania Avenue, one of several broad thoroughfares that sliced diagonally through the street grid, connecting the various departments of government and affording a breathtaking "reciprocity of sight."

Unfortunately, what L'Enfant possessed in architectural skill he lacked in social grace. Arrogant, impetuous, and utterly incapable of subordinating himself to authority, the Frenchman immediately ran afoul of his superiors, a

In his place Washington appointed Major Andrew Ellicot, "a gentleman of superior astronimal abilities," according to a 1791 Georgetown newspaper. He condensed and simplified L'Enfant's original plan.

As a concession Congress offered L'Enfant $2,625 and a plot of land in "a good part of the City," but the Frenchman, apparently insulted, refused. L'Enfant accepted a number of commissions after designing Washington, but never enough to keep him from poverty in his latter years. "Daily through the city stalks the picture of famine, L'Enfant and his dog," wrote fellow architect Benjamin Henry Latrobe. The ill-fated

commission of prominent landowners who were authorized to oversee the project. In one particularly damning incident, L'Enfant demolished a house owned by Daniel Carroll, newphew of one of the commissions, because it stood in the path of one of his streets.

Hearing of L'Enfant's behavior, Thomas Jefferson advised Washington that the architect "must know there is a line beyond which he will not be suffered to go." The president, perhaps feeling a bit paternal, first tried to appease the commissioners. But when L'Enfant demanded that the landowners be dismissed, Washington put his foot down. Less than 12 months after he started, L'Enfant was discharged.

Frenchman became a common sight in the capital, his shabbily dressed figure pacing the avenues of a city that, only years before, existed solely in his imagination. He died in 1825 and was buried in a modest Maryland cemetery.

It wasn't until the early 1900s, when the city's original plan was revived and expanded, that L'Enfant's talent was properly acknowledged. In 1909, the architect's remains were reinterred in Arlington National Cemetery at a site overlooking the capital, his greatest and most embittering achievement. ❏

ABOVE: Georgetown's busy waterfront at the end of the 18th century.

George Washington

There could have been no more reluctant person to serve as the country's first president than George Washington. "I greatly fear that my countrymen will expect too much from me," he wrote soon after his inauguration in New York in April 1789. By then he had already been elevated to the status of a minor god. Lavishly praised by his countrymen, he was known as "our Savior," "our Redeemer," and "our star in the east."

More a monument than a man, Washington, with little in his early life to prepare him for future greatness, labored under the burden of his own celebrity to lead the new nation when there was no precedent for such leadership.

Born in 1732 to a Virginia tobacco farmer and his second wife, Washington grew up near Fredericksburg, Virginia. After his mother died when he was 11, his older half-brother Lawrence encouraged frequent visits to his own estate, Mount Vernon, near present-day Washington. From one of his wealthy neighbors, Washington learned surveying and signed on to explore what was then the colonial western frontier.

For 12 years he was an active-duty soldier, much of that time served unsuccessfully. In 1754 he ambushed a French patrol on the frontier, an event that triggered the French and Indian War. During the Revolutionary War, he fought nine battles, losing or fighting to a draw six of them.

Lacking military genius, he also did nothing to distinguish himself intellectually. He had no appreciation for music or literature and, though he could hold his own in a social setting, he wasn't particularly gregarious. Even Jefferson, his close friend, could not avoid remarking on Washington's shortcomings. His talents, as Jefferson put it, "were not above mediocrity."

What Washington was, however, was scrupulously decent, his real source of power concentrated in the nobility of his character. When he was a boy, he copied 110 maxims of civility from an etiquette book, indicating his obsession with proper behavior. These values were exemplified in the compassion he showed for the men who served under him and his willingness during the Revolution to defer to a civilian leadership. Perhaps his most virtuous act was his resignation in December 1783

RIGHT: George Washington (1732–99), regarded even in his own lifetime more as a monument than a man.

as commander-in-chief of the American forces when he was at the height of his powers and could easily have declared himself king. But Washington wished only to retired to his beloved Mount Vernon, a modern Cincinnatus, the Roman who returned to his farm after saving his country.

Washington's strength, as first president, lay in his embodiment of people's aspirations and fears, and his ability to hold together a collection of former colonies that had no love for one another. Any impression that the colonies collectively united and rose up against the British is pure myth. Writing in 1765, Boston lawyer James Otis said: "Were these colonies left to them-

selves tomorrow, America would be a shambles of blood and confusion."

The Revolutionary War lasted 8½ years and, when it ended in 1783, America, though a name on a map, was hardly a country. It was into the ensuing fractiousness and chaos as America struggled to define itself and establish a constitution that Washington, coming out of retirement, stepped in to lead for two full terms as president.

Whatever native abilities he may have lacked, Washington, to his lasting credit, understood that the American experiment in which he was a central player could not be allowed to fail. As he wrote as he embarked on his first term as president, "For with our fate will the destiny of unborn millions be." ❏

BUILDING A NATION

Just as the city gained a toehold in the marsh, the British destroyed most of the major public buildings. It didn't come into its own until after the Civil War

In the summer of 1800 John Adams, second president of the United States, traveled by coach from Philadelphia to the nation's new seat of government, Washington City. Surveying the unfinished capital from a crest above the Potomac, the urbane Bostonian might well have felt a mixture of wonder and disappointment.

Not that the setting was unpleasant. The Potomac flowed gently below; dense woods and tobacco fields spread from the riverbanks; farmhouses were tucked into the valley, and tobacco boats sailed across the water. But in its incomplete state, the nation's capital must have seemed a remote and primitive place. The major government buildings, the great structures that were to embody the highest ideals of the young republic, were little more than rough skeletons surrounded by workmen's shacks and scattered debris.

Out of the mud

Pennsylvania Avenue, the grand boulevard that was to connect the Capitol Building and the White House, was a tangle of elder bushes and tree stumps. A few rows of boarding houses were complete, but the streets, the few that existed, were little more than muddy lanes. Compared to the great European capitals Adams had visited in the past, Washington was little more than an unkempt backwoods town.

Several months later Adams's wife, Abigail, put into words what her husband was perhaps too politic to express: "Washington City was not a city at all." Although she was pleased with the White House's riverfront location, Abigail complained that "not one room or chamber is finished of the whole." Moreover, the nearby port of Georgetown was "the very dirtyest hole I ever saw."

Ever the pragmatic New Englander, she used the White House's "great unfinished audience room" in which to hang the President's laundry. Other new arrivals concurred with the First Lady. Sir Augustus John Foster, a British diplomat, found the new capital "scarce any better than a swamp." Congressman Gouverneur Morris quipped: "We want nothing here but houses, cellars, kitchens, well-informed men,

amiable women, and other little trifles of this kind, to make our city perfect."

By 1809, Thomas Jefferson's last year as president, Washington was still a frontier town, criss-crossed with rutted trails, littered with debris and surrounded by wilderness. Work had proceeded at a snail's pace over the intervening years and neither the White House nor the Capitol building were complete. The nation's 106 delegates huddled in the House of Representatives – known derisively as "the oven" because of its poor design – while construction continued on the remainder of the building.

When President James Madison led the coun-

LEFT: a meeting of Congress in 1822.
RIGHT: constructing the Washington Monument, 1860.

try into a second war against the British, in 1812, Washington was still so undeveloped that many people were unconcerned, even though enemy warships were cruising Chesapeake Bay. Washingtonians had great confidence in Secretary of War John Armstrong , who assured the president that the British would not waste their time attacking the "sheep walk on the Potomac" when Baltimore, a far richer target, was also within easy reach.

When 4,500 British soldiers landed on the Patuxent River about 35 miles (56 km) south of the capital, however, it became tragically clear that Armstrong had misjudged the British

Navy. As residents and government workers fled into the countryside, American militiamen rushed to the capital's defense at Bladensburg on the District's border. But when the Redcoats attacked on August 24, 1814, the Americans beat a hurried retreat back to the city – one of the less glorious episodes in America's early history and known derisively as the "Bladensburg races."

By evening, the British, commanded by Admiral George Cockburn, were torching many public buildings, including the Capitol and the White House. Only hours before, the president's wife, Dolley Madison *(see page 124)*, had

THE BRITISH ARE BACK: THE WAR OF 1812

Early in the 19th century, while Britain and France under Napoleon battled it out on the high seas, American merchants devised an ingenious plan to protect their shipping interests as they carried goods between Europe and the Caribbean while still retaining American neutrality. By putting into American ports on their way back and forth across the Atlantic, they reasoned, the French and British goods in their holds became technically American, and therefore safe from seizure. Between 1803 and 1806 the annual value of foreign products "re-exported" from the US rose from $13 million to $60 million. Then the French and British instituted a series of blockades that practically shut down American business.

Compounding matters was Britain's annoying habit of impressing American seamen, snatching as many as 5,000 between 1803 and 1812 and forcing them to serve in the Royal Navy. The British also seized over 500 American ships, with Napoleon taking 200 more. The public clamored for war but President Jefferson, an isolationist at heart, negotiated the Embargo Acts instead. Little improved for American business as a result, and in 1812, James Madison, Jefferson's successor, declared war. The War of 1812 ended in a second stunning defeat for Britain in America, this time in 1815 at the Battle of New Orleans. Compared to Britain's 2,100 casualties, the Americans lost only eight men.

managed to save a few prized possessions from the White House, including Gilbert Stuart's famous portrait of George Washington.

When the British pulled out two days later, the capital was in ruins. "I do not suppose the Government will ever return to Washington," wrote Margaret Bayard Smith. "All those whose property was invested in that place will be reduced to poverty… The consternation about us is general. The despondency still greater." With the city in shambles and the government in dis-

THE CITY THAT HAS IT ALL

According to an 1822 directory, Washington was home to "a city hall, a theatre, a penitentiary, a masonic hall, four banks, 14 houses for public worship, and a circus."

of the Post Office, the only public building left undamaged, and voted to stay in Washington.

"A great[er] benefit could not have accrued to this city than the destruction of its principal buildings by the British," wrote Benjamin Henry Latrobe, one of Washington's leading architects after the War of 1812.

From the city's blackened ruins came a renewed determination to make Washington the national showcase that had been envisioned by Jefferson, Washington and L'Enfant. While President

array, a visiting Virginian feared that the republic itself might also crumble: "The appearance of our public buildings is enough to make one cut his throat," he wrote. "The dissolution of the Union is the theme of almost every private conversation."

But with American forces rallying against the British outside Baltimore and later at New Orleans, anxiety over the capital's and the country's future began to fade. With news of Yankee victories arriving daily in Washington, Congress reconvened in the cramped quarters

LEFT: British troops burn the city in 1814.
ABOVE: Georgetown as it looked in 1864.

Monroe took temporary quarters at the nearby Octagon House, the private residence of a wealthy Virginia planter which had somehow escaped the fires, the original architect, James Hoban, was called in to supervise the reconstruction of the White House.

At the other end of Pennsylvania Avenue, Benjamin Latrobe worked to restore the Capitol, following the design of original architect William Thornton. Construction began on several new federal buildings as well, including the magnificent neo-classical Treasury Building, Patent Office and Post Office, designed by Robert Mills, a student of Thornton's and the first American-born architect.

Up from the ashes

Washington was expanding in other areas, too. The short-lived Chesapeake & Ohio Canal, thought to be the key to Washington's commercial success, opened for business in 1830, but was quickly superseded by the Baltimore & Ohio Railroad. Georgetown University, founded in 1789, was joined by Columbian College (later George Washington University) in 1822. A grant bequeathed by an eccentric British scientist, James Smithson *(see page 143)*, in 1829 was finally discharged for the use of the Smithsonian Institution, housed in a medieval-style "castle" on the Mall. And in 1859, W. W. Cor-

coran began building the handsome Corcoran Gallery to house his coveted art collection.

Unfortunately, the city's problems were growing as fast as its buildings. Washington's notorious alleys were already filled with shacks and shanties and crowded with an ever increasing population of poor people. Vacant lots, many of them squeezed between the city's finest buildings, were heaped with trash and overgrown with weeds. Sanitation, never very sophisticated, was woefully inadequate.

Even the Washington Canal, the "miasmatic swamp near the Presidential Mansion," was choked with sewerage, silt and garbage, as were other marshy areas around the White House. An active slaughterhouse near the Mall added to the stench. As one army wife put it, "I was never in such a place for smells."

Crime was also becoming a major concern. "Riot and bloodshed are a daily occurrence," a Senate Committee reported in 1858. "Innocent and unoffending persons are shot, stabbed, and otherwise shamefully maltreated, and not unfrequently the offendor is not even arrested." And violence was hardly limited to the streets. More than once, congressional debates degenerated into bare-knuckle brawls.

Dueling pistols

The abolitionist Charles Sumner was nearly caned to death by a South Carolinian delegate in 1856. Congressman Daniel Sickles was acquitted for the admitted murder of his wife's lover in 1859. And political squabbles were regularly settled with dueling pistols. Add to this the usual round of gambling, drinking and whoring in which members of Congress rou-

FRANCIS SCOTT KEY'S NATIONAL ANTHEM

Francis Scott Key, born in Frederick County, Maryland, in 1779, was a lawyer and sometime poet who lived and practiced in a small house along the Potomac in Georgetown. He happened to be in Baltimore the night the British invaded on September 13, 1814, trying to save a friend from being impressed onto a British warship. In the course of the negotiations, Key, too, was detained, and along with his friend, watched as the British bombarded Fort McHenry that night, firing as many as 1,800 shells.

At dawn, when the American flag was still flying, though much the worse for wear, Key hurriedly penned the words that would become the National Anthem. Soon

the poem was being sung across America to the tune of "To Anacreon in Heaven," a British drinking song:
Oh, say can you see by the dawn's early light
What so proudly we hail'd at the twilight's last gleaming…

Key, who died in 1843, did not live to see his poem immortalized. By presidential proclamation, Key's poem finally became the nation's song in 1931 when in the grip of the Great Depression, Americans turned to the words for strength and inspiration.

In 1948, Key's home was torn down to make way for the Whitehurst Freeway and for the bridge that is named in his honor and links Arlington, Virginia, to Georgetown.

tinely engaged, and it's not difficult to see why one righteous soul denounced the "unparalleled depravity of Washington society."

But for all of Washington's urban problems, the nation's capital still had not come of age as a major American city. It was still a "sleepy country town," still tentative and only partially realized. During his visit in 1842, Charles Dickens expressed his delight at the casual grace of the nation's political elite, but couldn't help detecting a certain civic immaturity.

Washington "is sometimes called the City of Magnificent Distances," Dickens wrote, "but it might with greater propriety be termed the City

liberty and the pursuit of happiness. "Their are many kinds of hunters engaged in the Pursuit of Happiness," he wrote. Some "take the field after their Happiness equipped with cat and cartwhip, stocks, and iron collar, and to shout their view halloa!... to the music of clanking chains and bloody stripes."

Although some Washingtonians tried to shut their eyes to it, slavery was impossible to ignore. At the time of Dickens's visit, there were approximately 3,000 slaves in the District of Columbia. Prominent slave traders advertised in local newspapers; public slave auctions were held on a regular basis; and slaves could

of Magnificent Intentions." Its leading features were "spacious avenues, that begin in nothing, and lead nowhere; streets, milelong, that only want houses, roads and inhabitants; public buildings that need but a public to be complete; and ornaments of great thoroughfares, which only lack great thoroughfares to ornament."

Like other visitors, Dickens was especially disturbed by the presence of slavery, which he found all the more loathsome in the capital of a nation supposedly built on and dedicated to life,

LEFT: a bumpy ride at the Capitol.
ABOVE: the junction of 15th and Pennsylvania in 1865, the year of Abraham Lincoln's assassination.

be seen on the streets marching in chains or being held for sale in slave pens across from the Smithsonian Institution.

In 1849 a little-known Illinois representative, Abraham Lincoln, introduced legislation outlawing slavery in the District of Columbia, but the bill was quickly defeated. Instead, Congress adopted Henry Clay's Compromise of 1850, which, among other measures, abolished the slave trade in the District of Columbia but did not outlaw slave ownership.

At best, Clay's compromise was a stopgap measure, slowing but not stopping the movement toward a major confrontation. Abolitionists were becoming more influential in the

North and secessionists more militant in the South. The nation was headed for war with itself and the battle lines were being drawn in Washington.

"A house divided"

Sectional tensions escalated in Washington throughout the 1850s. Congress battled furiously over the Fugitive Slave Law and the extension of slavery into the territories of Nebraska and "bleeding" Kansas, so-called when pro- and anti-slavery forces came to

A PARADOXICAL PLACE

"Washington is the paradise of paradoxes," said the *Atlantic Monthly* in 1861. "It is a great, little, splendid, mean, extravagant, poverty-stricken barrack for soldiers of fortune and votaries of folly."

electoral vote and on a minority of the popular vote. South Carolina immediately seceded, and six other states followed.

On March 4, 1861, Lincoln was inaugurated under the watchful eyes of sharpshooters patroling nearby rooftops. In his inaugural address he spoke directly and bluntly to the newly formed Confederacy: "In your hands, my dissatisfied fellow-countrymen, and not in mine is the momentous issue of civil war. We are not enemies, but friends… Though passion

blows in 1856 resulting in the slaughter of some 200 settlers. Northerners were inflamed by the Supreme Court's pro-slavery Dred Scott decision, which effectively denied Congress the power to ban slavery in the territories by declaring that slaves were not citizens, but rather chattel. And Southerners were equally incensed when John Brown, an instigator in the Kansas affair, tried unsuccessfully to seize the federal arsenal at Harper's Ferry, Virginia, for the purposed of inciting a slave insurrection. By the following year, 1860, the conflict had reached boiling point.

On November 8, 1860, Abraham Lincoln was elected president without a single Southern

may have strained, it must not break our bonds of affection!"

But Lincoln's appeal fell on deaf ears. About a month later, on April 12, Confederate cannons fired at Fort Sumter. Virginia seceded two weeks later, following Arkansas, North Carolina and Tennessee. Robert E. Lee, formerly superintendent at West Point and never a supporter of slavery or secession, nevertheless declared allegiance to his home state, Virginia, and assumed the command of the Army of Northern Virginia. After an agonizing debate, Maryland, Washington's other slave-owning neighbor, agreed to remain neutral.

Washington was in a perilous situation. It

was a Northern capital in a Southern city, nearly half of its population hailing from, or sympathetic to, the Confederate states. As Southern soldiers resigned their posts and joined the Confederate Army, Lincoln called for 75,000 volunteers to come to the Union's defense. Thousands of troops poured into a capital that was little prepared to receive them.

By the end of April 1861, soldiers were being quartered in the Capitol, Treasury, Patent Office, and at Georgetown University. The unfinished Corcoran Gallery was commandeered by the Union's quartermaster; a bakery was established in the basement of the Capitol;

troops looming just beyond the district's borders, a massive effort was made to ring the city with fortifications and secure the surrounding countryside. Soldiers continued to flood the city, swelling the population from 61,000 to more than 1 million in a single year.

As the fighting raged at the Second Battle of Bull Run, Antietam, Fredericksburg and Chancellorsville, some 50,000 wounded men filled makeshift hospitals in churches and public buildings scattered throughout the capital. At night, carts loaded with dead soldiers rumbled through the streets, many of them headed across the Potomac River to the

Army horses, mules and other livestock were corralled in the mall.

When a showdown between Union and Confederate troops developed at Bull Run, Virginia, Washington's society people turned out in their finest carriages, fully expecting a glorious Union victory. The horror and humiliation of the Union defeat, and the panicked retreat of soldiers and civilians back into the capital, convinced even the most cocksure that this was to be a long and bloody fight. With Confederate

LEFT: since Washington's streets were a sea of mud, washday was a serious occasion.
ABOVE: the burning of Richmond in April 1865.

former estate of Confederate commander Robert E. Lee, which is now part of Arlington National Cemetery.

Troubled by crime before the war, the city was now hopelessly overrun. "Every possible form of human vice and crime, dregs, offscourings and scum had flowed into the capital and made of it a national catch-basin of indescribable foulness," one newspaperman reported. Aptly named neighborhoods such as Murder Bay were rife with gambling dens, whiskey joints, gangs and crooks. But with so many soldiers in town, the boom industry was prostitution. Captain Joseph Hooker tried to limit the trade to a red-light district called

"Hooker's Division," inadvertently lending the women his name.

In April 1862, President Lincoln freed slaves in the District of Columbia and then, in January of the following year, issued the Emancipation Proclamation, abolishing slavery in the Confederate states. In July 1864, Confederate general Jubal Early sent Washingtonians into a panic with a brief but terrifying assault on the District of Columbia, stopped only by the timely arrival at Fort Stevens, near the Capitol, of troops sent by General Grant from Petersburg, 100 miles (160 km) to the south. From an unprotected parapet, President Lincoln stood

at Atlanta. Richmond fell several months later, isolating the bulk of the Confederate Army. Surrounded by Union forces, General Robert E. Lee surrendered to General Ulysses S. Grant at Appomattox Courthouse on April 9, 1865.

Lincoln's assassination

Five days later, Lincoln reluctantly agreed to attend a play, *Our American Cousin*, at Ford's Theater on 10th Street. While Lincoln sat in the balcony, John Wilkes Booth, a well-known actor and Confederate conspirator, crept behind the President, shot him, and then leapt to the stage below, crying "*Sic semper tyrannis!* The

watching the battle, much to the dismay of the Union officers who could not convince their tall commander-in-chief that he made a tempting target for Early's rebel. Finally, when a medical officer standing near the president fell to a rifle shot, Oliver Wendell Holmes, then a lieutenant colonel and later a Supreme Court justice, grabbed Lincoln and shouted, "Get down, you fool!". It was the Confederates' last stab at the capital. After Union victories at Gettysburg and Vicksburg and the loss of Confederate power in the West, the tide of war turned decisively in the Union's favor.

In September 1864, Union general William Tecumseh Sherman delivered the *coup de grâce*

South is avenged!", breaking his leg in the fall. Fatally wounded, Lincoln was rushed to a nearby boarding house and died the next morning. A manhunt was organized, which turned up Booth two weeks later in a Virginia barn. The barn was set on fire and Booth was killed.

As Washington mourned, the assassination seemed to take on a sense of inevitability, as if Lincoln sacrificed his own life for the life of the Union. The country was at last one entity, but deeply scarred. Washington, and the nation, lay in ruins. ❑

ABOVE the Civil War ends as General Robert E. Lee surrenders to General Ulysses S. Grant in 1865.

Tracing Your Roots

The National Archives are best known for their original copies of the Declaration of Independence and the Constitution, kept in helium-filled bronze and glass cases that can be lowered instantly into a fireproof and shockproof safe. But Americans can also trace their family history here, sifting through more than 3 billion documents for priceless ancestral footprints.

Passenger lists for immigrant ships, the national census back to 1790 (which includes the names of heads of households), military and naturalization records, passport applications, pensions and land grants – the Archives contain the country's principal database for family histories and other uses. There is a free booklet available which is basically a guide to the collection; the pamphlet does, however, contain the warning phrase "the staff is unable to make extensive searches but, given enough identifying information, will try to find a record about a specific person."

What this means in practice is that family research has to start at home, so you should arrive at the Archives with a list of ancestral names, approximate dates of birth and places of residence and military details, as far back as you can go. Look in the attic, interview older relatives and check out the boxes which have been in storage for years. The more information the better. Armed with these clues, the Archives staff can be extraordinarily helpful – given that many of the records have no index organized by name.

A serious search really requires an investment in the National Archives' paperback book *Guide to Genealogical Research*, which offers a step-by-step procedure on how to go about research on the flimsiest of details, like "great-great grandfather Abner Smith who moved to Indiana sometime after the Civil War."

The point is that even in pre-computer days, it was not easy to inhabit the US without leaving a paper trace in the Archives. Military records (including the Confederate forces), seamen's certificates and pension files and land grants are useful, but the heart of the collection is the census data.

For ancestors born before 1861, it is much easier to start with the (free) males of the household. The 1790–1840 census lists give the head of household only; other family members are listed

without age or sex. For the 1850 and 1860 censuses, separate schedules list slave owners and the age, sex and color (but not the name) of each slave. These same lists do, however, give details of each free person in a household.

The 1880 census is even more useful, since there is a microfilm version of a card index with entries for each household that included a child aged 10 or less, including details of every household member. The indexes are organized by state, then alphabetically by the first letter of each last name, then by the first letter of the Christian name of each household head. Most of the 1890 census was lost in a fire in 1921; only 6,160 names remain.

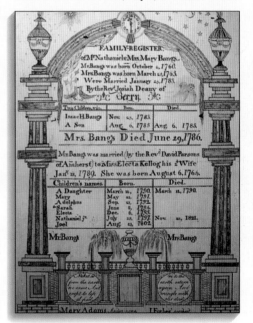

If your 18th-century ancestor came from Delaware, Georgia, Kentucky, New Jersey, Tennessee or Virginia, you may be unlucky. The census data for these states was burned by the British in the War of 1812. The Virginia government has since painfully rebuilt the census for about half the state's population in 1790, using mainly tax and property lists.

Ironically, since they lost their land as the white immigrants poured through the country, some of the best data concerns Native Americans. The school class lists from 1910 to 1939 are one place to start, but there is a near-census quality data bank for those Indians, mainly from the Cherokee, Chickasaw, Choctaw and Creek tribes, who moved west between 1830 and 1846. ❏

RIGHT: family trees provide vital details.

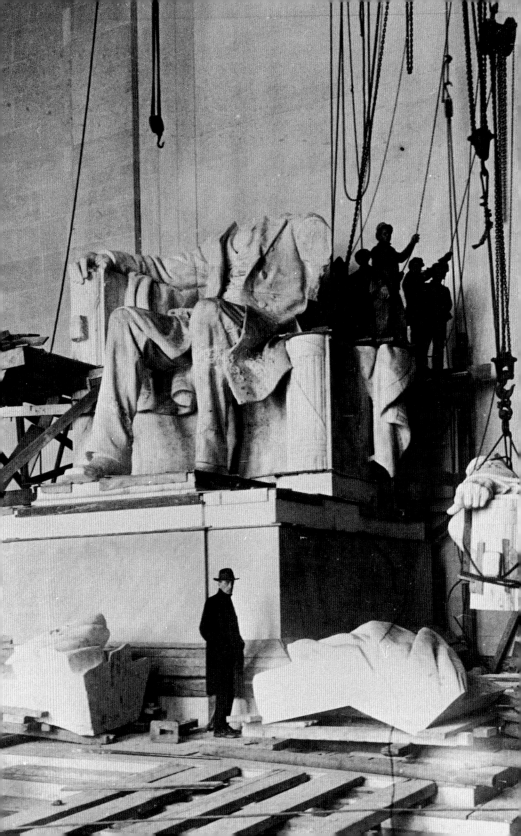

PRESIDENTS AND PROTESTS

Revitalization in the decades following the Civil War boosted Washington's prestige and made it the place where Americans gathered to voice their concerns

The Civil War left Washington a less provincial, less Southern and, all things considered, a less agreeable place to live. The Army of the Potomac had turned public buildings into a shambles and the streets into a "sea of mud." The city's slums were swollen with former slaves and discharged soldiers. Disease and squalor were evident in every neighborhood. "Crime, filth and poverty seem to vie with each other in a career of degradation and death," the Police Chief reported. A relief worker described "a cloud of darkness, poverty, rags, hunger, cold and suffering" in Washington's alleyways.

As before, there were those who wanted to relocate the capital to a more suitable place, a western city perhaps, where the country was still young and growing. But Congress was not yet ready to give up on Washington and, more importantly, Washingtonians were not ready to give up on themselves.

The Union had been spared, and the capital of the Union, where power was now concentrated, was Washington. By winning the war, the city had redeemed itself and was for the first time seen as the place where the destiny of the nation would be determined. The bureaucracy exploded, and Americans thronged to the city to take up federal jobs and to be at the center of power. In the decade following the Civil War, the population doubled to 175,000, overtaxing city services, such as they were, to the point that the nation's capital, as it struggled to define itself, was turning into, as the *Washington Post* put it, "a malarial joke."

Big spender

The first effort to rebuild Washington started in the early 1870s with the appointment of Alexander "Boss" Shepherd to the newly constituted Board of Public Works. A single-minded, energetic bull of a man, Shepherd ran herd over the local blue bloods and the District

government in his efforts to make the "Federal City worthy of being in fact, as well as in name, the Capital of the nation." Shepherd's swaggering, big-spending style made him as many enemies as friends in Washington, but despite his heavy-handedness, he got things done.

Shepherd also managed to drive the city

directly into bankruptcy, tallying up a $20 million deficit, more than three times the city's debt in its entire 70-year history. Although exonerated of any criminal wrongdoing by two Congressional probes, Shepherd's "negligent, careless, improvident [and] unjust" management of public funds finally forced him from office. Turning his attention next to a silver mining scheme, Shepherd took off for Mexico where he made his fortune before returning as a minor hero to the city that had once scorned him. Spendthrift or not, few could argue that Shepherd had literally brought the nation's capital out of the mud.

Shepherd's civic facelift ushered in a period

LEFT: Lincoln's memorial nears completion.
RIGHT: Ulysses S. Grant, president from 1869 to 1877.

of good feelings. Despite the national depression of 1893–94 and the unexpected arrival of the "Army of the Unemployed" – some 3,000 jobless men seeking government relief – Washington society managed to recapture the "extraordinarily easy and pleasant life" of the antebellum years. Tired of war, weary of politics, Washington became, in the novelist Henry James's words, "a city of conversation," blithely attending the business of society balls and teas and letting the government take care of itself. "The social side of Washington was to be taken for granted as three-fourths of existence," Henry Adams wrote. "Politics and reform became the detail, and waltzing the profession."

A second wave of civic improvements was launched at the turn of the century by Michigan Senator James McMillan, who commissioned some of the leading lights in architecture, including Daniel Burnham, Charles McKim and Frederick Law Olmsted, Jr, to create "a well-considered general plan covering the entire District of Columbia." At the heart of the McMillan Commission's proposal was a plan to sweep away the run-down buildings, roads and railways that had cluttered the Mall for years, to build a bridge connecting the Mall to Arlington National Cemetery, and to erect two new

ALEXANDER "BOSS" SHEPHERD

Alexander "Boss" Shepherd, the chief and strong arm of the city's Board of Public Works, never took "no" for an answer. In November, 1872, after being rebuffed by the president of a local railroad whose tracks interfered with Shepherd's new street grading and beautification plans, the Boss simply hired a crew of 200 men and ordered them to work through the night, ripping up the tracks. The next morning the railroad president was livid when he discovered what had happened, though he took no action against bulldog Shepherd, aware of whom he was up against.

Between 1871 and 1874, "Boss" Shepherd oversaw the installation of 123 miles (198 km) of city sewers, 3,000 street lamps, and almost 200 miles (320 km) of sidewalks, not to mention the more than 25,000 trees he had planted along city streets.

Despite his efforts at bringing civilization to a capital city that was barely an outback, Shepherd today is little more than a footnote in the city's history. In a city crowded with monuments, not one memorializes him, although until 1980 there was a statue of Shepherd on Pennsylvania Avenue at 14th Street. For reasons never fully explained, then-mayor Marion Barry had the statue removed to an abandoned car lot along the Anacostia River where the Boss now gazes across the water into Maryland.

monuments, one to Abraham Lincoln, the other to Thomas Jefferson.

It was an expansive time for Washington as well as for the nation. The Spanish-American War came to a swift end, catapulting Teddy Roosevelt and his Rough Riders into the national spotlight and giving the US its first colonial possessions, the Philippines, the Pacific island of Guam, and Puerto Rico. Roosevelt joined President William McKinley's re-election bid as the vice-presidential candidate,

ing Machine Company, forerunner of IBM) came too, drawn by the intellectual and commercial opportunities offered by the Smithsonian Institution, Patent Office and technology-minded legislators. Electric lights made an illuminating debut in the 1880s, and by the late 1910s motor cars were chugging through the streets and terrorizing unsuspecting pedestrians.

America's entry into world politics brought new responsibilities to Washington. When war broke out in Europe in 1914,

and found himself in command of the Oval Office after McKinley's assassination only four months later.

With the industrial revolution in full bloom, a new breed of "social Darwinists" gravitated toward the capital, housing their families in extravagant Victorian mansions in the West End, particularly on Massachusetts Avenue (now "Embassy Row"). Inventors such as Alexander Graham Bell, the Wright Brothers and Herman Hollerith (founder of the Tabulat-

LEFT: President Woodrow Wilson addresses Congress on going to war with Germany in 1917.
ABOVE: the Roaring Twenties reaches Glen Echo.

the Allies looked to the United States to exercise its newfound military muscle. Although President Woodrow Wilson was an early advocate of "peace without victory" in Europe, Germany's persistent attacks on American ships forced his hand toward war. In a special address to Congress on April 2, 1917, Wilson dubbed Germany's U-boat campaign "warfare against mankind." Congress took its cue and declared war against Germany four days later.

Almost overnight, the "city of conversation" was transformed into the hub of an international war machine. "Life seemed suddenly to acquire a vivid scarlet lining," Helen Nicolay, a contemporary writer, remembered. "Old prejudices gave way to passionate new beliefs. Old precedents were wrecked in an endeavor to live up to the duty of the hour." As thousands of Americans passed through the capital en route to the trenches of the Argonne, the city was swept into the war effort. Military advisers, scientists, intellectuals, industrialists and clerks poured into the city; the federal government expanded dramatically; government housing was hastily erected; and boxy "tempos" were thrown up along the Mall to use as office space. Once a "sleepy country town," Washington suddenly became a major player in global politics.

The US military presence tipped the balance

likes of F. Scott Fitzgerald, Eugene O'Neill or Al Capone, there was no shortage of bootleg whiskey, suffragettes, shiny new Fords or extravagant parties. The nation's outlook seemed so rosy, in fact, that President Herbert Hoover could confidently announce at his 1929 inauguration: "The poorhouse is vanishing among us. We in America today are nearer to the final triumph of poverty than ever before in the history of the land."

It was a particularly unfortunate prediction. On Black Thursday, October 24, 1929, only seven months after Hoover uttered those words, the stock market crashed, the economy took a

of power in Europe and within two years peace negotiations were underway at Versailles. In 1919, Washingtonians cheered as President Wilson and General John "Black Jack" Pershing led triumphant American soldiers down Pennsylvania Avenue. The "war to end all wars" was over, a new decade was on the horizon and Americans were ready to enjoy the fruits of victory.

Good times, bad times

Prohibition kicked off the Roaring Twenties on a rather dreary note, but somehow the taboo on liquor made the good times seem even better. And although Washington couldn't boast the

nose-dive, and the Great Depression had begun. The hardships of the Depression were slow to reach Washington, but when they hit, they hit hard. City-wide income was slashed by nearly half; unemployment soared to 25 percent, 50 percent among blacks. People lost their jobs, their homes, their dignity. Ramshackle "Hoovervilles" sprang up in vacant lots, and breadlines stretched along sidewalks. Hundreds of "hunger marchers" trickled into the city seeking relief from the federal government, and protests erupted daily in front of the White House and the Capitol.

In the spring of 1932, some 20,000 World War I veterans marched into Washington from all parts of the US demanding early payment of a cash bonus due in 1945. The self-proclaimed Bonus Expeditionary Force, or Bonus Army, was allowed to occupy several condemned buildings on Pennsylvania Avenue and to set up a large shantytown – known as Camp Marks— on the Anacostia Flats. "The arrival of the bonus army seems to be the first event to give the inhabitants of Washington any inkling that something is happening in the world outside of their drowsy sun parlor," author John Dos Passos noted wryly. As far as President Herbert Hoover was concerned, however, the Bonus Army was strictly a "local problem." He

troops that bore down Constitution Avenue included a tank unit commanded by Major Dwight D. Eisenhower and an infantry unit under General George Patton. Leading the column was General Douglas MacArthur, decked out in full military regalia.

After the soldiers cleared out the condemned buildings on Pennsylvania Avenue, they marched on Camp Marks and burned it down. The bonuseers suffered seven casualties; the movement was thoroughly crushed. "Every drop of blood shed today or that may be shed in days to come as the result of today's events can be laid directly on the threshold of the White

promised to veto any bonus legislation and refused to meet with Bonus Army spokesmen.

As spring turned into summer, the situation escalated. The veterans' protests became more urgent; clashes with police became more frequent, and rumors circulated that the Bonus Army was infiltrated by subversives. When two protestors were shot during a confrontation with local police, Hoover seized the opportunity to call out federal troops and drive the Bonus Army from the capital. The column of

House," declared Sergeant Waters, the Bonus Army spokesman, and the American public seemed to agree. Herbert Hoover was decisively trounced by Franklin D. Roosevelt in the 1932 presidential election.

The New Deal

Roosevelt's New Deal brought some relief to Washington by creating hundreds of federal jobs. Workmen were hired to scrub the Washington Monument and to manicure parks; artists were commissioned to decorate public buildings; librarians and archivists were given work tending government documents.

But the Depression's grip on Washington

FAR LEFT: ice-skating on the Reflecting Pool.
LEFT: Franklin D. Roosevelt, architect of the New Deal.
ABOVE: the poor remembered on the FDR Memorial.

didn't truly ease until the US entry into World War II. On December 7, 1941, Japan attacked the US Navy at Pearl Harbor, and Washington was once again swept into war. Anti-aircraft guns were mounted on rooftops; guards were posted at reservoirs, bridges and railways to protect against saboteurs.

A secret bunker was constructed for the president beneath the White House. Bullet-proof glass and black-out curtains were installed in the Oval Office. The city's great monuments were left unlit. There was even a plan to paint the Capitol dome black in order to protect it from air attack.

secret Manhattan Project for the rapid development of the atomic bomb.

Tragically, Roosevelt never saw the outcome of the war. After leading the country out of the Depression and into the greatest war of the century – and transforming Washington into an international center of power – Roosevelt died of a stroke. Less than a month later, Roosevelt's former vice-president, Harry S. Truman, declared victory in Europe. Truman also authorized the deployment of the world's first nuclear weapons – the first on Hiroshima on August 6, followed by another on Nagasaki three days later. Japan capitulated. World War II

Faced with the enormous task of building and coordinating a 7-million-man military, the federal government grew faster and larger than ever before. Hundreds of "government girls" arrived in the capital to take up posts as secretaries and clerks. The newly built Pentagon (completed in 1943), the largest office building in the world, quickly filled with military personnel. The Office of Strategic Service, forerunner of the CIA, was formed in order to handle espionage and other covert affairs.

In 1942, President Roosevelt received a letter from Albert Einstein advising him of groundbreaking experiments in atomic fission. Some time later, Roosevelt authorized the top-

was over. Washington and the world had been transformed. And a far more complex and insidious conflict was already under way.

The obsession with Communism

Despite peacetime demobilization, the federal government continued to expand its new role as a world power. With fascism dismantled in Europe, the Truman administration turned its attention to the "containment" of Communism, indirectly spawning a new generation of think tanks, consulting firms, political-interest groups and bureaucrats dedicated to the pursuit (and some would say the perpetuation) of the Cold War.

While American troops fought a bloody "police action" against Chinese-backed troops in North Korea, Senator Joseph McCarthy manipulated Communist paranoia at home in order to advance his political career. "I have here in my hand a list of 205 [people] known to the Secretary of State as being members of the Communist Party and who nevertheless are still working and shaping the policy of the State Department," McCarthy announced during one of his inflammatory speeches.

Although his accusations were groundless, the mere insinuation was enough to sully reputations and ruin careers. Blacklists, red-baiting

try can do for you, but what you can do for your country") and promising a "new frontier" of social reforms. Abroad, Kennedy escalated the American military presence in South Vietnam and played a white-knuckle game of nuclear brinkmanship with Soviet Premier Nikita Khrushchev over the placement of Soviet missile bases in Cuba.

At home, Kennedy was increasingly preoccupied with the explosive issue of desegregation, and with the growing urgency of the civil rights movement and its most prominent leader, Dr Martin Luther King, Jr.

In the early 1960s, Washington itself was still

and loyalty oaths became the order of the day in Washington as McCarthy targeted first the State Department and then other branches of the government and military in his hunt for "Communist sympathizers."

The election of John F. Kennedy in 1960 brought a welcome change of atmosphere to the city. Although an ardent supporter of the Cold War, the young charismatic president forged an idealistic national agenda stressing individual activism ("ask not what your coun-

very much a segregated city. District schools had been openly segregated until the Supreme Court's 1954 Brown vs Board of Education of Topeka decision. Not long before, black congressmen were barred from whites-only bathrooms on Capitol Hill. In a televised address, Kennedy informed the nation that blacks had "twice as much chance of becoming unemployed... one-third as much chance of becoming professionals... [and] about one-seventh as much chance of earning $10,000 a year" as whites. He had only to look outside the White House to see the grim reality these statistics represented.

While Kennedy was trying to push his civil rights bill through Congress in early 1963,

LEFT: World War II ends, officially.
ABOVE: demonstrators came to Washington from all over the US to protest against the Vietnam War.

plans were already being drawn up by Martin Luther King and other civil rights leaders for a massive March on Washington. On August 28, some 250,000 people converged on the capital to voice their support for civil rights legislation. As the nation watched, whites and blacks locked arms and marched to the Lincoln Memorial, chanting slogans, carrying placards and singing "We Shall Overcome," the unofficial theme of the civil rights movement.

Dream on

With an enormous crowd gathered around the Reflecting Pool, King delivered what is per-

stalled in the House of Representatives three months later when, on November 22, 1963, the news reached Capitol Hill of President Kennedy's assassination in Dallas.

Crisis of confidence

If anything, Kennedy's successor, Lyndon Baines Johnson, deepened the federal government's commitment to civil rights. He dispatched national guardsmen to Selma and Montgomery, Alabama, to protect "freedom riders" from local police. He pressured Congress to pass the civil rights bill in 1964, and developed comprehensive social programs,

haps the most famous speech by a black American. "I have a dream," King sang out. "It is a dream deeply rooted in the American dream. I have a dream that one day this nation will rise up and live out the true meaning of its creed: 'We hold these truths to be self-evident, that all men are created equal.'"

For many, it was the high point of the civil rights movement, an enormous upwelling of support, a symbol of interracial harmony and a confirmation of King's philosophy of nonviolent social change.

And yet, the March on Washington seemed to do little to speed Kennedy's civil rights legislation through Congress. The bill was still

including a massive "war on poverty." He also strengthened the US's involvement in the Vietnam War, increasing the number of American soldiers from 20,000 to more than 500,000 and stimulating a backlash among the already active antiwar movement.

It was a long way from Capitol Hill to the poorest neighborhoods, however, and by the mid-1960s racial conflict was approaching the flashpoint in cities throughout the country. Riots broke out in Los Angeles, New York, Detroit, Philadelphia and elsewhere, underscoring black frustration at the slow pace of change. In the spring of 1968, Martin Luther King organized a second March on Washington to drive home the need for economic equality as well as equal rights.

But King never made it to Washington. On April 4, 1968, while members of King's Poor People's Campaign awaited his arrival at Resurrection City – a small shantytown located on the Mall – Martin Luther King, Jr was assassinated in Memphis, Tennessee.

The reaction in Washington's poor black neighborhoods was immediate. As the news of King's death spread, an angry crowd began gathering at the corner

MONSTER PROTESTS

Demonstrations against the Vietnam War centered on Washington. In 1971, more than 7,000 people were arrested on one single day as the downtown area reeked of a mixture of marijuana and tear gas.

eral workers in Washington early that afternoon, creating a massive traffic jam that served only to hamper police, firefighters and emergency medical teams.

The situation continued to spin out of control, and police were unable to slow it down, much less stop it. Entire city blocks were engulfed in flame, belching thick black smoke over Capitol Hill and the White House; shops and supermarkets were gutted by looters and set on fire; clashes between the police and rioters grew increas-

of 14th and U streets, while groups of youths went to neighborhood shops demanding they close in deference to King's passing *(see page 45)*. By the following day, the crowds had turned into unruly mobs, and the scattered outbreaks of arson and looting had blossomed into a full-scale riot. Hoping to defuse unrest throughout the nation's cities, President Johnson asked Americans of all races to "reject the blind violence that has struck Dr King who lived by nonviolence." Johnson released fed-

ingly brutal. After consulting with city officials and civil rights leaders, Johnson called out the national guard and regular army – dispatching over 5,000 soldiers into the most troubled neighborhoods.

Gradually, as soldiers secured the streets, the violence wound down. In 36 hours of rioting, the city had sustained 12 deaths, 1,000 injuries and some $27 million in property damage.

Sit-ins, tent-ins and think-ins

But the Washington riots were only a symptom of a more pervasive and deep-seated political disaffection. Protestors had always come to Washington to air their grievances, but starting

---

LEFT: Martin Luther King delivers his historic "I have a dream" speech in 1963.
ABOVE: the aftermath of King's assassination in 1968.

in the mid-1960s they seemed to come in ever-increasing numbers and with a burning sense of urgency. They staged sit-ins, tent-ins, think-ins and peace-ins. They blocked traffic, occupied federal buildings, petitioned legislators and practiced passive resistance. The more radical among them threw rocks, bottles and firebombs and purposely engaged police in combat.

Although the protestors represented all sorts of special interests – the Black Third World Task Force, the Student Mobilization Committee, gay liberation, women's liberation, Vietnam Veterans Against the War, flower children, labor unions, the American Indian Movement,

and many others – the overwhelming focus of their demonstrations was the Vietnam War. The protest movement culminated during another March on Washington in which an estimated 750,000 antiwar demonstrators descended on the capital, making it one of the largest protests (if not *the* largest) in Washington's history.

The Watergate scandal

Disenchantment with the "establishment" seemed to crystallize in 1972 when Vice President Spiro Agnew came under investigation for a variety of charges including tax evasion and accepting bribes. Agnew resigned in October 1973 and was later convicted of tax eva-

sion. By that time President Richard Nixon and several key members of his staff were also being investigated for their part in the cover-up of a break-in at the Watergate building, which housed the Democratic Party's headquarters. Nixon vehemently denied any knowledge of the attempted burglary and wiretapping, promising Americans that he was "not a crook."

"I want you to know," he said in his State of the Union address to Congress and the nation in January, 1974, "that I have no intention whatever of ever walking away from the job that the American people elected me to do for the people of the United States."

A special prosecutor was appointed to look into Nixon's involvement in the scandal, starting a highly publicized tug-of-war over Nixon's secret Oval Office tape recordings. The president eventually surrendered the tapes, although an incriminating eight-minute silence remained a mystery.

With his credibility fatally undercut, and impeachment proceedings already underway, Nixon finally admitted what the press and public already suspected: that he had stopped an FBI investigation of the Watergate break-in that would have implicated members of the White House staff. Nixon announced his resignation on August 8, 1974. He was the first president to give up his office.

Nixon's resignation set off a crisis of confidence that would take the better part of a decade to mend. The remainder of Nixon's term was assumed by President Gerald Ford, who stirred up yet another controversy by pardoning Nixon of any crimes he might have committed.

Democratic president Jimmy Carter took office in 1976 promising to move beyond the malaise and disaffection of the Vietnam and Watergate years, only to be politically paralyzed by a major recession, a Mideast oil crisis, a revolution and seizure of Western hostages in Iran and a botched rescue attempt.

When Ronald Reagan took office in 1980 after a landslide victory for the Republicans, Washington's conservative political establishment was on the eve of a major resurgence, but Washington itself – the city beyond the marble monuments – was a city split increasingly apart by an ever widening economic divide.   ❑

LEFT: campaign buttons commemorate some of the men who made it to the White House.

Race Riots

While the 1968 riots that nearly destroyed much of downtown Washington are probably the most remembered, the city has had an unfortunate history of racial tension that has erupted into violence. In 1835, an alleged assault by a slave on Mrs William Thornton, widow of the architect of the Capitol and Octagon House, touched off one such event. The Epicurean Eating House at 6th Street and Pennsylvania Avenue was a popular restaurant owned and operated by Beverly Snow, a black man. When rumors spread that he had made disparaging remarks about Mrs Thornton, and generally about the wives of certain white mechanics who worked in the nearby Navy Yard, a crowd of whites looted and all but destroyed Snow's restaurant. The mob went on to vandalize several black homes, business, and churches, and completely destroyed several black schools.

With the help of friends, Snow escaped to Canada, where he remained. Despite a new city ordinance that forbade blacks to operate their own business, Snow's reopened a year later, probably because white customers wanted it to, but this time under the proprietorship of Absalom Shadd, another black man who kept the establishment going for the next 20 years.

While the situation for blacks in the ensuing years were always precarious, the situation ignited again during the so-called Red Summer of 1919. The District was one of two dozen or more cities throughout the country that experienced racial rioting after black soldiers returned home from having served in World War I. There were 70 known lynchings in the year after the war, and several of the victims were soldiers in uniform.

Anger among blacks over the treatment of black veterans grew, and in DC the issue made headlines that became only more sensational when reports surfaced that white servicemen had begun to hunt for blacks in the southwest section of the city. Blacks armed themselves against the white mobs who entered their neighborhoods and federal troops had to be called in. A heavy rainstorm finally brought an end to the five days of street violence that left 30 dead and hundreds injured. Disgraceful as the situation was, blacks received little or no restitution and

RIGHT: a woman is arrested in front of the White House after a 1963 anti-racism demonstration.

the issue of racism was left to smolder until the fateful spring of 1968.

Before that year, the 14th Street corridor, especially near U Street, had drawn a substantial population of Southern rural immigrants and had been a thriving black commercial center. By the 1960s, though, the area had fallen to crime and drugs. Shortly before 7pm on Thursday, April 4, 1968, the rage erupted with the news of the assassination of Dr Martin Luther King in Memphis.

Looting began at the Brookland Hardware store and was followed by fires that burned out of control along 14th Street, 7th Street and H Street, NE. After three days of rioting and 12 days of occupa-

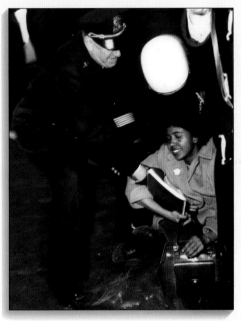

tion by federal troops, tensions finally subsided, leaving 12 dead, more than 1,000 injured, and $27 million worth of property damage.

Before the riots, 14th Street businesses grossed between $75 million and $100 million annually. By 1970, business receipts totaled less than $4 million and 90 businesses had either been destroyed or closed. Federal funding to restore the area never materialized and it wasn't until the mid-1980s that the area started to turn around. Today, the picture looks brighter, and small businesses, restaurants, and several theater companies have opened, giving the neighborhood a reputation as an up-and-coming theater district. ❑

WASHINGTON TODAY

Racism, drugs, and poverty make up Washington's seamier side, contrasting dramatically with the grandeur normally associated with the city

It's often said that Washington is really two cities. The Washington most people know is a city of diplomats and public servants, press conferences and cocktail parties. The other Washington is a city of run-down neighborhoods, intractable poverty, street crime and crack houses.

But there's a third Washington, too, less concrete than the others. As the nation's capital – the stage on which national and international politics are played – Washington occupies a symbolic space. What happens here is emblematic of what happens in cities throughout the country. And what's happening in Washington is enough to make anybody frightened about the future of America's inner cities.

The story of violent crime is all too familiar. Drug gangs move into a neighborhood and innocent people suffer. This is a poor state of affairs to happen in any city. But to occur at such an alarming rate in Washington, DC – a monument to the nation's highest ideals of democracy, freedom and equality – is chilling.

Millions of tourists visit Washington each year, but few see this side of the city – the poor, predominantly black areas "behind the marble mask" of statues, museums and government buildings.

Murder capital

The 1980s brought mixed fortune to the city. A real-estate investor with money in a plush downtown property, or a young professional riding the crest of Reaganism, probably did very well indeed. The catch-phrase of the 1980s – "greed is good!" – was as valid in the capital as on Wall Street. And although a few wheeler-dealers ended up on the business end of a subpoena, there were always more winners than losers, especially among the big-time players.

For Washington's lower classes, however, things didn't go well. Ronald Reagan's much-

LEFT: kidding around at the Capital Children's Museum (on 3rd Street behind Union Station).
RIGHT: the city's murder rate soared in the 1980s.

touted trickle-down theory – which stated that wealth at the top of the economic scale flows naturally to the bottom – apparently did not take into account the many ways cash is diverted on its downward journey. The economic boom of the 1990s only compounded the problem. People on the lower rungs of the

economic ladder have slipped through the cracks and the fall-out is painfully obvious. Today there are an estimated 6,000 to 10,000 homeless people in the city, many camped out on ventilator grates, in doorways, overcrowded shelters and welfare hotels.

Despite the metropolitan area's considerable wealth (an average annual income of close to $50,000 per household, well above other cities), one in five of Washington's residents, according to the latest census report, live below the US poverty line, and most of them are black. The average income of black households is half that of whites. Infant mortality is nearly double the national average. Public school drop-out rates are among the worst in the country. And signs of improvement are on the distant horizon.

By the late 1980s, Washington was also suffering from a drug problem of epidemic proportions. And with drugs came violence. Between 1985 – the year crack cocaine was introduced – and 1988, the number of yearly homicides in Washington jumped from 148 to 372, giving Washington the highest per capita murder rate in the country (60 per 100,000) and an undisputed claim to being the murder capital of the nation. In 1990, the number leapt again, to far more than one murder every day.

As one journalist remarked around that time, the US capital was "statistically more dangerous than Belfast or the West Bank," outdoing both the Israeli-occupied territories and the whole of Northern Ireland in the number of violent deaths. The majority of victims and perpetrators, he continued, were young black men. "Homicide is… by far the most common cause of death among black males between 15 and 34 in Washington."

For a few years in the late 1990s the situation improved and the murder capital stigma was passed on to other southern cities such as New Orleans and Atlanta, but murder in DC is on the rise again, approaching 250 victims by the

THE HOMELESS

Out of mind, perhaps, but seldom out of sight, Washington's homeless are a persistent presence. You'll see them clustered near heating grates, asleep on park benches, and panhandling in the streets of Georgetown. A disproportionate number are mentally or physically disabled, or are alcohol and drug abusers.

A growing number, however, are the working poor. According to Edward Orzechowski, president of Catholic Charities in Washington, more than 100 of the 600 people recently served in his program for the homeless were actually in employment but were unable to afford the high cost of housing.

Most are not a threat, but it's hard to know who may or may not be dangerous whenever you're approached for a handout. Whether you respond or not is a personal decision, but if you want to help but aren't sure how wise it is to give money directly to someone who may be a drug addict, or possibly even a con artist, here are some organizations where you can send a check and be assured that your money will be put to good use.

So Others Might Eat: 70 O Street, NW, Washington, DC 20001; tel: 202-797-8806

Catholic Charities: 924 G Street, NW, Washington, DC 20001; tel: 202-772-4300 end of 2002. In less than an hour on a single night in November 2002, for example, four people were gunned down, all in the poorer sections of the city and with no witnesses to aid police in solving the crimes.

Snipers at large

A month earlier, the region was transfixed and terrorized by the random killings carried out in the surrounding and wealthier suburbs by the pair known as the Washington snipers. For the three weeks that the snipers were on the loose, police across the region were mobilized and news coverage was ceaseless.

Throughout the continuing crisis in DC, municipal and federal officials seemed more interested in passing the buck than in solving the problem. Three-term mayor Marion Barry, an outspoken veteran of the civil rights movement, complained about the lack of federal dollars. But the charges of corruption that plagued 12 city officials in his administration, and the drug charge that finally landed him in jail did nothing to make his case and loosen Congress's grip on the purse strings.

More recently, mayor Anthony Williams, soft-spoken, bow-tied, and well educated, has fared no better where the federal government

The November homicides in DC, however, received comparatively little police attention or news coverage. And the reason may have to do with motive. In the case of the snipers, no one could pinpoint the motive; the shootings were terrifyingly random. In the case of DC's November homicides, however, the motive, especially in the poorer neighborhoods, is presumed to be no mystery. Whatever the details of each shooting, the main culprit is likely to have been drugs.

LEFT: Washington is a focal point for lobbyists.
ABOVE: Marion Barry, the mayor who backed the black underclass, "the least, the last and the lost."

is concerned. While nothing as scandalous as Barry's drug charge, Williams managed to discredit his administration by having forged thousands of voters' signatures on the petition required to put his name on the ballot during a mayoral election.

For this he was fined $250,000, the largest fine ever imposed by the Board of Elections, which did nothing to endear him to the powers that be on the Hill. For instance, funds that would have been used by Williams to build thousands of units for the homeless were denied by a Republican-controlled and paternalistic Congress, convinced that the city is incapable of ruling itself.

Home rule

The ordeal of the Barry scandal put issues of race and racism into high relief. With a 70 percent black majority, Washington has long been considered a center of black political leadership – a fact that has vexed the perennial battle over District home rule and possible statehood. As far back as 1865 – when DC citizens didn't even have the right to vote in federal elections – it was generally accepted that home rule meant some degree of black rule.

Today, of course, the situation is quite different. Washington residents vote in local and federal elections; the mayor and city council

initiate local legislation. But District government is still not autonomous. Under the Home Rule Act of 1974, the US Congress has the power to review and amend District legislation, including the budget. Supporters of limited home rule cite Washington's history as a federal district and Congress's obvious interest in the management of the capital. The charge is still being made, however, that Republican factions in Congress and in the city are simply unwilling to hand over the reins of power to black leaders.

The irony of the District's political status is difficult to miss. The District of Columbia is not only the capital of the US, it is purported to be the capital of the free world. So why is it

that DC has no representatives in the US Senate and only one non-voting delegate in the US House of Representatives?

The answer is simple: because the District of Columbia is not a state. It has enough people to be a state (the District's population is higher than that of Vermont's, Alaska's or Wyoming's), and its residents certainly pay enough federal taxes. But as supporters of statehood often point out, a new state of Columbia would be entitled to two seats in the Senate and one voting seat in the House of Representatives. And considering the District's demography, those seats would most likely be filled by black Democrats.

As Senator Edward Kennedy, a longtime supporter of DC statehood, put it, Washington suffers from the "four toos": "The District of Columbia and its residents are too urban, too liberal, too Democratic and too black." Washington, the US government's monument to itself, has become a national proving ground for the limits of black political power.

The big challenge

What the future holds for Washington is anybody's guess, but the city's social problems are undeniably daunting. The District's substantial black middle class, frightened by urban violence, has fled to the suburbs, taking their substantial tax dollars with them. The city itself is almost bankrupt. Putting Washington back on an even keel, and managing the historic tug of war between City Hall and Capitol Hill, is an uphill challenge.

Washington is the nation's capital, the place where America and other parts of the world look for direction. It's also a black-majority city and a flagship of black political leadership. Although its problems are considerable, few cities can draw on the pool of talent and resources available to the national capital. Optimists say that Washington has already got a hold of its bootstraps and that it's only a matter of time before it pulls itself up: one indication is the opening of a Metro station in Anacostia; another the revitalization of downtown and of Southwest. Optimistic signs are needed, because Washington doesn't just belong to Washingtonians, it belongs to everyone. ❑

LEFT: a protest march to legalize drugs.
RIGHT: Japanese-American students, part of the city's widening ethnic mix.

LIFE IN WASHINGTON

As the permanent population swells, the city is becoming more distinctive and dynamic, influenced by factors other than the changing fortunes of politics

Washington, according to John F. Kennedy's jibe, is one of those cities that mixes "Southern efficiency with Northern charm." Times have changed since that remark, but this almost laughably small place – a diamond-shaped area of modest residential streets – is one of the most contradictory capitals in the developed world.

For a start, it is described as a town – and, despite the glittering presidency of John Kennedy, a pretty provincial one at that. The foremost argument that engaged residents and visitors at that time was whether it was a Southern town or a Yankee one. Now the question of tourists who stroll along the Mall marveling at some of the world's finest museums is more likely to be: does anyone, apart from that man in the White House, actually live here? The city appears devoted to official buildings, official business and people who visit. It seems detached from the nation it governs. But to its inhabitants it is very much an established city – in fact, it would be more true to say that it is four cities.

The four faces

There is the Washington that is most generally conjured up by the name – the administrative city that governs the vast military and bureaucratic machine that Washington has become. This is the city defined by the White House, the Pentagon and the Capitol, and the legions of local inhabitants who make the machine work.

Then there is social Washington, hovering not so discreetly behind the closed doors (to anyone who does not clutch an engraved invitation) of the exclusive salons of Georgetown, Kalorama and Embassy Row. Its purpose is to woo, soothe, encourage, coerce and promote useful relationships among the politically influential. For this is not a city like New York or

San Francisco that has grown up around the more usual physical and cultural needs of a socially integrated community. Its crème de la crème are not drawn together by vibrant local theater, innovative restaurants, imaginative grocery stores, fashionable, witty or daring style. The thing that counts most is power. It

was at the exclusive Georgetown salon of Pamela Harriman, later made US Ambassador to France, that Bill Clinton was first veted, then promoted to run for president.

The third Washington is the city that is three-quarters black and known as a drug and murder capital. But there is a fourth Washington, and it is this Washington that is finally forcing the capital into becoming a coherent, normal place to live, functioning beyond the shadow of the Capitol. It is the Washington that lies outside the District of Columbia line.

In Chevy Chase, Bethesda, Arlington and the nearby environs of Maryland's Prince George's County plus the wealthier counties of Mont-

---

PRECEDING PAGES: rendering the national anthem at a Redskins football game; the concierge every visiting VIP dreams of finding.
LEFT: keeping in touch on Capitol Hill.
RIGHT: the late Pamela Harrison, influential hostess.

gomery in Maryland and Fairfax in Virginia – are the suburbs where these days you will find burgeoning business and residential Washington. While demolition teams scrape away downtown DC to erase the old ghettos still clinging to the skirts of Capitol Hill, out beyond the DC line a vibrant new Washington has been growing.

The face of the future

DC, kept deliberately unattached to either of its neighboring states in order not to show favoritism, is probably the last American city under renovation and construction in the classical quasi-European style. Yet while old

and social care provoked a government drive to increase manpower in these professions. The result was a surge in staff levels at the departments responsible for overseeing them.

Unable to adequately accommodate these new bureaucrats in the old federal buildings and state departments in town, some government offices made the move into the suburbs. Major international banking and communications institutions followed, spreading outwards in Virginia in the direction of Dulles International Airport and making the suburb of Reston a burgeoning city. The effect is almost a reversing of the power process, with the activity of

columned buildings receive face lifts and once-problem areas become gentrified, what is now recognized as the standard modern American city mushrooms on the other side of the District line. Pushing outwards beyond the Beltway (the 8-lane traffic nightmare of a highway that girdles the city) are the high-rise office buildings, the shopping malls, the freeway strips that signify the city of the future anywhere in America.

This new Washington is the city that has grown out of President Lyndon Johnson's grandiose program in the 1960s for the Great Society. This dream for a nation in which citizens would have the best in education, medical

the arteries in the suburbs helping the central body function. Increasingly, what takes place on the outskirts of town supports, justifies and shores up the activities and continued existence of the business of the center.

What pushed, with explosive suddenness, the process of transforming Washington from a sleepy Southern town into a forceful power city were the race riots of 1968 that followed the assassination of Dr Martin Luther King. What these fires began, the demolition teams continued, and the political process took over. In 1970 the redevelopment of Washington began, including the idea of pulling it down and rebuilding it afresh. Requests for building and development permits were made and fulfilled, and the process of cleaning up the city was underway with a vengeance.

Marion Barry, who took office as mayor in 1978, used his powers to push the construction program forward with even greater fervor. Whole areas were razed while others were restored. To see the contrast between old and new, stand on Dupont Circle, where Massachusetts Avenue crosses Connecticut Avenue. The Connecticut Avenue to your south is new and high-rise. The block of Connecticut to your north is jumbled, low-rise and on a more human scale.

themselves to Washington only for the duration of the President who appointed them.

Politicians and shrinks

These days politicians who once would have returned home after the end of their term of office now choose, in increasing numbers, to stay on to become PR executives, lobbyists or consultants in the mushrooming think-tanks, trade associations, and law firms. You can understand their desire to stay put. Washington is an attractive and lucrative city in which to play power games.

These are not the only group of people who

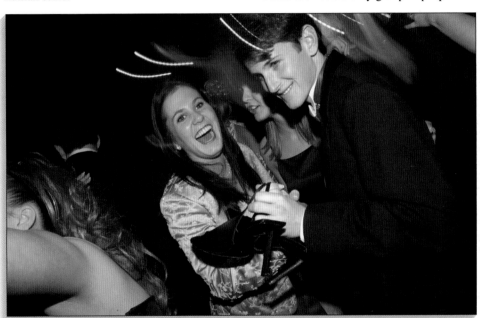

Permanent residents

With this transformation, Washington began to attract people with firm commitments to putting down roots. This was unlike the intentions of the original founding fathers, who sited the capital in an uncongenial spot because they hoped this would discourage politicians from spending too much time here gathering power. While the backbone of government bureaucracy is necessarily run by a permanent staff, people in high political office have traditionally committed

Left: good ol'-fashioned entertainment after George W. Bush's inauguration in January 2001.
Above: post-inauguration parties are energetic affairs.

now make a point of establishing at least one of their homes in Washington. The growth in government, banking, and institutions has drawn to the city a positive plethora of lawyers, to the point where they have become the butt of caustic jokes.

There are over 55,000 of them working in practices large and small, but over 100,000 people in Washington actually have law qualifications. This is more than the number of doctors in the city, despite the fact that in the Maryland suburb of Bethesda sits the vast compound of the National Institute of Health, the prestigious research center set up by President Nixon's government to find a cure for cancer. But government needs lawyers to construct and implement the law and to protect itself, and businesses need lawyers to confront the government. Now litigation appears to have become a way of life among the residents, too.

Washington has also become a major employer of specialists in emotional problems, with more therapists, counselors and psychiatrists per capita than any other American city. Perhaps the pressures of power are just too great for some. Despite this growth, however, Washington is still a comparatively slow-paced, provincial city, although its polish increases

tory theaters. There are also concerts, lectures, classes, and art exhibits of every type throughout the city in small galleries and halls, and of course at the Smithsonian, whose prestige attracts the finest performers, artists, writers, and historians as instructors and lecturers for its wide offering of classes.

While there is a constant carousel of night-clubs and music cellars for students and the yuppie community, there is no glitzy night life for the power people. This is part cultural: Washingtonians have to get up far too early to party all night long.

Political Washington is a city that closes

daily. While its New York neighbors may find it, as the actress Barbra Streisand did, "a stuffy city," it is no longer the backwoods hick town.

Culture central

There are more than 5,000 restaurants in the city, where Washingtonians can dine on the national cuisine of practically any country in the world. There are good theater companies, much relished by indigenous Washingtonians, including the world-class Shakespeare Theater, plus the Kennedy Center, home to the National Symphony Orchestra, the National Theater, which hosts many of the Broadway companies on tour, and several outstanding smaller reper-

down early. Wilbur Mills, the all-powerful chairman of the Congressional Ways and Means Committee who leaped into the fountains in some disarray with fan dancer Fanne Foxe, would agree that it doesn't pay to be seen in enthusiastic party mode, unless the spectators are part of the party. Politics dictate Washington's night life. Fund-raising dinners are more valued than gourmet occasions, and these days most events are run by PR consultants on behalf of lobbyists.

Although the late Mrs Pamela Harriman, wife of deceased Ambassador Averell Harriman and herself once an ambassador to France, will always be remembered as one of Washington's most prominent hostesses, the days of a coterie of influential women who could make or break a political career have passed.

Nevertheless, old money, new habits and/or general behavior moves in and out of fashion and depends entirely on who's in the White House. A case in point is the Clinton administration, when social Washington lost a good deal of glamour and pizzazz. Clinton's "teenage" advisors (as they were disparagingly, sometimes enviously, referred to),were rarely seen behaving with the kind of flamboyance usually expected of their age group.

After the terrorist attacks of September 11,

acceptable in one administration will no doubt be out of favor with the next.

Security concerns

Since the terrorist attacks, Washingtonians have become noticeably more security-conscious. Unlike many foreign capitals, where bomb scares, searches, and other inconveniences are a regular way of life, Washington, since the Civil War, has had little experience with violence directed at it from an outside source. With barely a check of identification by the guards, public buildings have been largely accessible to most. And until the attacks, delivery trucks

2001, which killed 189 people at the Pentagon, the atmosphere in the Bush White House, though still "Texas casual," became definitely serious. When national security adviser Condoleezza Rice, for example, addressed the press, laying out a White House strategy for war with Iraq, or explaining how the new Eastern European members of NATO could help in halting terrorism, the atmosphere was rarely lightened with a smile. The point is that each president brings his own personality to DC, and what is

bringing printed matter and other supplies necessary for the business of Congress drove unimpeded to the open loading docks of the buildings on Capitol Hill. Free and open to the public, Washington seemed to operate on faith that the democratic principles that guide the nation would somehow keep it from harm.

Though inexperienced with this sort of wholesale and spectacular violence, Washingtonians nevertheless rose to the occasion. On the morning of the attacks, workers left their offices and, without panicking, calmly headed home to their families. On that blue-sky and crisp autumn day, streams of people walked quietly across the city, northward into Mary-

LEFT: influential CNN talk show host Larry King.
RIGHT: the damage to the Pentagon after terrorists crashed a hijacked plane into it in September 2001.

land, and across the bridges into Virginia. They waited patiently for trains at the Metro stations. They were aware of what had happened, but not yet sure why, and the atmosphere all over the city was eerily silent.

But by the next morning, everyone had returned to work, and the business of government went on as usual, though in the following months federal workers went about their jobs with an increased sense of mission. In the weeks after the attacks, air patrols flew day and night over the city.

Many of the attractions that draw tourists were closed until security could be increased.

Reagan National Airport was closed and was the last airport in the nation to re-open. With the tourists staying away, there were suddenly no lines at the museums, and suddenly locals had a choice of tables at the city's best restaurants. Most of the sites have re-opened and the crowds are back, though the everyday routine has been affected.

Security was tightened all over the city, with checks conducted at every tourist site and government building. Activities that seemed even remotely suspicious were no longer ignored, but reported to the police. The White House temporarily suspended tours, the Pentagon said it had no plans to allow anyone inside who wasn't there on official business, and there was even talk of installing a system of surveillance cameras across the city, much to the annoyance of those who still believed that Washington, if no longer entirely safe from foreign threat, must itself remain free and open.

Life changed in Washington, and daily life, as a result of the attacks, undeniably become less convenient. Washingtonians, though, are an intrepid breed, driven by the sense of duty that led them into public service in the first place. No matter what happens next, one thing is certain: Washington won't shut down.

Civilized and violent

To anyone familiar with other American cities, Washington is perhaps the most handsome. Any architectural appetite can be satisfied here, from Egyptian and Greek to Victorian and the post-Modern. Gazing up and down the Mall, this would seem to be the most civilized of cities. Yet, despite appearances, and rather incongruously for a city so taken by surprise by 2001's terrorist attacks, it also has one of the highest murder rates in the world *(see page 48)*. Almost 100 percent of these crimes occur in the North-east and Southeast, areas that are predominantly African American and Hispanic.

American apartheid may legally have ended in 1964, but Washington is an apartheid city, with middle-class whites living west of the 16th Street line that divides the city from the north down to the White House. Unemployment and violence are part of the daily lives of those who live in the east.

The motives for this state of urban warfare in the east part of Washington have vexed and perplexed sociology departments in learning institutions throughout the nation.

Because the place still in flux and development, it is difficult to define. Its crime and welfare problems call for real and sweeping efforts in order to resolve them. But this focus on one part of Washington should not obscure the others. This is, after all, four cities. Desperate, magnificent, stylish and provincial, Washington is also a town which possesses a rare, distinctive quality. It's a place where, in the words of Ralph Waldo Emerson, "an insignificant individual may trespass on a nation's time." ❑

LEFT: veteran satirist and columnist Art Buchwald.
RIGHT: peace and love on the Fourth of July.

BLACK WASHINGTON

*The level of crime among young black males drove even middle-class blacks
into the suburbs. But there are some signs that things are improving*

For years, Washington's black neighborhoods were known collectively as the "secret city," the city behind the marble edifices and the pomp and politics of official Washington, the city segregated from white Washington. Within these cloistered avenues existed a vital and thriving social and economic culture whose citizens overcame the barriers of discrimination, realizing their full potential and in the bargain enriching American life. Like other American cities, DC continues to undergo the changes that result in gentrification, shifting demographics, and the flight to the suburbs. But, block by block, Washington is reclaiming itself, tearing down the old and putting up the new, and with each passing year black Washingtonians are finding a viable place for themselves, contributing to the resurgence of the city.

High-ranking supremos

Black Washington is more than just the figure of Marion Barry, whose behavior when mayor garnered him many national headlines. In recent years there have been a number of high-ranking supremos, including Sharon Pratt Kelly, the city's first black woman mayor, General Colin Powell, Secretary of State, Eleanor Holmes Norton, District delegate to the House of Representatives, and Condoleezza Rice, George W. Bush's national security advisor.

But within just a few blocks of the White House and the Capitol are the weed-strewn lots and crack houses that cluster just north of Massachusetts Avenue around 14th Street. Open-air drug markets flourish where Quincy and Newton cross 14th Street, spilling over to the basement areas where addicted prostitutes offer sex for $5 and less, the price of a vial of crack.

These are the mean streets. The arrest of Mayor Marion Barry in 1990 on drug charges (most were subsequently dropped) confirmed the international image of the city as one vast

LEFT AND RIGHT: more of the city's black population now enjoy middle-class prosperity, but poverty and homelessness are still intransigent problems.

ghetto surrounding a besieged core of public buildings and government offices which a largely white bureaucracy visited nervously by day, fleeing home at night to the suburbs.

Yet Washington is also the home of a prominent black middle class. The city's Dunbar High, on the corner of 1st and N streets NW,

was the first black high school in the country, and produced generations of black leaders, such as Benjamin David, the first black US Army general; Charles Drew, the medical scientist who discovered blood plasma; and Edward Brooke, who grew up in the District but returned to represent Massachusetts as the first black senator of the 20th century.

The relatively new University of DC and the long-established Howard University, often called "the black Harvard" with its 18 schools, 12,000 students, 8,000 employees and $500 million annual budget, have created a strong intellectual foundation. While the city's public school system has one of the country's highest drop-out rates, DC still has the highest percentage of black college grads of any major American city.

Along Upper 16th Street, known locally as the Gold Coast, is a suburb of expensive homes for the city's black elite. On Sunday mornings, the streets around the fashionable black churches of the Gold Coast and Shepherd Park are thronged with luxury cars and well-dressed families. But this prosperous and growing middle class tends increasingly to behave like its white counterparts, and flee the problems of the inner city. Head north-east of the DC boundary and you reach Prince George's County, which in 1990 became the first suburban county in the country to have a black majority population.

Deep roots

The roots of Washington's black community go back to the founding of the city in the established white South. Located south of the Mason-Dixon line, the old boundary between the slave and the free states, the nation's capital was born in the curious contradiction between slavery and black genius.

When the city's architect Pierre L'Enfant was dismissed by George Washington for insubordination, his initial sketches were reconstructed

DUKE ELLINGTON

In the decades following the Civil War, the prime address for respectable and hard-working middle class blacks was T Street, NW, also home of famed jazz pianist and composer Duke Ellington, who grew up at no. 1212. Edward Kennedy Ellington was born in 1899. His father worked as a butler and caterer, then as a technician at the Navy Department. At his mother's insistence, "Duke," as he was known for his impeccable dress and fastidious manner, studied piano. While working as a soda jerk after school, Duke composed his first piece, "The Soda Fountain Rag," and started sitting in with local bands. By the 1920s Ellington's "Washingtonians" was the city's leading society band. In 1923 he left

for Harlem, and by the 1930s he and his band became a fixture at the famed Cotton Club.

A prolific and masterful arranger, Duke also composed more than 6,000 songs, including "Satin Doll," "Mood Indigo," and "Don't Get Around Much Anymore." In 1943, he was the first non-classical musician to perform at New York's Carnegie Hall. After World War II, his popularity waned, but his performance at the Newport Jazz Festival in 1956 revived his career. He died at the age of 75 in 1974. The Calvert Street Bridge was renamed in his honor, and the Duke Ellington School for the Arts keeps his spirit alive by providing education for the city's most musically gifted.

and made into a viable design plan by a free black surveyor, Benjamin Banneker. Son of a former slave and an English indentured servant girl, Banneker was born in 1731 and raised on a farm near Baltimore. The wife of the farmer taught him to read, but Banneker taught himself astronomy and mathematics. When Major Andrew Ellicott, L'Enfant's replacement and a neighbor of Banneker's, needed a surveyor the much admired black intellectual was immediately recruited.

entrenched, although small schools and churches were founded, and some black artisans, such as masons and blacksmiths, were permitted to open small businesses, that is when their skills were to the advantage of the whites.

On the eve of the US Civil War in 1860, there were 11,000 free blacks in the city, and more than 3,000 slaves. After the Civil War and the abolition of slavery, the Freedmen's Bureau launched such great institutions as Freedmen's Hospital and Howard University, established in 1867, which

Banneker, however, was the exception. Fully 80 per cent of the black population living in the city that he was surveying was enslaved and after the British destroyed the city in 1814, much of the physical effort of rebuilding fell to those slaves. As early as 1808 a $5 fine was imposed on any black person found on the street after 10pm; two years later, a jail term was added. A series of laws known as the Black Codes increasingly restricted blacks. By the middle of the century, slavery remained

would become the pioneers of black progress. Beginning in 1866, all adult men were given the vote, regardless of race, and by 1870 blacks were legally permitted service in restaurants and other public places, though these laws were not widely enforced. By 1950, there were still no more than a handful of restaurants across the city where a black could order a meal.

An interesting reminder of these hopeful days after the emancipation of the slaves is the Frederick Douglass Memorial Home and Museum, at 1411 W Street SE, in the heart of what is now the ghetto of Anacostia. An accomplished journalist and eloquent abolitionist, Douglass *(see profile, page 237)* might be downcast by the

LEFT: a mural commemorates Duke Ellington, who led the city's leading society band in the 1930s.
ABOVE: art for sale in the Eastern Market.

levels of crime around his old home. But as an escaped slave he would relish the fact that the city government of Washington is now run by the great-grandchildren of slavery.

It has, however, proved a bitterly slow and uneven process. Early in the 20th century, blacks held few jobs in the government of the city or in the federal government, and most of those were at the lowest levels. In 1922, when the Lincoln Memorial was dedicated to the Great Emancipator, a separate stand was erected for black dignitaries, a symbol of the segregation that still governed the city's schools, restaurants and theaters. Only the trol-

leys, the buses and the stands of Griffith Stadium, the field where the famed Washington Senators baseball team played, were integrated, but not the field itself.

The problem of Washington's brand of racism was pointedly demonstrated in 1939 when the famed black contralto Marian Anderson was denied permission to sing at Constitution Hall, the concert venue owned and operated by those denizens of white American respectability, the Daughters of the American Revolution. Eleanor Roosevelt, the outraged wife of the president, stepped in and arranged for the singer to perform in front of the Lincoln Memorial. Anderson's outdoor concert drew tens of thousands

and her glowing performance, while it did nothing to change the Daughters' stance, brought the problem of segregation, once again, to the nation's attention.

The 1920s saw the city's black population fall to its lowest ebb, no more than 25 percent, and the demographic tide which makes today's city three-quarters black began to flow only with Roosevelt's New Deal and the government's explosive growth during World War II.

But the black character of modern Washington was shaped by two crucial events. The first was the Supreme Court's order to desegregate the public school systems in 1954, which dramatically accelerated the white flight to the suburbs. And, 14 years later, the assassination of Martin Luther King sparked off the racial riots and the looting and burning which reached within four blocks of the White House. The white flight was followed by the steady drift of shops and service industries and jobs to the suburbs, and created the social condition which became known as the inner city. The 1968 riots and the fires created the conditions for a rebirth.

Democracy is extended

The riots reinforced the case for giving the national capital enough democracy to elect its own form of civic government. After a century as a fief of Congress, the city was able to vote for its own school board in 1968, and for its own (non-voting) delegate to Congress in 1971. In 1974, the city finally won the right to elect its own mayor and largely black city council. This helped spur the growth of the black middle class, as they appointed sympathetic police chiefs and blacks rose in the civic bureaucracy.

The destruction of so much of the city center created a golden opportunity for property development, new office buildings to house the growing federal bureaucracy and the lawyers and lobbyists and consultants and corporate offices that began to grow around it. But that opportunity has been seized upon only in the past few years. Fourteenth Street, the black commercial center, has nearly completed its transformation, while neighborhoods further east and into the Southwest section are still under construction. Slow though the going may be, new building still means construction and service jobs, and a revenue source in the form of income and property taxes that will only benefit the city.

By 2002 there were more than 49,000 civic

employees in this city of 600,000 people, the highest proportion in the country. Mayor Barry created a political machine, a vast patronage system which mobilized an electorate already solidly loyal for the Democrats and which under the later administration of Anthony Williams showed no signs of abating. The District, for better or (some would say) for worse, was the only electoral community in the entire US to vote against the national landslides for the successive Republican presidents, Richard Nixon, Ronald Reagan, George Bush and George W. Bush.

The rate dropped during the Clinton administration's years of property, but then began to rise again, forcing the District's substantial black middle class into the suburbs. A survey in the early 1990s found that around 42 percent of young black urban Washingtonians were either in prison, on probation, on parole, awaiting trial or being sought by the police. Three out of four people in District jails were inside for drug-related offences. Their share of the prison population had tripled in 10 years. Their recidivism rate averaged 70 percent. Little has changed.

America had seen this kind of ethnic machine politics before, with the 19th-century Irish in Boston or the Italians in New York in the 1930s. What helped turn Washington so sour was crack cocaine. From a peak of just over 200 murders in 1981, the murder rate dropped steadily until 1985's low of 153. Then came crack, and the killings soared annually to more than 500 murders a year after 1990, with the fastest growth among young black males.

LEFT: can a black girl make it to the White House?
ABOVE: Condoleezza Rice did, joining George W. Bush's team in 2001 as national security advisor alongside Colin Powell as secretary of state.

Signs of hope

This is the bad news. The better news can come in unexpected ways, such as the springtime riots of 1991, when DC's Hispanic population in Mount Pleasant spent three nights burning and looting in a protest against police "racism." The point in this case was that the police were black – which made an intriguing change from the black attacks on racist white police in the 1968 riots. And in the District Courts on Indiana Avenue, for all the depressing parade of black defendants in the dock, the black lawyers and black judges and black reporters on the press benches suggest a more hopeful society emerging, however slowly. ❑

THE NEW IMMIGRANTS

Washington's new communities are diverse and vibrant, adding culture and culinary variety. But there's tension between them and the the city's poor blacks

A popular reinterpretation of the CIA's initials among savvy Washingtonians is "the Culinary Institute of America." This is a dig at the extraordinary coincidence of the sudden surge in new ethnic restaurants that comes with the ebb and flow of political events around the globe. Democratic American voters – and Republicans, too – may object to some of the meddling in the internal affairs of small and distant countries, but they can always look forward to something new to eat at the end of the revolution.

Country cooking

After the fall of Saigon, Washington was suddenly introduced to the delights of Vietnamese cooking. And not just with one single restaurant. The whole extended family fled too; uncles, aunts, cousins, in-laws established Vietnamese eating places all over town. Wilson Boulevard in Arlington, Virginia, on the other side of the Potomac, is fondly know as Little Saigon. In the early 1980s, Ethiopian cooking, influenced by the Italians who occupied the African country in the 1930s, became popular. And, with the Soviet invasion of Kabul, an assortment of Afghani restaurants sprang up. More recently, Latin American cooking became all the rage.

Inevitably these refugees are go-ahead and determined individuals, possessing the drive to make things work in their new life. Their bent for enterprise and sense of initiative is high.

The United States has been the first home of political or religious refugees ever since it was born. Indeed, that is why the nation *was* born. So it can be no surprise that for today's survivors of domestic revolt, America should still appear as the welcoming new motherland, willing to offer comfort and succor to all in need – or a goodly portion of them, at any rate.

Every nation has its own immigrant influx, but while West Indians and Asians gravitate towards Britain, Algerians towards France and

Turks towards Germany, everyone gravitates towards the United States.

Melting pot

The original "melting pot" that America held itself to be was based on a mix of European races. These days, as America's political influ-

ence stretches throughout the entire world, immigrants come from every nation. Although it has remarkably stringent immigration laws, it does turn a relatively kind eye on political refugees, particularly those from a nation in which the United States has been overtly or covertly involved. It would be hard not to look favorably upon the immigrant applications of that country's people, particularly when so many of them used to run the pro-American faction back home.

Once here, the new Americans tend to gravitate towards the areas of the city in which their fellow countrymen are already established and to the jobs their fellow countrymen already

---

LEFT: the city has a substantial Asian community, which has helped widen the range of cuisine (**RIGHT**).

hold. So that a large number of Washington's cab drivers are Iranian and Ethiopian. Filipino women become housekeepers and cleaners, while the Chinese congregate in Chinatown on H Street NW and join the catering trade. With their flamboyant shop signs, street names in Chinese, touches of Chinese influence in the architecture including the huge red and green pagoda-inspired Friendship Archway that greets Metro riders at the entrance to the Chinatown station, and stores packed full of products and goods

AMERICAN DREAM

New arrivals are drawn to the region as the local economy expands beyond the federal government to offer jobs and opportunities in the private sector.

identifiable only to the aficionado, they have succeeded in turning a small tract of central Washington into a country within a country.

There is also a large Russian community that reflects several generations of its own nation's turbulent history, from the elder generation of titled White Russians who came to Washington to escape the revolution, through the influx that fled the horrors of Stalin's purges, to the newest arrivals of political and Jewish dissidents. Their focus is the Russian Orthodox churches, St John the Baptist on 17th Street NW, and St Nicholas on Massachusetts Avenue NW. For the Russian Orthodox Christmas on January 7 they offer a midnight mass, and on

Easter Sunday they emerge from church rejoicing *"Kristos Voskreseniye"* – Christ is Risen.

Some of these new Americans become skilful small-time entrepreneurs, making the most of their contacts with their original homelands by organizing sales in hotel rooms, advertised on local radio, of leather goods, jewelry or fabrics brought in by visiting family. Others stand at tables along the streets of Georgetown and the busy downtown corners along K Street and Connecticut Avenue selling everything from rain gear to sunglasses to flowers to jewelry made by Asian, Indian or Oriental village women.

Land of opportunity

The enlarged migration to Washington began after World War II, as America became a superpower. The city was deluged by an influx of foreigners accompanying the tangible expressions of the rest of the world's concern or excitement – depending upon the government back home – at the US's new political powers.

This was certainly the capital in which to have a footing. Embassies opened, Washington became home to the World Bank, the International Monetary Fund and countless other international financial corporations and political institutions. The foreigners made plans for long-term postings. All this was new to a capital that had deliberately isolated itself on an unwelcome square of marshland.

But with the Golden Years of the 1950s, everyone wanted to stay on. The "other superpower," after all, was a less congenial posting. So the infrastructure grew to support the tastes and requirements of all these foreigners. With citizenship more easily acquired then, staff attached to foreign businesses became American and opened shops to supply their ex-countrymen with the specialist foodstuffs, clothing and reading matter which they desired. Washington's foreign community established itself.

By and large, each community of new Americans gets on fine with each other because most are equally balanced and relatively small. Besides, they recognize that they are still the newcomers. They are committed to making a success of their new status that will benefit their children for generations to come. One does not abuse one's host. The situation changes when it comes to single ethnic groups who are themselves American but forced economically to settle in one part of town, or when a single ethnic group expands greatly beyond the rest.

Ethnic rivalries

Then the potential for mass disaffection, rioting and protest arise; witness the race riots of 1968 and the black versus Hispanic uprising during the spring of 1991. Apart from the disturbing suspicion and rivalry which has grown worse between the growing His-

> ### TOWER OF BABEL
>
> In Fairfax County, a close-in Virginia suburb, classes in the public schools are offered in 22 languages to serve the needs of the immigrant population.

unrest. Much of this optimism lies with the opening of a Metro stop in deprived, strife-ridden Anacostia. The station's 1,300-car parking lot just off a major interstate is designed to draw commuters from the prosperous Prince George's County, and real estate investors are already moving in. Commercially zoned land not far from the station has risen to a value 10 times its price less than a decade ago, and the Metro stop area has been billed as a "regional center for economic development."

panic community and the African-American community, the dividing line in Washington is less between the individual ethnic communities, than between white Americans and everyone else.

Of all the murders that take place regularly in the capital, almost all occur in the non-white sections of Washington, particularly the east. In 1991, more murders took place within a stone's throw of Georgia Avenue than in the entire British Isles. There is hope, however, in combating the spiraling crime rate and racial

Plans and schemes for Anacostia are ambitious. But so far, that's all they are: ambitious plans. Only time will tell whether these schemes will work, however, for money has been pumped into Anacostia before, and it still remains a place of great unquiet.

But Washington's ethnic mix has made for a city that has lost its provincial edges by embracing a multitude of cultures. Any day of the week you can hear on the streets of downtown four or five different languages. There is almost no culinary speciality that is impossible to find. Local art galleries regularly mount shows these days from countries whose work previously remained at home. ❑

LEFT AND RIGHT: Washington's ethnic mix has helped the city to dilute its traditional provincial attitudes.

ARTS AND ENTERTAINMENT

The choice is wide, embracing a world-class symphony and opera company, clubs and theaters, ethnic eating places and the exuberant Cherry Blossom Festival

Washington takes its time out seriously. Among the people who run the city and for whom the city is predominantly run, culture comes with a capital C. Dress is often black tie and the most sought-after invitations are for strictly formal events. But though the diplomats and the power brokers may set their chins for the fund-raising dinners or the theatrical season, there are some others who enjoy themselves in a more lighthearted way. The young residents of Adams Morgan, the African-American, Hispanic, Asian and student neighborhoods have a range of cultural activities which might well remind any junior embassy official of the exoticism of home.

While mainstream Washington makes a point of leaving its dinner parties promptly at 11pm (lights in suburban homes are out by 10.30), in Adams Morgan, or downtown on Connecticut Avenue, or along southwest Washington's waterfront and up Georgia Avenue, the beat goes on. Bars and nightclubs line the streets.

Concrete egg box

Power Washington is happy to focus its cultural life on the Kennedy Center, with its six theaters housed in a rectangular concrete egg box that overhangs the Rock Creek Parkway along the banks of the Potomac. The Opera House and the Concert Hall, home of the National Symphony Orchestra, are the sites of the most glittering of Washington's cultural events, drawing international stars and companies. Washington's opera season, under the artistic direction of famed tenor Placido Domingo, is a magnet to the city's social elite. Imaginative and sometimes daring productions – from its celebrated production of Maria Ewing singing Strauss's *Salome* and writhing through the dance of the seven veils completely naked, to the very esoteric and rarely heard *King Arthur* by the 17th-century English composer Henry Purcell – draws a packed audience ready-dressed in its cocktail best.

LEFT: Blues Alley veteran Charlie Byrd (1925–99).
RIGHT: Placido Domingo, the big noise in local opera.

Culture does not have to come all dressed up, however. One lovely concert hall accessible to the general public is the Wolf Trap Farm Park's Filene Center, off the Dulles Toll Road in Vienna, Virginia. Although it's necessary to get there by car, this open-air auditorium has sweeping lawns in front, on which you can picnic

before a performance. The repertoire includes music, ballet, jazz and pop. In Washington itself, there are several smaller stages where a black tie would be positively out of place.

Foremost among these are the Arena Stage, the Studio Theatre, the Woolly Mammoth Theatre Company, the Source Theatre, and Ford's Theatre, where Abraham Lincoln was shot. Their drawback is that they are not always well served by the capital's theater critics, who in overpraising less worthy productions do not force the companies to strive for the high standards appropriate to a capital city.

Washington boasts no fewer than two dozen theaters where on any day of the week you can see live performances, maybe not always to New York standards, but always entertaining. No less than Broadway musical legend Stephen Sondheim has a hand in operating Arlington's Signature Theater, known for its experimental musical productions.

One outstanding venue is the Shakespeare Theatre at 450 Seventh Street, NW which puts on Shakespeare's plays and other classics, sometimes in updated settings (such as a production of Shakespeare's *Much Ado About Nothing* set in 1920s England). The company often casts leading actors from the film world in key roles. Stacey Keach, who became famous

Washington's social crème de la crème see that this is earned through fund-raising. Washington's most respected public figures, hostesses and senior members of its diplomatic community spend a good deal of their social lives at fund-raising events dedicated to keeping the symphony orchestra and the opera offering up-market entertainment for yet another season.

Fund-raising, an entertainment in itself, has produced an aristocracy of its own, every bit as exclusive and powerful as the titled salons of old Europe. Every day of "the season" (September until June) these people's social calendars are filled with appointments to dine at

as TV's tough private eye Mike Hammer, impressed audiences with his intimately evil Richard III and film actress Kelly McGillis gave a fine performance in *The Duchess of Malfi*, John Webster's complex tale of intrigue and betrayal.

Fine arts fundraising

The National Symphony Orchestra under the direction of Leonard Slatkin is probably the greatest pivot for the melding of Washington's cultural with its social life. Though supported by the National Endowment for the Arts, the orchestra needs millions of dollars more each year to keep going and a doughty team of

meals that they will have paid for themselves. The meals don't come cheap and if they do come free, beware – they don't always come sponsored by the pure in heart. The private dining rooms of Washington hotels have been the scenes of breakfasts hosted by shady Savings and Loans officials eager to soften the hearts of congressmen and senators, or luncheons thrown by banks that have set the world money markets reeling with horror.

Favors are floated, temptations subtly intimated, power increased, territory acquired, briefings given, at generous meals in hotels such as the Hays-Adams, the Ritz-Carlton, the Mayflower, and the Hotel Washington. Dinners follow a similar vein, whether actual funds are involved or not. While venerable Washingtonians congregate at four-figure-dollar plate dinners for their favorite causes, foreign diplomats along Embassy Row and in Georgetown bring together influential political figures over private dinners to express among the chink of champagne flutes the kinder, more humane version of their nation's political viewpoints.

Jazz and blues

Apart from the private dinners for the glitterati that take place in Georgetown, that area of the capital has diminished as a center for night-behind M Street. Blues Alley is where top-name entertainers perform, attracting to its dim and smoky room people like Nancy Wilson, Quincy Jones and Pieces of a Dream. Wynton Marsalis and the late jazz singer Eva Cassiday have been among the jazz greats who have recorded albums here. Otherwise, there is little left of the Washington jazz boom of the 1940s and '50s. What survives can mostly be found at clubs such as Bohemian Caverns (2001 11th Street, NW), and at Wolf Trap and Kennedy Center concerts.

You'll find plenty of other types of popular music performed live, though. Try the 9:30

time fun and games. Some night clubs remain, but the commercial heart of Georgetown today is two main thoroughfares lined with over-priced boutiques and overstretched restaurants. These are appreciated most by people under 30 years of age and over $30,000 in salary who have a strong shouting voice. These are *noisy* restaurants. The young bloods of town love a good shout at the end of a hard day's work.

What Georgetown does have is Blues Alley, located at 1073 Wisconsin Avenue in an alley

Club at 815 V Street N.W. for rock that's innovative and, if nothing else, loud, The State at 220 North Washington Street in Falls Church in Virginia for everything from '60s vintage rock bands (Leon Russell) to blues (John Mayall), and the Birchmere at 3701 Mount Vernon Avenue in Alexandria for everything from a Hank Williams tribute to a sing-a-long with the folk song group The Kingston Trio.

Washington is also home to a number of excellent smaller symphonic orchestras including the National Gallery Orchestra, which offers Sunday evening concerts in the stately marbled garden court of the West Building of the National Gallery of Art. The concerts are free

---

LEFT: Leonard Slatkin conducting the National Symphony Orchestra.
ABOVE: *A Wonderful Life* at the Arena Stage.

and have featured some outstanding performers, including jazz greats Billy Taylor and George Shearing. Dumbarton Oaks, the famous Georgetown mansion, also holds concerts for intimate audiences, as do a number of the smaller museums in the city, such as the Phillips Collection at 1600 21st Street, N.W. Musical choral groups also have a presence in the city, especially the Choral Arts Society of Washington, 190 voices strong and performing in the city for 38 years.

Then, of course, there's the Smithsonian. It's the world's largest museum complex, and besides paintings and sculptures, it offers an on-going series of popular lectures, concerts,

performances, classes, and excursions around and beyond the city.

Film Festival

Filmfest DC, usually held in early May, has developed into a showcase for premiering dozens of American and international films. With first-time screenings in cinemas across the city, exhibitions, and much publicized celebrity receptions, Washington's film festival has become one of the country's top film markets.

If you're after something a little more exotic, Adams Morgan, a "gentrified" neighborhood of South American and Third World restaurants and clubs, is the place to go. In fact, you will probably eat as well in Adams Morgan as you will anywhere in Washington, where so often the menu reads better than the food tastes.

Honest cooking

Neither the restaurants in Adams Morgan nor those in Arlington, Virginia, just the other side of the Potomac, suffer from that Washington dining disease of pretentiousness, because their cuisine is genuinely ethnic – in Arlington predominantly Vietnamese and South East Asian – and produced by chefs who learned their honest cooking back home.

Washington likes to think of itself as a culinary outpost in the civilized Western world. European gourmets may sneer that it still has a long way to go, particularly for those without the luxury of an expense account. Power Washington is admirably served by a range of top-line restaurants presenting excellent French, Italian and European food. The disappointing experiences come in that gap somewhere between the fast food emporia and the $50-a-head meal.

This is where the ethnic food restaurants come into their own. Most are reasonably priced and the dishes are accurately presented because of the abundance of Third World spices, pulses and vegetables found in shops along Columbia Road and in Arlington. Restaurants striving to serve modestly priced European fare are less fortunate in their supplies, being bound by laws that restrict the import of dairy and meat products essential in provincial European dishes.

Street parades

Washington's Hispanic, Asian and African-American communities have all developed their own styles of entertainment. Along with a generous supply of specialty restaurants, one community or another appears to organize a street party or a festival practically every weekend. Usually announced ahead of time in the weekend pages of the Washington newspapers (check especially the "Weekend" section on Fridays in the *Washington Post*) , these are exuberant events, with their beauty queens, their calypso, reggae, mariachi bands, single-stringed instruments, tin bells, cymbals and drums, decorated floats, vivid dancers and booths of strange and exotic foods. The most colorful is Adams Morgan Day, traditionally held in September, when the streets are packed with enthusiastic Washingtonians.

One of Washington's most public events is the annual Cherry Blossom Parade, which attracts up to 200,000 visitors. These decorative trees along the banks of the Potomac came about because of a symbolic blossoming of US–Japanese relations almost 100 years ago.

It began with a remarkable American botanist, David Fairchild, founder of the Foreign Plant Introduction office at the Department of Agriculture, who is credited with introducing some 75,000 new varieties into North America. His name is now accursed by many farmers and private gardeners because he brought in the *kudzu* vine from the Orient as a livestock food

A thoughtful Japanese Ambassador cabled back to Tokyo that US–Japanese relations could be happily cemented with a gift of 2,000 young trees, which were then shipped across the world at considerable expense. The trees would last at least 30 years, said the flowery note from the Japanese emperor, which would symbolize the enduring friendship between the two countries. (He was not far wrong. The friendship ended when war broke out after the Japanese attacked Pearl Harbor, exactly 31 years later.)

This first shipment of 2,000 trees was burned when US Customs found the trees diseased. Undeterred, the Japanese tried again, and in the

and as an excellent soil restorative during crop rotation. But the vine spreads so rapidly and has proved so tenacious that it has become known as the killer weed.

In his travels in the East, Fairchild fell in love with the Japanese cherry blossoms, and brought back some seedlings which he grew successfully at his private estate in Chevy Chase, just outside Washington. In 1910 the president's wife, Helen Taft, decided that Japanese cherry trees would be just the thing to cover a local eyesore by the Potomac river.

LEFT: in full swing at the Blues Alley nightclub.
ABOVE: a Cherry Blossom Festival beauty contest.

spring of 1912 a new consignment of 3,000 healthy trees were planted along the banks of the Potomac river.

In 1935, the city inaugurated the annual Cherry Blossom Festival as a way of attracting tourists. These days there is a three-hour Cherry Blossom parade, with bands and floats and a Miss Cherry Blossom contest. The date shifts with the weather and the peak time of blossoming, but is usually in the first two weeks of April. Only a few dozen of the original 1912 trees still remain, most of the rest having succumbed to old age, but the stocks are constantly renewed through exchanges with the city of Tokyo, which continue to this day. ❑

SPORT

Power politics may be Washington's first sport, but football runs a close second.
Except for baseball, the city hosts every other major league sport

Everyone works out in Washington. Health clubs are jammed with the sweaty and the spandex-clad, while the calorie-burning trails of the C & O Canal and the George Washington Parkway are crowded with bicyclists and joggers zooming along from "zero-dark-thirty" (local 24-hour clock-speak for pre-dawn) until

The city's grid of circles and angled streets form a system of cross-town mini-parks that give it more green space than probably any other major American city, though it's Rock Creek Park that is the city's jewel. This four-mile forested stretch in the heart of Washington is laced with trails perfect for walkers, hikers,

well into the evening. Even during the worst of the summer heat, which can reach a suffocating 100°F (38°C) and stay there for days at a time, heartier Washingtonians routinely lace up their running shoes at noon for a heat-stroke inducing lap or two around the Mall. Exercise seems the way Washingtonians choose to bleed off the tension that results from a steady diet of high-stakes politics and international intrigue.

For those who can't afford the expensive club memberships, or who just want a change from all that pedaling and running, DC also offers 93 recreation centers, 156 tennis courts, and 45 public pools, all free to residents, plus access to three public golf courses for nominal fees.

in-line skaters, and even equestrians (there are stables here, and you can sign up for a guided ride). What's best is that thickly shaded Rock Creek can be as much as 10 degrees cooler than the rest of the city in summer, a much needed bit of relief for the weary and overstressed. Just as they take everything else, Washingtonians take their recreation very seriously. And there's nothing they're more serious about than the local football team, the Redskins.

Football frenzy

Regardless of the huge social and economic gaps that often fragment Washingtonians into stubbornly separate camps, the city is brought together by the Washington Redskins. Ever since the team first reached the football Super-bowl finals in 1973, won it in 1983 and again almost 10 years later, the team has been taken to that part of the city's bosom normally reserved for expense account lunches. Though the team's record the past few years has been dismal, the fans' enthusiasm has yet to wane.

The home games are played at the new FedEx Field in nearby Laurel, Maryland, which seats more than 86,000. The team's owner, Dan Snyder, an ambitious advertising mogul who likes to refer to himself as "a guy's guy," paid $800 million for the Redskins, a record for a North American sports franchise. Pugnacious and temperamental (in 13 months he fired three head coaches), Snyder was determined to make his investment pay off.

For a mere $205 million he sold the privi-lege of naming the Redskins' stadium to the package delivery giant Federal Express. And fans were asked to pay more than the support-ers of any other team in the National Football League. Snyder charges a whopping $220,000 per season for one of his stadium's clubby suites where fans can watch the game in luxury. Even individual season tickets in the nosebleed section are pretty pricey, though so far nobody's complaining. "Fans love me," Snyder has quipped, and evidently they must. Despite the cost, the waiting list for a season ticket is now more than 10 years long. As for a ticket to a single game, forget it – no such thing exists.

To have any chance of attending a game, and having the privilege of eating overpriced luke-warm hot dogs while your neighbors tune in to the radio commentary and you peer at the midget activity on the field below, you'll have to find a kindly local with a spare ticket who will invite you along. If you're rich, powerful, famous or beautiful enough, you might be invited into the owner's private box. Failing that, do what the rest of the city does and watch the games on television on Sundays.

You're out!

When football season's over and April's here, discussion turns again, and with considerable longing, to baseball. If the 'Skins cross every

barrier of race, class, politics, and sex to unite Washington in a unique way, it may be because the city has no baseball team to get behind.

Twice it has lost teams, once to Minneapolis and then to Texas, and now has to make do with the Baltimore Orioles, an hour's drive away. Named after the Maryland state bird, the Ori-oles play at the 48,000-seat Camden Yard, a comfortable stadium of traditional design in a park setting and worthy of Baltimore's most famous sporting son, the legendary Babe Ruth. But it is still in Baltimore. To the humiliation of the capital, even the expansion of the Major League didn't help Washington's chances.

But talk of bringing baseball back to Wash-ington hasn't died. Four different groups of wealthy businessmen, including one made up of Black Entertainment Television founder Robert L. Johnson and Redskins owner Dan Snyder, are bidding for teams. So far Major League base-ball officials have shown no confidence in the city and its ownership group to put together an acceptable deal to bring baseball back to the nation's capital. Several sites around the city have been considered for the new stadium that would be required to entice a team to Washing-ton, but critics of the plan point out the cash-strapped city can ill-afford the outrageous $500 million price tag. This plus the city's history of

LEFT: two Washington Redskins move in on a Dallas Cowboy.

RIGHT: the 'Skins attract loyal fans.

crime and its dysfunctional government, hamstrung further by Congress's hold on the purse strings, doesn't exactly put Washington on the baseball executives' shortlist.

The last team to play here was the Washington Senators who made Griffith Stadium, named for the team's owner, their home in 1903. Negro league teams, including the Homestead Grays and the Black Senators, also played here. The tradition of the president throwing out the first ball of the season was begun at Griffith Stadium with William Howard Taft in 1910. In 1953, fans at Griffith Stadium watched Mickey Mantle hit his famous home run, a dis-

tance of 565 ft (172 meters), which was so long that it cleared the left field bleachers.

Despite the city's love for the Senators, the team played its last game in September 1961 and four years later the stadium was demolished to make way for the expansion of Howard University's hospital center.

If you want to see a game without driving to Baltimore, catch the Orioles Class AA farm team, the Bowie Baysox, who play at the Prince George's County Stadium in Bowie, Maryland. The Carolina league affiliate of the Cincinnati Reds, the Potomac Cannons, also play their games nearby at the Richard Pfitzner Stadium in Manassas, Virginia.

Basketball

Basketball fans find it much easier to attend a professional game. Both men's and women's professional teams play downtown at the MCI Center, along with Georgetown University's Hoyas. The men's professional team, the Washington Wizards (known formerly and rather unfortunately, considering the city's high murder rate, as the Washington Bullets) haven't had a good season in decades. But things are threatening to get interesting with the team's newest president, basketball's great Michael Jordan who regularly comes in and out of retirement to help boost his team on the court.

Whatever the men may lack, the women more than make up for with the Washington Mystics, the women's professional team which draws the biggest crowds of any team in the league. And of course there are the city's golden boys, the Georgetown Hoyas where champion players Patrick Ewing and Alonzo Mourning got their starts.

From October through April, the MCI Center also serves as home for the Washington Capitals, the city's National Hockey League team which may just win a Stanley Cup yet.

Soccer, tennis and golf

Because of its cultural diversity, the DC area is as soccer-friendly as you can get in the US. The adult leagues tend to be made up of teams from embassies or groups of expatriates from the same country, but out in the suburbs, where the sport has gained favor with middle and high school students, "soccer moms" spend their afternoons running mini-van jitneys from soccer field to soccer field. When the Redskins moved out of Robert F. Kennedy Stadium, their former home, soccer moved in. It is now the home of the major league soccer team known as the DC United, made up of both top US players and foreign stars.

As for tennis, the Legg Mason Tennis Classic has made DC a required stop for professionals on the tour. Veterans Andrea Agassi and Pete Sampras, plus newcomers such as James Lake play in this tournament named for its sponsor, the Baltimore investment firm of Legg Mason. The Fitzgerald Tennis Center at 16th and Kennedy streets is located pleasantly enough in a national park, but the matches, for reasons known only to the tennis executives, are played in the unmerciful heat of August.

Come May, when the weather is actually pleasant, golf fever spreads when the Kemper Open comes to the Congressional Country Club in close-in Bethesda, Maryland. Once the pros leave town, though, the ugly-pants set takes back the links and golf in the private clubs becomes an exclusive game for political insiders, especially male political insiders. For the past 80 years the men-only Burning Tree Club, also in Bethesda, has been a favorite golfing spot for Republican congressmen and senators, and more than a few presidents.

But while women may occupy seats in the House and the Senate, they'll never get into

to either to give up membership or force the club to change its policy.

Paddling the Potomac

Fortunately, no such restrictions apply when it comes to the Potomac River. Everyone can enjoy the city's most important geological feature, which winds its way down from the mountains of West Virginia and through the piedmont, emptying into the Chesapeake Bay a good 75 miles (120 km) south of Washington. Nearby Georgetown and Alexandria were both thriving commercial seaports when Washington was founded and now, along with the

Burning Tree, which enforces Saudi-like rules when it comes to females. Women are not even permitted as guests. The club's regressive code has become such an embarrassment for the city that preaches freedom and equality to the rest of the world that even George W. Bush's administration, not generally regarded for its social sensitivity, acknowledged the hypocrisy and declared the club off-limits for its officials.

The rest of the Republican leadership, however, persists in its unenlightenment, refusing

LEFT: the Washington Wizards in action.
ABOVE: former Olympic medallists on the Potomac.

Southwest waterfront, harbor all manner of pleasure craft for weekend boaters and sport fishermen who catch mostly large-mouth bass and rockfish.

The water is brackish from Georgetown down to the Bay, and dangerous above Georgetown among the rocks at Great Falls, so it's not good for swimming, but it's ideal if you like to kayak or canoe. Along with a boat, you can also rent a river guide who'll show you the city from the more relaxing perspective of the water. On a clear summer evening with the moon rising over the monuments, Washington may be at its most beautiful when you're adrift in the middle of the Potomac. ❏

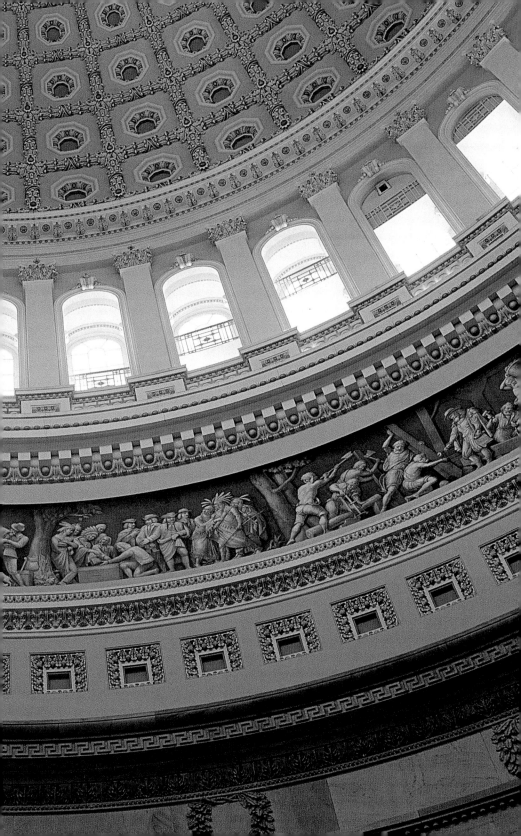

POLITICS AND OTHER POWERS

*Plenty of good food, political contributions, and a steady stream of lobbyists keep
Washington's power players energized and at the top of their political game*

Washington is the most political of cities, designed and built to be the capital of what has become the last superpower. But while the stone and marble structures which house and embody government power are plain to see, the reality of power remains elusive. An American president may find it easy to despatch aircraft carriers and troops around the globe, but he can have trouble getting his budget, or his nominee to run a major department, through Congress. Congress may pass a law, only to find the White House defies it, or the Supreme Court redefines it.

The power and the story

For the visitor, therefore, the power structure of Washington which is on show in Congress and the White House can be deeply misleading; it is part of the reality of power, but only a part. There are two maps of the power process to bear in mind. One is public, and made up of the great and imposing buildings from which power is exercised. The other is private, secret and composed of a series of subtle and personal links through which power is wielded and enjoyed.

The public face of power is etched in a series of straight lines, running as true and potent as the legendary ley-lines of ancient Britain. One axis of power runs along Pennsylvania Avenue, from the White House to the US Capitol and the Supreme Court just behind. To the south of this great artery is the Federal Triangle, housing the great departments of government from Justice to Agriculture to the National Archives, and the Federal Trade Commission and the Internal Revenue Service. Another axis runs to the Pentagon, just across the river. A third heads out to the Virginia suburbs in McLean, where the CIA has its vast headquarters.

Each of these power centers has its own suburb. The Pentagon has spawned its own bureaucratic military-industrial complex. And the US

Congress has become a veritable city in its own right, with its own subway system to ferry the senators and representatives and their staffs around the complex of office buildings which surrounds the Capitol.

To the north are the Dirksen and Hart and Russell buildings, each named after a powerful senator, and each containing the vast and formal committee rooms where so many public hearings and so much history has taken place. Joe McCarthy's witch-hunts for Communist sympathizers in government, J.W. Fulbright's Foreign Relations Committee hearings against the Vietnam War, Sam Ervin's probes into Richard Nixon's complicity in Watergate – these marble halls witnessed all these dramas.

Less well-known are the grandiose office suites where each of the 100 senators have more than 30 staff members whose salaries are paid by the taxpayer. Much of their time is taken up processing the 25 million letters a year Congress receives from the voters.

On the House

To the south of the Capitol stand the Cannon, the Longworth, the Rayburn, O'Neill and Ford buildings where the 435 members of the House of Representatives reside in slightly more modest splendor. House members average about 14 staff members, but their committee rooms defer not at all to the Senate's self-importance. The basement of the Rayburn building contains the House gym, one of the few places where the legislators can get away from voters, journalists and lobbyists and hang around with each other. Celebrities can be squeezed in, but only to be useful, like the way Arnold Schwarzenegger was recruited to give advanced tuition in weight-lifting.

The B-2 Stealth bomber was almost killed in this gym, for it was here that the liberal Congressman Ron Dellums and the conservative Republican John Kasich struck up the weight-lifting friendship which let them realize they both had good reasons to oppose the expensive warplane.

PRECEDING PAGES: inside the Capitol's dome.
LEFT: lobbyists in the lobby of the Willard Hotel, where the term is thought to have originated.

It is by no means a conspicuously luxurious place. The swimming pool is a modest 60 ft (18 meters) long, and it contains a basketball and paddle ball court, weights, stationary bicycles, two treadmills and a stairmaster. The Helene Curtis company provides free soap, and there are constant complaints that the towels are too small.

Because of the cramped locker room, it is a single-sex gym. A smaller women's gym has been opened upstairs. But the House gym is getting too crowded for serious workouts; when a noted Speaker of the House decided to lose 100 pounds, he went to the gym of the University Club on 16th Street instead.

Executive suburb

One of the fastest-growing suburbs is the White House itself. The Treasury huddles close alongside to the east, but beyond the western wing, which Richard Nixon transformed from a swimming pool into a press room, stands the evidence of the growth industry of the presidency: a large 19th-century building of gray stone and pillars, the Executive Office Building.

It used to be sufficient to house the entire civil service, the Navy and Commerce and the State Department. These days, it cannot even house the White House staff, whose more than 2,300 members have spilled over into a redbrick New Executive Office Building on the far side of Pennsylvania Avenue. It is more luxurious than it looks – $350,000 was spent just to redecorate the gym.

The EOB has the best vending machines and automats in town. At 3am, White House workaholics can get cash, pay their bills, shuffle their bank accounts, buy stamps, send off a last-minute anniversary present to the spouse they never see via Federal Express, and get hot macaroni and cheese or cold baloney sandwiches along with a health drink. They can do all this without ever seeing another human being.

The bowels of the White House are the center of the discreet power structure which really runs the political side of the city. The White House mess is one of the most exclusive places in town – like the White House tennis court, even presidential assistants have to go on a waiting list.

Most of them settle for the spartan cafeteria in the EOB, where staffers openly wonder how much of the $2.4 million that the taxpayer spends each year on subsidizing the 22 exclusive dining rooms in the Capitol and Pentagon and White House ever filters down to their plates. (The hospitality prize goes to the Pentagon, which averages $100,000 in revenues for every $450,000 in subsidies).

Breakfast of champions

The serious political sessions, when the fundraisers confer with the Congressmen, take place in private dining rooms in the big hotels, or in the corporate suites and the think-tanks. Powerful people have to eat, and on any given day, they will be breakfasting and lunching and having drinks in a series of discreet places. There are the private rooms of the National Press Club, the top floor of the arms controllers' think tank on Dupont Circle, or the Heritage Foundation on Massachusetts Avenue NE.

The stretch of Massachusetts Avenue between 17th Street and Dupont Circle offers

---

POLITICAL PARTIES

Except for the occasional third political party, such as the Know-Nothings of the 1850s, the Mugwumps of the 1880s, or Texan Ross Perot's United We Stand of the 1990s, America's political system is comprised of two parties. The Democratic Party is the oldest continuous political party in the world. It was founded in the 1780s in opposition to the Federalists, who believed that the country's interests ought to be put before those of the states. Thomas Jefferson was the first president under the Democratic-Republican Party, as it was known then. Over time the party's principles have shifted to embrace social reform, the redistribution of wealth through taxation, and the protection of the "common man." The Republican Party, on the other hand, is typically less interested in the disenfranchised, although it drew its initial strength from the anti-slavery movement. It was founded in 1854 and its first president was Abraham Lincoln. Since then, its objectives have become more closely aligned with big business and the wealthy.

The symbols of the parties were created by 19th-century cartoonist Thomas Nast. His Democratic donkey was seen as representing stubbornness in protecting the rights of the working class, and his Republican elephant signified that party's ponderous resistance to change. Nast himself was a Republican.

the private rooms of the Brookings Institution and the seventh floor of the School of Advanced International Studies of Johns Hopkins. CIA and State Department experts, academics and the occasional journalist gather here for utterly off-the-record seminar suppers on Middle East affairs, and domestic policy-makers meet the Brookings economists over no-alcohol lunches.

This personalized power structure has an occasional public face, seen at a handful of restaurants. Fashions change all

> ## WORK FORCE
>
> Over 300,000 people living in the metro area are employed by the federal government and serve in every capacity from cabinet officers to park rangers and museum curators.

D Street, NE) and La Colline (400 North Capitol Street, NW), especially popular now and where senators go when they get tired of the famous pea soup in the Senate dining room.

Apart from food, the other two commodities that bind the Washington power structure together are gossip (witness the Monica Lewinsky scandal, which consistently remained the top news story for months and overshadowed Bill Clinton's last year as president) and, especially, money,

the time. In the Kissinger years, Sans Souci (now gone) and Maison Blanche (1725 F Street) were the places to be seen. In the 1980s, Nancy Reagan made the reputation of the Jockey Club (2100 Massachusetts Avenue). Though his detractors joked that President Clinton's favorite watering-hole was Wendy's, the hamburger chain, his taste in food expanded to include the elegant Philomena's (1063 Wisconsin Avenue) and the Bombay Club (815 Connecticut Avenue). Then there are the Capitol Hill restaurants such as the Monocle (107

ABOVE: it can be lonely at the top, as this shot of Ronald Reagan, president from 1981 to 1989, tries to show.

Fund capital of the world

Among those with the most clout in Washington are the fund-raisers, people who are keen to understand the problems a politician faces in affording the advertising time on television that is necessary to get elected. Washington is the fund-raising capital of the world. In the 2002 mid-term non-presidential elections alone, the Republican Party raised over $289 million (compared to the Democrats' measly $127 million). Much of this money was contributed through the PACs (Political Action Committees) of such influential groups as the National Rifle Association, the National Association of Realtors, and the American Medical Association. Having gained access, the lobbyists representing these organizations press their cases, cruising the halls of Capitol Hill in astonishing numbers, and effectively setting the legislative agenda, much to the chagrin of the American electorate. The term "lobbyist" derives from the 19th-century influence peddlers who congregated in the lobbies of public buildings, most especially the Willard Hotel near the White House. Today there are an estimated 23,000 special interests

telephones a Congressman, he is unlikely to be put on hold. And a journalist from the big papers, the *New York Times* or *Los Angeles Times*, the *Wall Street Journal* or the *Washington Post* or *USA Today*, will also be put straight through. The same applies to the main TV networks and CNN, and to the top diplomats of the main allies. So the main media offices and embassies, and the Georgetown and Cleveland Park homes of the top diplomats and journalists, are also part of the secret map of political Washington.

There are a handful of occasions when all the elite of Washington congregate in a single place, and the politicians, the fund-raisers, the diplomats and media come together as one great herd. An example is the annual ball for the Washington Opera, usually thrown by the French ambassador. The summer garden party held in honor of the Queen's birthday by the British Embassy is another. Such occasions have one thing in common apart from the guests; the DC police force calls them "limo-locks," from the traffic jams of the sleek, long limousines.

Inauguration Day

There is only one event which needs so many limos that reinforcements have to be driven down interstate 95 from New York City. This takes place every four years in January, on the president's inauguration day. The weather is often very cold. When John F. Kennedy took the oath in 1961, the temperature was below freezing. In 1985 it was so cold they held Ronald Reagan's inauguration indoors.

A chilly driving rain and the presence of a small but very vocal crowd of protestors who lined the parade route marred George W. Bush's inaugural in January 2001. Convinced he had stolen the election from his Democratic opponent Al Gore, the crowd jeered the president, who stayed out of sight and locked behind the tinted glass of his limo.

Whatever the climate, atmospheric or political, the limo-lock that grips the streets just behind Pennsylvania Avenue, gives access to the windows and the rooftops from which the fortunate nibble their canapés and sip champagne as the president and his wife wave their way back from

groups in Washington with as many as 90,000 people engaged in lobbying efforts. The pharmaceutical industry, whose excessive profits have been the source of much recent discussion, has 623 lobbyists alone on Capitol Hill, a rather striking imbalance considering there are only 535 Congressmen.

The most generous contributors to individual candidates usually include the Teamsters trade union, the American Medical Association, the National Association of Realtors, the National Education Asssociation, the United Auto Workers, the Association of Trial Lawyers of America, and the National Rifle Association.

If any official of these lobby organizations

the swearing-in on the Capitol steps on their triumphant parade to the White House.

The limos lock again in the evening, for the inaugural balls. These now take place all across the city, in the vast and echoing Building Museum, in the marbled hall of Union Station, and at the big hotel ballrooms. The political power brokers of all 50 US states, the fund-raisers and the donors, the far-flung members of the new White House family and the new Cabinet all gather for a glimpse of the president and his

WORLD HEADQUARTERS

Besides serving as headquarters to the World Bank, the International Monetary Fund, and business giants such as AOL Time Warner, Marriott Hotels, and MCI, the city also hosts 175 foreign embassies.

power was aphrodisiac enough for his escorts to be some of the most beautiful women in the country. But sex is dangerous in Washington – for proof, see the next chapter, "Political Stings and Scandals".

Gore Vidal, the novelist of Washington and its ways (and a relation of Al Gore and his father, a former senator from Tennessee), observed that the city had become the modern Rome, and it was therefore to be expected that it would develop its own scandals to rival those of the Caesars.

spouse as they take the floor for a token waltz, and then glad-hand their way out of the door to the next ball.

On the House

If this does not sound like much fun, that is because the real pleasure of Washington is not in the lavish and sleek way this city entertains itself, but in the erotic grip of power. Henry Kissinger was never a handsome man, but his

LEFT: TV news anchors such as Judy Woodruff can gain considerable access to the corridors of power.
ABOVE: President George W. Bush was protected by bullet-proof glass at his 2001 inaugural.

The difference is that the power of Rome's emperors came to overwhelm the Roman republic and the institutions of law and senate which had built its greatness. And in spite of that vogue phrase "the Imperial Presidency," the democratic institutions of the United States have proved to be of sterner stuff, and the concept of the American empire is more metaphor than a replica of Roman imperialism.

The secret of political Washington is the way the private and the public maps of power interact to diffuse that power and spread it through the Congressmen and corporations and media to let it flow back to the 50 states which find their focus in this modern Rome. ❏

How the Federal Government works

The American experiment in democratic self-government entered its third century in 1991 with the bi-centennial of the Bill of Rights. These rights are now codified in 10 amendments to the Constitution and were drafted shortly after the Constitution itself to settle lingering anxieties about the limits of the state's powers over the lives of individuals.

Educated in the social theories of such European Enlightenment thinkers as John Locke, Thomas Hobbes, and Jean-Jacques Rousseau, the founders shared a deep mistrust of any concentration of power, having recently overthrown the rule of a despotic English king. The system of Thomas Jefferson, James Madison and Alexander Hamilton was designed to ensure that no institution or faction could seize control of the apparatus of government.

The system is a federal one, so the national government's powers are limited. Most criminal law, for example, is written by state legislatures. When Americans refer to the Federal government they mean Washington, DC. The government has three branches, each able to limit the powers of the other two. This is known as the separation of powers and the system of checks and balances. The three branches are organized by function: the legislative, executive, and judicial branches.

The legislative power is vested in Congress, which consists of two bodies, the House of Representatives and the Senate. Both are elected directly by the people and act according to majority vote. They are organized into committees, which consider, amend and send to the entire membership bills which win a majority of committee members. Bills which fail in committee can be resurrected by a majority of the entire membership of the House or Senate.

Membership in the House of Representatives is apportioned according to population. The total number of Representatives is fixed at 435, and they are allocated to each state's proportion of the total national population. Each decade a census determines the official population and congressional seats are adjusted accordingly. Bills can be introduced in either body of Congress, but only the House can initiate appropriations of funds, the real business of policies.

The Senate is considered the senior body. Senators represent larger constituencies and are elected to six-year terms. They are meant to operate on a broader scale than House members, thinking more of the nation as a whole. Each State has two Senators, so a large state such as Texas has no more Senators than a tiny one such as Rhode Island. This was another means of balancing things out. The original small colonies feared dominance by the large ones.

The Senate is nominally run by the Vice President of the United States. But his (or her) power in the Senate is limited to voting only when there is a 50–50 tie. The majority and minority leaders,

elected by their memberships, exert the real power, as do the committees which conduct the detailed analysis and the behind-the- scenes politicking that actually run the government.

When a bill passes majorities of both bodies, it is sent to the President for signature and can then become law. Frequently a bill has been amended by each body in different ways, so it is then referred to a "conference" committee in which Representatives and Senators hammer out the differences and reintroduce the bill.

The Executive power resides in the Presidency. Presidential and Vice Presidential candidates always run as a team, so there is no practical way they could be of different parties. They are the only officials elected by all the people, and are thus presumed to have a mandate to govern along the lines of their campaign programs. While the people vote for presidential candidates by name, they are voting for a slate of electors pledged to the candidate.

Each state has electoral votes equal to the sum of its two Senators and all its Representatives. This is called the electoral college. The electors vote not in proportion to the popular vote in their states, but all vote for the candidate receiving a majority. Consequently, Presidential elections focus on closely contested states, especially on the large states. Presidents and

Vice-Presidents are limited to four-year terms, and each can be re-elected once for a total of eight years in office.

The President proposes a program of legislation to Congress and Congress are expected to act on it. When the same party controls both branches, this can go smoothly.

The President's best-known role is in the foreign policy area, negotiating treaties, and in moments of international tension, as the Commander-in-Chief of the Armed Forces. The Congress, however, must ratify the treaties and has

ABOVE: the president delivers his annual State of the Union message directly to Congress.

the power to declare war, so as international relations have become very complex there has been confusion which branch can actually do what.

The Executive is charged with enforcing laws and implementing legislation. The President appoints members of the Cabinet. They are the leaders of the executive departments such as the Departments of State (the foreign office), Justice, Treasury, Agriculture, Labor, Commerce; and some new departments created by changing times, like Energy and Transportation. Each appointment must be approved by the Senate.

When Congress passes legislation to the President for enactment, it can be signed into law or vetoed. A veto stands unless the Congress can then muster a two-thirds vote in each house to override it.

The third branch is the Judiciary. Members of the Supreme Court, the Appeals Courts, and the lower-level federal courts are appointed for life by the President but require confirmation by the Senate. The courts interpret the Constitution and laws passed by Congress and the President.

A by-product of the split between the parties has been that much legislation emerges from revision and compromise with ambiguous meaning and ends up being adjudicated by the Supreme Court. Since the Constitution and most of its amendments were written and ratified a long time ago, there is always a debate on their proper application in the very different economic, technological, and cultural environment of today. The court can invalidate a law by determining that it violates the Constitution. But Congress and the President can usually enact the main parts of such laws by amending them to delete the parts found unconstitutional.

There is another *de facto* branch consisting of semi-autonomous regulatory groups, charged with watching over securities markets, foods and drugs, interstate commerce, telecommunications, and other fields. These groups are established by Congressional acts and their leaders are presidential appointees requiring Congressional confirmation.

The American system is complicated and not as efficient as a more authoritarian one might be. It requires much balancing and compromise, and it is often frustrating to the citizenry. But in a modern, complex, multi-racial society, it is essential to have broad consensus on major issues, and this is what the system, regardless of its other problems, usually achieves. ❏

POLITICAL STINGS AND SCANDALS

In Washington, sex, money and power form a strange and delicate balance
that tests even the most resilient, and sometimes topples the mighty

Scandals are system-specific; they reveal a lot about the nature of the society in which they take place. Britain tends to be fascinated by sex scandals, the French take more interest in financial scams, but Washington is obsessed with power and its abuses.

Occasionally sex alone can make a Washing-

former paramour Monica Lewinsky. Investigation of the scandal, headed by special prosecutor and Republican lapdog Kenneth Starr, resulted in Clinton's impeachment. He managed to cling on to his office, but the scandal tied up the nation's collective attention in pages of press and hours of air time for more than a year.

ton scandal, like the oddly innocent occasion when chairman of the House Ways and Means Committee Wilbur Mills was found frolicking in the Washington fountains in the early hours with his stripper mistress Fanne Foxe. And US Senator Gary Hart lost his bid for the 1988 Presidential race when he challenged the press to follow him to verify that he was not being unfaithful to his wife. The *Miami Herald* watched his back door, and proved him to be a liar.

But it would be hard to top former president Bill Clinton, who lied before an audience of millions of TV viewers, declaring outright that he did not "have sex with that woman," as he referred to the young presidential aide and his

A sex scandal sinks that much lower when it's linked to drugs. Witness the celebrated 1990 "sting" operation,when Mayor Marion Barry was arrested in a hotel room smoking crack cocaine with a former mistress. Everything was captured on film, including the Mayor's half-hearted and unsuccessful attempt at seduction. During the trial that followed, sympathetic Washington blacks sported T-shirts that read, "I saw the tape – Bitch set him up."

Power generates more dramatic scandals than sex and drugs because of the US political system which divides power among the executive, the legislature and the judiciary branches of the government. Inevitably, they fight for the spoils.

But the post-war prominence of Washington as a superpower capital has raised the stakes of the power struggles, and added the juicy new ingredients of spy wars. It's a poor scandal these days that doesn't include the CIA.

The FBI tries its best, however, and unfortunately has succeeded. The betrayal of his country by Robert Hanssen, self-described loyal FBI employee and devout Catholic, did nothing to enhance the credibility of the agency that would be drawn to the center of a controversy in the weeks following September 11, 2001. Turf wars and mismanagement of information, both internally at the FBI and in exchange with intelli-

legendary first FBI director J. Edgar Hoover is said to have kept his job for so long because of the secret files and tape recordings he kept on everyone else in the power game. Hoover's excuse was the FBI's charter to run counter-espionage inside the US, while the CIA was restricted to operations overseas. Looking for elusive evidence of King's Communist Party links, the FBI instead collected tapes of Martin Luther King's bedroom activities.

Hoover's collection of evidence about the sex life of President John F. Kennedy and Attorney-General Robert F. Kennedy, including their affairs with Marilyn Monroe and the Presi-

gence chiefs at the CIA, admittedly not very sexy as scandals go, nonetheless had tragic consequences when thousands died during the terrorist attacks on New York and Washington.

Hoover's snoopers

Enforcer of the law and standard bearer of all that was considered to be moral and just in America, the FBI has repeatedly been the victim of its own hypocrisy and bad behavior. The

LEFT: a news camera picks up an early encounter between Monica Lewinsky and Bill Clinton.
ABOVE: FBI chief J. Edgar Hoover, center, kept files on both John and Robert Kennedy.

dent's dalliance with Mafia molls, ensured that the FBI director was unsackable. President Lyndon Johnson put it succinctly: "I'd rather have him inside the tent pissing out than outside pissing in." After Hoover died in office, the FBI scandals began to leak out, including the largely illegal COINTEL Progamme, in which the FBI organized and authorized break-ins and burglaries against Black Panthers and activists protesting the Vietnam War. None of it, though, could compare to reports that surfaced after his death that the chaste, no-nonsense and unimpeachable Hoover had a secret, too: a penchant for dressing in women's clothes.

Beyond making interesting headlines, this

sort of thing amounts to nothing more than fool-ishness, hardly a threat to the republic, although the financial scandals that erupted after George W. Bush came to power certainly challenged its principles. Substantial campaign contributions made by Texas-based former energy giant Enron Corporation to both political parties, but especially to Bush's presidential campaign, may have made it possible for shady accounting practices, in which enormous corporate losses were hidden while massive profits were declared on paper, to slide right by investiga-tors at the Securities and Exchange Commis-sion, the government's watchdog agency

charged with overseeing business. Enron's cor-porate reach was extensive and the fallout over its manipulation of the energy market while the Bush White House stood by was widespread. The situation created an energy crisis in the state of California and in several other western states, Wall Street dipped 3,000 points, and millions of middle-class Americans, who had put their faith in American business, lost their savings.

Enron officials, forced to testify before Con-gress and explain their actions, were also put on trial. As for Bush's appointee at the SEC, Har-vey Pitt agreed to be the president's fall guy and tender his resignation as the agency's top watch-dog, which the president quickly accepted.

Nothing new

But such scandals are nothing new. Take, for example, the Teapot Dome affair, back in 1923, during the presidency of Warren Harding. Albert Fall, the Secretary of the Interior, persuaded the Navy Secretary to transfer government oil reserves to the control of the Department of the Interior, which then leased drilling rights to pri-vate investors. Fall earned $100,000 from the Doheny oil corporation, and $300,000 from the Sinclair oil company for leasing the vast Teapot Dome oilfield. When found out, Fall resigned, and went to work for Sinclair. He was later con-victed of taking bribes worth, in modern terms, at least $10 million.

This was peanuts compared to the sums involved in the greatest Washington scandal of the 19th century, the $4 million in 1870s money siphoned off from the Internal Rev-enue Service by the Whiskey Ring. This was a group of distillers in St Louis who were closely involved with Colonel Orville Bab-cock, the chief aide to President Ulysses S. Grant. They provided him with a beautiful blonde courtesan known as The Sylph, whose red underwear became legendary, and cigars wrapped in $1,000 bills.

Babcock was twice acquitted of wrongdoing, after President Grant gave evidence to the court on his behalf. Babcock, who had been General Grant's military aide in the Civil War, almost managed to win the island of Santo Domingo for the US flag through bribing its President, Buenaventura Baez, but the Senate refused to ratify the deal. The feats of "Orville the Incred-ible" had everything – sex, bribes, tax evasion and secret international deals by a military hero working inside the White House. He provided the first of the classic Washington scandals, and even got away with it.

Washington's best scandals never die, but rumble endlessly on through the courts and media, and then through books and the yellow-ing copies of old newspaper files, until they finally pass into legend and develop a new kind of life as footnotes to history. Ex-president Richard Nixon, for instance, found the 1990s opening with a best-selling reminder of the 1972 Watergate scandal in a book called *Silent Coup*. The book claimed that the real Water-gate conspiracy had been hatched deep inside the Pentagon in order to discredit Nixon for his detente policies towards the Kremlin.

Watergate's impact

Watergate remains the classic scandal because it led to the fall of a president. It all began in the election year of 1972, when a team of dirty tricksters attached to the Nixon re-election campaign were accidentally caught red-handed when they tried to bug the headquarters of the opposition. The offices of the Democratic National Committee were in Washington's prestigious Watergate building (then owned by the Vatican). Bob Woodward

FIRST RESIGNATION

Richard Nixon offered this 17-word resignation to Secretary of State Henry Kissinger on August 9, 1974: *Dear Mr. Secretary: I hereby resign the office of President of the United States. Sincerely, Richard Nixon*

grace. There was a theory that the CIA deliberately discredited Nixon to avoid an inquiry into its own (illegal) covert operations in dealing with protesters against the Vietnam War. The Soviet newspaper *Pravda* claimed all along that "forces of imperialism and the military-industrial complex conspired against a peace-loving President who had reached new understandings with the Soviet Union." The Chinese suggested that the KGB had helped topple Nixon because he had

and Carl Bernstein, two cub reporters on the *Washington Post*, were in the police court when the bugging team came up for trial – and a White House phone number was found in their possession. So began the long and tangled trail which finally connected Richard Nixon to the former CIA agents in the bugging team. Two years after the Watergate break-in, Nixon was finally forced to resign.

But the theories of what really lay behind Watergate continued long after Nixon's dis-

become too friendly to Beijing. *Silent Coup* was one of many conspiracy theories.

One of the villains in this revamped version of the ultimate Washington scandal was Bob Woodward, the *Washington Post* reporter who helped break the original Watergate story. Before getting into journalism, Woodward had been in Naval Intelligence (true) where he had briefed senior Pentagon staff including the future White House aide General Alexander Haig (unproven). Haig later became the *Washington Post*'s Deep Throat (unproven), the inside source whose leaks to Woodward from the White House sank the Nixon Presidency (true).

Even while *Silent Coup* made the best-seller

LEFT: the Watergate complex.
ABOVE: Richard Nixon in 1973 accepts responsibility on nationwide television for the Watergate break-in.

lists, Washington was awash with other reverberations of the scandals of the previous decade. A crack in the CIA's wall of silence over the long-dormant Irangate scandal (in which the Reagan administration covertly backed the right-wing Contras in Nicaragua) widened into a breach which brought a new round of criminal charges in the affair which simply refused to pass into history. Five senior CIA officials were formally placed under scrutiny by the Irangate Special Prosecutor after Alan Fiers, the Agency's head of operations in Central America from 1984 to 1986, turned state's evidence in order to avoid further charges.

own spymaster and Director of Central Intelligence, the late William Casey.

The most dramatic of these cases is the "October Surprise," the allegation that the Reagan campaign conspired with Iran in 1980 to prevent the release of the hostages from the US Embassy in Tehran before Reagan had successfully challenged Jimmy Carter for the presidency. William Casey, a member of the Reagan campaign team, is said to have held secret negotiations with Iranian officials in Paris and Madrid in 1980, designed in effect to help Ronald Reagan steal the election.

Mr Casey was also a shadowy presence in

This revival of the Irangate affair, and the return of the time of troubles for the CIA, gave an eerie new life to issues that seemed long buried, from the Irangate scandal, to the October Surprise, to the trial of Panama's Manuel Noriega which explored the CIA's readiness to wink at drug trafficking by its sources. What they all had in common, including the way that the BCCI international banking scandals threw up links to the CIA, was the way the Reagan presidency gave its spies and covert operators a free-wheeling license to operate, with scant regard for law and morality, in the name of national security. The single name which linked all the cases together was President Reagan's

court in the trial of Manuel Noriega in Miami, where the defense produced bank statements which showed that Noriega was on the payroll of Casey's CIA as a regular informant, and that his dealings with Cuba and with Colombian drug-runners were at the CIA's behest.

In the Irangate affair Casey was again a central figure. Congress was appalled by Irangate because it had legally banned the US government from financing the Nicaraguan Contras. The Reagan administration got around this by using proceeds from the covert sale of arms to Iran, a ploy which Casey claimed to have inspired in a bizarre deathbed confession to the *Washington Post* journalist Bob Woodward (there's that name again). But Irangate contained all the crucial ingredients of a really juicy Washington scandal. First, Congress could get its own investigating committees involved, and second, the law had been broken so there was a court process with leaks from lawyers and from the grand jury. Third, there was some bizarre black humor which helped fix the scandal in the public mind.

The tabloid media which helps fix the image of any scandal likes an oddity. Take the Pumpkin Papers, at the heart of the scandal which gave us the word McCarthyism. Whittaker Chambers, an ex-Communist who began testifying in the late 1940s about Communist sympathizers in government before the hearings of the then Congressman Richard Nixon and Senator Joe McCarthy, owned a farm in Maryland. It included a pumpkin patch. One of the highlights of his testimony against the State Department official Alger Hiss was that some secret papers had been hidden in one of his pumpkins.

The Irangate affair

The Irangate affair also began bizarrely, when White House aides took a cake and a Bible to the Ayatollahs in Tehran, to start the relationship which led to the transfer of $150 million in US arms and spare parts from Israel to Iran. Irangate became special when the villain of the case became a hero. Marine Colonel Oliver North, a zealous military aide in the National Security Council of the White House, stood before the US Congress in his uniform and his medals and announced that his only crime was patriotism. Finally, there was a beautiful woman in the case, Oliver North's secretary Fawn Hall, who smuggled some of the crucial documents out of the White House in her underwear.

North was later convicted of obstructing the course of justice by shredding White House documents before the investigators came to collect evidence. The popular President Reagan rode out the scandal by suggesting that he had not been paying attention at the time. This claim was widely believed after the White House admitted that by 6pm on an average day he was usually in pyjamas for a quiet evening

dozing in the front of the TV. But Irangate returned to the headlines when the Special Prosecutor began probing into the work of the CIA in central America when the Irangate funds were steered to the Nicaraguan Contras. (Oliver North, in the meantime, became a high-profile DC politician.)

This focus on the CIA's role not only placed William Casey in the spotlight, it also turned up the heat on Casey's deputy director, Robert Gates, President George Bush Sr.'s choice to be the new director of Central Intelligence. Mr Bush had other reasons for disquiet at the way the ghosts of the scandals were crawling from

their graves. As the vice-presidential candidate in 1980, and named by some witnesses as a possible participant in the October Surprise meetings in that year, the President had a personal interest in the matter. As a former director of the CIA at the time when General Noriega was apparently recruited as a CIA agent, George Bush had a professional interest in that case too. But he probably never got to know him well enough to investigate Noriega's red underpants, worn to fend off the voodoo spells of his enemies. The legendary Sylph of President Grant's scandalous years doubtless had other, more wholesome, motives for her choice of scarlet underwear. ❏

LEFT: Dustin Hoffman and Robert Redford dramatized the Watergate investigation in *All the President's Men*.
RIGHT: Oliver North tells his side of the Irangate affair.

FOREIGN AFFAIRS

Diplomats from all over the world are posted to Washington, offering glitz and polish to the city's edgy political image

"**M**y dear," the State Department's legendary Chief of Protocol, Ambassador Joseph Read, was once overheard to announce at an embassy party, "you've no idea how many diplomats I have to expel each day! When I started this job I told them to ring me at any time day or night if they had a problem. These days, I only answer the phone for homicide or pillage."

Overseeing the manners of Washington's foreign embassy community is a taxing job, if the behavior of its other foreigners is anything to go by. There is a temptation among those representing their businesses *en poste* to a little informal delinquency – be it only in ignoring the parking rules, in chauvinistic comment, or in a somewhat aggressive response to the nation's speed limits. Diplomatic behavior, too, tends to get a little out of hand. After all, there is nothing so pleasant as pushing against the fences when one is an almost-protected species.

Diplomatic impunity

Protected these diplomats certainly are, with their traffic-snarling motorcades, their exclusive "S" license plates that exempt them from paying fines for parking anywhere they please or for blocking the narrow side streets of Georgetown with rows of limousines as they are delivered to and from their parties.

These pivotal events where policies, not recipes, are discussed and proposed, are generally held in the residences of the ambassadors. The formal bureaucratic work is done in the less glamorous chanceries. Washington's embassies are for the most part wonderfully pompous or exotic edifices, with columns, turrets, gargoyles, stucco and fluting and – considering their central situations – set in astonishingly large gardens. Most of the embassies line Massachusetts Avenue and 16th Street.

The fact that the embassies, though well protected, are accessible to the public, being placed

as they are so directly on the street, makes them vulnerable to protesters and pickets. This is not necessarily a bad thing. Though a law banning demonstrations within 500 ft (150 meters) of embassies was in force at the time, picketers during the mid-1980s at the South African Embassy, opposite the British Embassy, never-

theless got their anti-apartheid message across. The effect that their regular shouts, placards and arrests had on the traffic on busy Massachussetts Avenue helped to dramatize and change the way Americans looked at South Africa and the plight of black South Africans.

Cultural affairs

Aware of the effect that their presence has, embassies have been trying harder to gain wider support and solid public understanding. When you set foot in one of Washington's 175 embassies, you are literally standing on a piece of foreign territory. The impression you take away from that experience help form a sense

LEFT: lunchtime at the official French residence.
RIGHT: a maid takes a break in the Spanish Embassy.

of that country's character, a character that many diplomats are hoping reaches beyond the stereotypical. The Romanians, for instance, would like to be known for more than Dracula, the Swedes for more than Volvos. To that end, almost every embassy offers a program of cultural activities open to the public *(for more information, see page 213)*. Where else, as one *Washington Post* reporter observed, can you experience an Indonesian jazz violinist decked out in Tibetan headdress, singing scat? "You could go out to an embassy every night in this town," Jerome Barry, founder of the Embassy Series which organizes some of the embassies'

to pack up the champagne and wait until his flight back home leaves.

The embassies themselves generally survive, but not in the case of the Iranian Embassy of the 1970s, once the neighbor of the South African Embassy. During that decade, this was where some of Washington's most lavish parties were held. "The youngest champagne we serve is Dom Perignon '69," was the house motto. Journalists writing social columns could not be chastised for using that over-worked word "glittering." The guests were as exclusive and flowing as the champagne. But when the Shah was toppled, so was his ambassador in

cultural events, told the *Post*. Art exhibits, champagne tastings, opera, lectures, there's something for everyone with even a smidgen of interest in the world at large.

Of course, having an embassy doesn't guarantee keeping one. When an ambassador's government falls back home, the ambassador often falls too, particularly if his demise has been caused by a coup or revolution that has ousted his entire party of support. Some diplomats squat defiantly in place: the Panamanian Embassy sided with their American hosts and refused to recognize General Manuel Noriega, accused of drug-running and other wrongdoings. From one day to the next, an ambassador may be required

Washington. The Embassy was seized in 1980 by the United States and converted into offices for the State Department.

Conditions at the Iraqi embassy were even worse. After 1991, when Iraq and the US severed relations, the abandoned 1930s-vintage brick mansion on Massachusetts Avenue that once served as the embassy fell into disrepair, its leaky roof resulting in extensive water damage inside. Repairs, though, were slowly made, paid for by long-frozen Iraqi assets.

Social butterflies

Diplomatic life appears from the outside a pleasant, if repetitious, affair. There are the stylish parties, the useful luncheons, the important dinners, the drinks parties, the cultural events and National Day celebrations. At the British Embassy, a traditional brick manor house design by the Edwardian architect Sir Edwin Lutyens, dinner guests raise their glasses to the health of the Queen, which Anglophiles relish, though other Americans, less patriotic, may giggle.

FOREIGN FUN

Washington's 175 embassies are a hidden source of culture and entertainment, available to the public and a good way to meet people from other countries.

It is important, as social occasions tend to follow a regular pattern, to be able to offer an angle that captures the interest of Washington society.

ton loved the dogs, the parrot and the mynah bird that came with the household. He was recalled at exceptionally short notice back to the motherland.

Social success is a fragile thing. The Canadian Embassy, with its glamorous youngish ambassador and his wife, were the center of attraction for some time, until the wife, furious that a much-courted White House aide was not about to grace her party, slapped her social secretary in front of Canadian journalists, an unfortunate audience. Despite her

It can be something as simple as where one comes from: the embassies of Eastern Europe, so much in the news for a while, enjoyed a sudden, though passing, vogue. Later on, the Hungarian Embassy was the focus of Washington gossip with the arrival of a charming ambassador who had no previous diplomatic experience at all. *Le tout* Washington waited agog as rumors circulated of Hungary's displeasure at its ambassador's social success – too politically incorrect for them, perhaps, though Washing-

success creating a desirable social salon of opinion makers, power brokers, respected politicians and journalists, that slap swept it all into thin air and public disgrace.

Though the ambassadors of those nations playing a central role in world politics have a serious job to do, for many lesser nations and their senior diplomatic staff, the posting is a gentle passage in which to enjoy being part of a civilized city with a ready-made social scene. But those at the less prestigious end of the embassy lists will not receive quite so many invitations; ambassadors whose pay checks seem so often delayed are reduced to living on the canapés and snacks of the cocktail circuit.

LEFT: Washington's wealth of embassies makes for ethic diversity at the supermarket.
ABOVE: African diplomats at a UN reception.

For the diplomatic wives who don't work there are charities and fund-raising committees to join, art gallery lectures to attend, instructive courses to take and amateur theatricals to give one's all to. The Adventure Theater at Glen Echo for children is a popular repository of diplomatic talent, with embassy wives throwing themselves into dramatizations of fairytales.

No sneakers

But if nothing else needs to be borne in mind to keep one's social head above water, the absolutely crucial thing to understand on the dime-thick invitation cards is the dress distinc-

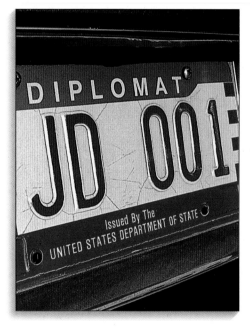

tion between "Informal" and "Casual." There is never a call for blue jeans and sneakers in the latter category. And a scribbled "P.M." in the lower corner does not stand for after lunchtime, but for *Pour Memoire* – "to remind you."

These wives' spouses may arrange small informal lunches to lobby mid-level Capitol Hill bureaucrats, or an off-the-record meeting ("but read my lips") with local journalists and foreign correspondents, while their ambassadors tackle the serious negotiations with Congressional heavyweights and State Department officials. Though these meetings may appear relaxed occasions, the ambassadors are unlikely to be discussing the latest exhibition at the National

Gallery of Art. They will be using this and any other opportunity to broach policy issues – their own and those of the United States.

Among the embassy community, as on Capitol Hill, working breakfasts are popular. Not only is this a useful time of day for a visiting politician to meet his own press for in-depth briefings, but in some cases the restricted size of these events makes them admirable occasions for dropping "exclusive" information. The need for embassy officials to create opportunity to make their nation's views understood has become such a competitive race in a tight but filled social year that public relations firms in Washington have been booming.

These experts in the arts of subtlety and the profitable use of hospitality devise all manner of social events in order to draw to their receptions for visiting musicians, traveling art collections or conservation activists, those American officials who can exert some real influence over their clients' particular foreign nation.

Foreign correspondents

Taking note of all this in the hope of making sense of it for the readers back home and, more importantly, reporting what is taking place on the Hill and in the White House are the foreign correspondents. Correspondents, too, need to build their contacts, in order to have to hand interpreters of the political scene prepared to be quoted in the foreign press. So they join in the round of the breakfast briefings, the lunches, the dinners at embassies, in diplomatic households and the homes of other news media hounds; everyone on the scent of the story of what is really going on.

There are over 1,000 foreign correspondents, including film cameramen, filing constantly for newspapers, journals, radio and TV stations. Many use the facilities of the National Press Building on 14th and F streets, NW. Unlike the senior diplomats with their established run-of-staff residences, the correspondents live where they choose, throughout the city and its suburbs. While in general a diplomatic tour in Washington lasts for three years in the larger embassies (though of any length greater for smaller nations), journalists are usually in town for the full span of a presidential term.

They come in, if they can, with the election campaign or the inauguration of a new president. There is a practical side to this timing.

Congressmen and senators, state officials and bureaucrats weighed down by their own work, may not find the time, once they have their feet firmly under their Capitol desks, for a correspondent newly arrived in the middle of a presidential term, particularly one who comes from a country that may not have much significance for them. It is easier to make and develop those crucial contacts when all parties are "new and eager boys" together.

The optimum time to make contacts with the administration is during the election campaigns, when staff members enthusiastically make themselves available to members of the press from home may be taken on the basis of a newspaperman's editorial or an embassy's briefing, the profits of many conglomerates depend on the Washington office analysis of what is taking place on the Hill, in the Pentagon budget, at the World Bank or the International Monetary Fund. Hence a watchful eye is necessary.

A quick read through the yellow pages of Washington's telephone directory gives a glimpse of what may be happening politically, economically and socially in the rest of the world. As the lists of Vietnamese restaurants stand as a reminder of America's involvement in that South East Asian war, the high presence

anywhere. They all use the same campaign planes, stay at the same hotels, and drink at the same bars.

Career opportunities

The foreign community also includes vast numbers of employees of foreign businesses, large and small. Like the journalists and many of the diplomats, they come with their children, settle into local neighborhoods, enroll their children in local schools and sign them up for baseball and soccer teams. As political decisions back

of Japanese firms confirms the growth of Japanese investment in the United States. The biggest investors of all, however, are the British. This creates very little stir, as their names hardly show up as foreign.

A posting to Washington may be wonderfully glamorous, but the focus on American and world affairs takes up so much time and energy that some members of the foreign community have been known to leave the capital without ever having bothered to know any Americans who didn't have a desk on the Hill. That attitude is changing, however, as cultural attachés realize the need to sell their own cultures more forcefully to the world's only superpower. ❏

LEFT: diplomats enjoy numerous privileges.
ABOVE: pumpkin time at the French residence.

PLACES

A detailed guide to the city, with principal sites clearly cross-referenced by number to the maps

"**D**riving around Washington, it seems like everybody knows where they're going but me," observed a resident of eight months' standing. In fact, even long-term Washingtonians with a fixed destination often find themselves navigating the city's traffic circles more often than they would like. Frustration is intense because, on a map, getting around the District appears manageable.

Architect Pierre L'Enfant's 1791 design was a masterpiece of city planning. Based on a diamond-shaped grid, with the Capitol as its central point, the city is divided into four quadrants: northeast, northwest, southeast and southwest. North and South Capitol streets form the border between east and west, while the Mall and East Capitol Street serve as the division between north and south. The quadrants are mirror images of each other: numbered streets run north and south, lettered streets run east and west. Once the alphabet has been used up, east/west streets have two-syllable names: Adams, Bryant, Channing, etc. When these run out, three-syllable names begin: Albemarle, Brandywine, Chesapeake, etc.

Unfortunately, this elegant system is then invaded by rogue elements, like diagonal avenues – named for American states – and other streets, which criss-cross in a disorganized fashion. Contemporary planners have added to the confusion by implementing one-way systems and changes of lanes during rush hours.

A simple way around all this is to walk, rather than drive. "Monument Washington" is perfect for pedestrians, although bear in mind that the historic appellation "The City of Magnificent Distances" still applies – things are further away than they look. Fortunately, Washington has more open spaces than almost any other town in America, so finding a place to rest footsore feet, either around the Mall or in areas like Georgetown or Dupont Circle, is rarely a problem.

Neither is finding a Metro stop, even outside the city in nearby Arlington or Falls Church. Beyond the suburbs are the green rolling hills of rural Maryland, where you can sail in the Chesapeake Bay or visit Baltimore and Annapolis. In Virginia are the cool, smoky Blue Ridge Mountains, perfect for hiking and camping, plus more sites of Civil War battles than any other state in America.

Washingtonians *do* appear as if they know where they're going, and are very determined to get there. With such a variety of pleasant places to choose from, who can blame them? ❑

PRECEDING PAGES: the White House, founded in 1792; the architecture for the federal city's institutions was based closely on classical models; DC has been called "the city of white marble."
LEFT: statue of Jefferson in the rotunda of the Jefferson Hotel.

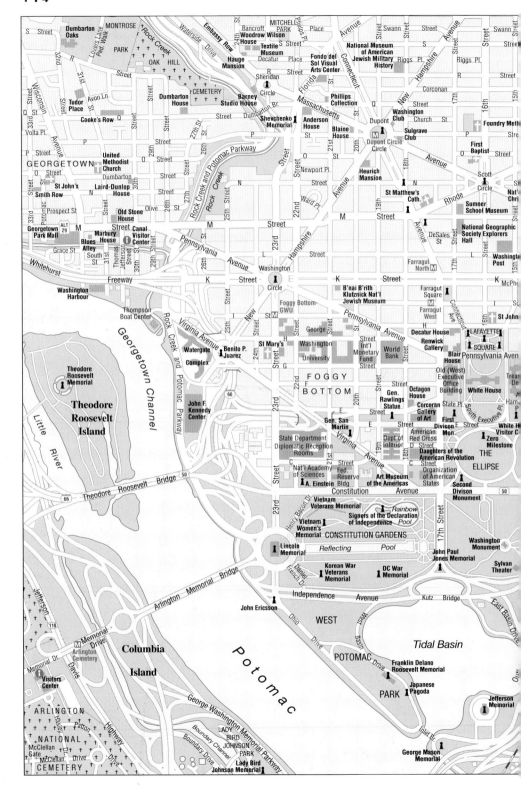

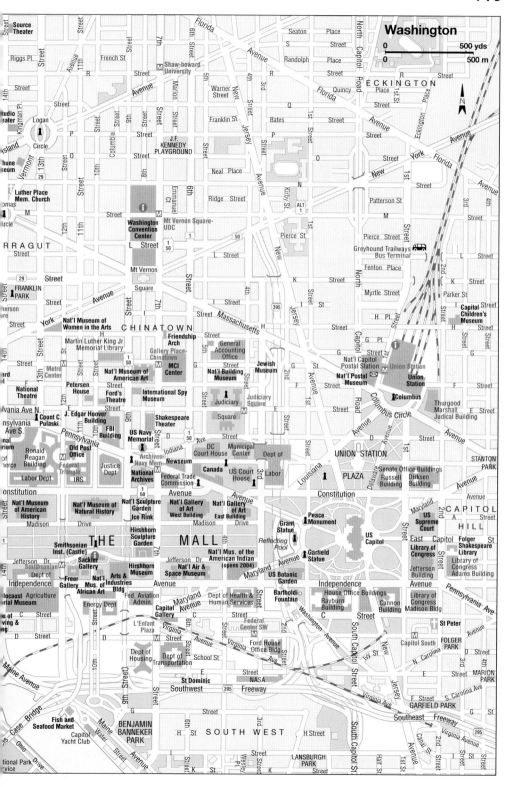

Map
on page
119

THE WHITE HOUSE

*The architect had to omit a planned third floor to save money,
but the relative modesty of the White House was meant to
make a deliberate statement about American democracy*

As official residences of world leaders go, the **White House ❶** is among the less remarkable, and in a city full of gracious mansions it is by no means the most magnificent. It is, however, the most important. Presidents come and go, but the White House endures, physical evidence of the founding fathers' intention that the president is first a citizen in service to his country, here temporarily and beyond the need of princely trappings. Oddly enough, the president who foresaw the need for an official residence and who was responsible for seeing to its construction, George Washington, was the only president who did not live here. He died before the house was completed.

Through its history, presidents have chronically complained of its inadequacies and undertaken to improve it, often with mixed results, and while it is stately, it is not an imposing fixture on the landscape. Nor is it so sacred. Once it was burned nearly to the ground, and twice structural engineers have pronounced it beyond the bother of saving and recommended that it be pulled down altogether.

It has even been vandalized. Rowdy celebrants at populist president Andrew Jackson's 1828 inaugural, convinced that the participatory democracy he espoused made the White House quite literally their own, destroyed what furnishings they didn't carry off as souvenirs.

Over the years the "people's house," as it is sometimes viewed, has become increasingly off limits to the very people to whom it stands as a symbol, owing chiefly to security concerns. East and West Executive avenues, which flank the White House, have both been closed for decades, Pennsylvania Avenue in front of the White House has been closed since 1995 when a gunman shot at the mansion from the sidewalk, and general tours of the White House itself have been sus-

pended. Of course, American democracy is not about a house. It's about an idea, and that idea has been the singular force that has shaped the White House into the unique residence that it is for the American presidency.

A sense of history

Since the terrorist attacks of September 11, 2001, the White House has been closed to the public (except for school tours, organized veterans' groups and limited special events). While you may not be able to tour it, you can still get an excellent sense of the house and its history if you stop by the **Visitor Center ❷** at 15th and E Streets, NW in the Department of Commerce building (daily 7.30am–4pm; closed Jan-

LEFT: the White House in 1850.
RIGHT: the White House in winter.

uary 1, Thanksgiving Day, December 25; tel: 202-208-1631, or visit www.white-house.gov, or for information regarding special events tel: 202-456-7041). The National Park Service rangers who staff the center are and helpful, and in lieu of a tour there are several exhibits, a 30-minute film entitled *Within These Walls*, and a gift shop.

Lines to tour the White House were traditionally long, much longer than the free 20-minute self-guided tour itself. There were no docents on duty to offer commentary or answer questions – a strange arrangement, considering the history and the stories associated with the house and everything in it.

Fits and starts and fire

The White House, the oldest public building in Washington, was designed by the architect James Hoban, who drew on the Georgian-style manor houses of his native Ireland for inspiration. His plan, for which he was awarded the $500 prize in an open competition, called for an elegant three-story stone structure with a columned portico and an eagle carved into the pediment. Construction began in October 1792, but a lack of funds and skilled stonemasons stalled the project. A year later, at George Washington's request, Hoban redrew his plan, omitting the third floor and reducing the overall dimension as a cost-saving measure. Work began, but at a snail's pace. When the US government officially moved from Philadelphia to Washington in November 1800, the house still wasn't finished.

John Adams, the second president, and his wife Abigail were first to take possession of the still incomplete house, grumbling all the while about the slow pace of Southern workmen and the lack of convenience. Jefferson followed the grumbling Adamses, but was no happier, displeased with what he regarded as the mansion's excess. It was, as he wrote, "big enough for two emperors, one Pope, and a grand Lama." Always the architect, Jefferson designed low-terraced pavilions for

BELOW: 19th-century elegance.

Map on page 119

either side of the house to add much needed utility and to give it a more graceful appearance. During his eight-year tenure he indulged in his other favorite hobby, landscaping, by planting formal gardens and installing a circular carriage drive to service the north entrance.

In August 1814, the house was torched by the British but spared total destruction when a violent thunderstorm suddenly erupted, dousing the flames. James and Dolley Madison, then in residence, moved to the nearby Octogan House, owned by the prominent Tayloe family, and called in Hoban to oversee the restoration. The walls of the White House still stood, but the stone had cracked in the intense heat of the fire followed by the sudden cooling of the rain.

In the two weeks that followed, 32 stonemasons set to work restoring the walls, which Hoban had ordered painted white to cover the charring from the fire and which unofficially gave the house its name. It took another two years to replace the roof. Although few of the rooms were inhabitable, President James Monroe, who followed Madison, was determined to move in and hold the traditional open house on New Year's Day in 1818, wet paint, wet plaster, and all. That day, 3,000 visitors showed up to admire the restored house which Hoban had rushed to complete, taking shortcuts by using timbers instead of brick in some of the bearing walls which would cause trouble later.

Home renovations

In 1901, when Theodore Roosevelt and his wife Edith moved in with their six children, they were at once struck by the house's inadequacies. Besides the fact that there were only two bathrooms, the State Dining Room, where the president was expected to entertain guests, was so small that it could hold only six people. To prevent collapse during large receptions, the floors had to be propped up from the cellar, and the elevator that had been installed for the president's convenience shot out

MODERN CONVENIENCES

The original White House had no plumbing, of course, and water for John and Abigail Adams, the first occupants, had to be hauled from half a mile away. Thomas Jefferson erected a cistern in the attic and installed an ingenious system of wooden pipes, which were connected to two water closets and replaced the Adamses' outhouses. In the 1830s, Andrew Jackson was able to enjoy the White House's first real plumbing system, which offered both hot and cold water delivered in iron pipes. There was still no central heat, though. That would have to wait a few more years until almost 1840 during Martin Van Buren's administration. At 340 pounds, William Howard Taft, who arrived at the White House in 1908, presented an entirely different sort of bathroom problem, solved by an a team of ingenious plumbers who fashioned a custom bathtub to fit the hefty president and which was big enough to hold four ordinary sized men.

During the 1840s, James Polk snuffed out the candles and kerosene lamps and lighted the White House with gas. In 1865, Andrew Johnson had a telegraph installed, and five years later Rutherford Hayes brought in a telephone and the typewriter. Electricity came with Benjamin Harrison in the 1890s. Harry Truman had the first White House TV set, and Jimmy Carter the first computer.

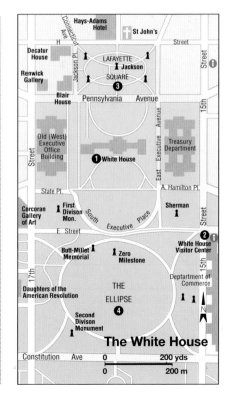

The White House

sparks when it wasn't broken down entirely. A team from the Army Corp of Engineers examined the house and suggested that it be torn down, but Roosevelt insisted on restoration instead and hired the architectural firm of McKim, Meade, and White. Out went the utilitarian looking greenhouses, which provided fresh vegetables and flowers for cutting but which had sprung up unattractively all over the lawn.

The architects also fashioned a new public entrance on the east side, and most importantly, created the West Wing, a presidential working space with a Cabinet Room and the president's private Oval Office, though that wouldn't be completed until President Taft's time.

Besides the addition of the third floor, part of Hoban's original plan and finally installed during the Coolidge administration, little was done to maintain the house over the next 40 years. Investigations into its structural integrity during Harry Truman's presidency revealed that it was again on the verge of collapse. The point was rather dramatically brought home in 1948 when the leg of a piano in daughter Margaret Truman's room broke through two floorboards and knocked the plaster down in the family dining room below. The engineers reported that the exterior walls could be salvaged (though as Truman put it, they were "standing up purely from habit,"), but the house itself had to be gutted.

The Trumans moved across the street into Blair House, while architect Lorenzo S. Winslow and his team set to work underpinning the original stone walls with concrete and erecting new steel framing at a cost of $6 million. They completely built out the walls, floors, and ceilings of each room, added Truman's famous south portico balcony, and finally completed the work in March 1952, at which point Truman led the prominent newscaster Walter Cronkite on a tour of the restored house to show it off to America. He even sat down and played the East Room's great Steinway piano for the cameras.

THE PRESIDENTS WHO LIVED IN THE WHITE HOUSE

(The first president, George Washington, 1789–97, died before it was completed)

John Adams (F)	1797–1801
Thomas Jefferson (D-R)	1801–09
James Madison (D-R)	1809–17
James Monroe (D-R)	817–25
John Quincy Adams (D-R)	1825–29
Andrew Jackson (D)	1829–37
Martin Van Buren (D)	1837–41
William Henry Harrison (W)	1841
John Tyler (W)	1841–45
James K. Polk (D)	1845–49
Zachary Taylor (W)	1849–50
Millard Fillmore (W)	1850–53
Franklin Pierce (D)	1853–57
James Buchanan (D)	1857–61
Abraham Lincoln (R)	1861–65
Andrew Johnson (R)	1865–69
Ulysses S. Grant	1869–77
Rutherford B. Hayes	1877–81
James A. Garfield (R)	1881
Chester A. Arthur (R)	1881–85
Grover Cleveland (D)	1885–89
Benjamin Harrison (R)	1889–93
Grover Cleveland (D)	1893–97
William McKinley (R)	1897–1901
Theodore Roosevelt (R)	1901–09
William H. Taft (R)	1909–13
Woodrow Wilson (D)	1913–21
Warren G. Harding (R)	1921–23
Calvin Coolidge (R)	1923–29
Herbert C. Hoover (R)	1929–33
Franklin D. Roosevelt (D)	1933–45
Harry S. Truman (D)	1945–53
Dwight D. Eisenhower (R)	1953–61
John F. Kennedy (D)	1961–63
Lyndon B. Johnson (D)	1963–69
Richard M. Nixon (R)	1969–74
Gerald Ford (R)	1974–77
Jimmy Carter (D)	1977–81
Ronald Reagan (R)	1981–89
George Bush (R)	1989–1993
Bill Clinton (D)	1993–2001
George W. Bush (R)	2001–

Parties: D = Democratic; D-R = Democratic-Republican
F = Federalist; R = Republican; W = Whig

Map
on page
119

Later, the interior of the White House underwent substantial redecoration, starting with the efforts of Jacqueline Kennedy, who formed a Fine Arts Committee to restore the rooms to their original style and grandeur. The project institutionalized the idea of the White House as a place to reflect American history through furnishings, and renewed the sense of the White House as a house of history. At his wife's urging, President Lyndon Johnson established the Committee for the Preservation of the White House, which provided for a permanent curator.

The grounds

Thanks to a permanent crew of arborists, landscape architects and gardeners, the White House grounds are a showcase of 500 trees, 4,000 flowering shrubs, and 12 acres of impeccably green lawn, though this wasn't always so. In 1800, the 17 or so acres (7 hectares) that comprised the White House grounds were mostly mud, not counting the piles of construction

materials, the shacks that housed the workers, and the barnyard animals that ambled about. During his term, Jefferson transformed the workers' shacks into sheds for the animals, built an exterior wine vault, and added wings for a coal house, a slaughterhouse, and a dairy, though these improvements were lost when the White House was burned.

In the 1820s, John Quincy Adams, an experienced gardener, transplanted hundreds of seedlings during his woodland walks when he occupied the White House, some of those seedlings now stately specimens. In 1877, Rutherford B. Hayes began the still surviving tradition of each president ceremonially planting a tree on the grounds.

The park in front of the White House, **Lafayette Square ❸**, where office workers enjoy their lunches in good weather and protestors routinely congregate for their marches, was originally part of the White House grounds until Pennsylvania Avenue was cut through in 1822. For

BELOW: demonstration of the Wright Type B airplane in 1911.

much of the 19th century the **south grounds** were open to the public and served as a park for picnickers. Spontaneous baseball games were held here, and in summer evenings the Marine Band entertained the crowd. In the late 1800s the grounds were permanently closed and the iron fence erected to protect the privacy of the president and his family.

Flanking the White House, **East Executive Avenue** was added in 1866 and **West Executive Avenue** in 1871, though for security reasons both are now closed. As part of the city's public works project during the 1870s, the canal to the south, basically an open sewer, was covered over and the **Ellipse** ❹ was installed. It's here every year that the president presides over the lighting of the National Christmas Tree, a ceremony accompanied by free live performances featuring operatic singers and military and popular bands. Performances continue throughout December; check the listings in the *Washington Post* for dates and times.

The famous **Rose Garden** adjacent to the Oval Office was originally planted on the White House grounds by Woodrow Wilson's wife Ellen in 1913. Landscape architect Rachel Lambert Mellon, at the request of Jacqueline Kennedy, redesigned it. Richard Nixon's daughter Tricia was married in this garden, which can accommodate 1,000 guests. On the east side of the White House, Lady Bird Johnson installed a companion garden complete with reflecting pool and named it the **Jacqueline Kennedy Garden**.

Improvements to the grounds since then have been relatively minor, mostly reflecting each president's personal preferences. Richard Nixon, for instance, removed Dwight Eisenhower's putting green, which Bill Clinton put back, and Gerald Ford installed a swimming pool.

Public spaces

Five public rooms are on show when tours are available: the **East Room**, the **Green Room**, the **Blue Room**, the **Red Room**

BELOW: children are checked out by the Secret Service before taking a White House tour.

FIRST PETS

Chickens, cows and pigs roamed the White House grounds well into the 20th century, the last barnyard animal being William Howard Taft's cow Pauline, who left when the president vacated the White House in 1913. Since then there has been no end to first family pets that have occupied the premises. Herbert Hoover's son Allan had a pair of alligators, which he tried but often failed to keep contained in a bathtub. Among Secret Service agents who guarded him, Franklin Roosevelt's faithful Scottie dog Fala was also known as The Informer. Whenever the president stumped the country by train, Fala, who had to be walked, was first off the train as soon as it stopped, thereby confirming to all that the president was on board. The Kennedy menagerie was probably the largest and included Pushinka, a pup of the Russian space dog Strelka given as a gift by Soviet leader Nikita Kruschchev, plus hamsters, canaries, lovebirds, ducks, a cat, a Welsh terrier, a police dog, and a pair of ponies known as Leprechaun and Macaroni. Perhaps the most unusual White House pets were a pair of tiny dogs whose breed was known as Sleeve that Commodore Matthew Perry brought back from the Orient and gave to President Franklin Pierce. The dogs were so little that they fit comfortably in a coffee cup saucer.

Map on page 119

and the **State Dining Room**, with the rest of the house comprised of private family rooms and offices which have always been off-limits. Tours enter through the **East Wing** and into the **Ground Floor Corridor** and past the **Library** where a **Gilbert Stuart** portrait of George Washington, painted about 1805, hangs over the fireplace. This room also features two portraits by **Charles Bird King** of Indian emissaries who visited President James Monroe in 1822 for the purpose of establishing peaceful relations with the Americans as the nation expanded westward. Perhaps foreshadowing the natives' ultimate demise, one of the representatives, Eagle of Delight, a member of the Oto Tribe, contracted measles during her visit and died shortly after she returned home.

Across the hall from the Library is the **Vermeil Room**, sometimes called the **Gold Room**, and used as a ladies' sitting room during formal occasions. The **China Room** and the **Diplomatic Reception Room**, where foreign dignitaries are offi-

BELOW: the East Room.

cially received and where Franklin Roosevelt delivered his "fireside chats" to the nation during World War II, are also off this corridor which features several official portraits of recent First Ladies. A Sheraton-style breakfront made in Baltimore around 1800 holds several pieces of official White House china, including a plate George Washington used, and some pieces of the service used by Lincoln. Flanking the stairway on the way up to the East Room are two porcelain busts, one of George Washington and the other of Benjamin Franklin.

The East Room

The gleaming white and gold **East Room** reflects the kind of elegance you'd expect considering the caliber of "public" that gathers here – the monarchs, the heads of state, the celebrities and politicians. Designed originally as a Public Audience Room, it normally contains little furniture and is used for the entertainments that follow state dinners and formal receptions.

The room's centerpiece is the **Steinway grand piano** supported by gilded eagles. The full-length Gilbert Stuart portrait of Washington that hangs here, the only object original to the White House, was rescued by Dolley Madison when the British set fire to the mansion *(see panel)*. In this all-purpose room, seven presidents have lain in state, including Lincoln and Kennedy; two presidential daughters, Alice Roosevelt and Lynda Bird Johnson, were married; and Richard Nixon delivered his resignation speech in 1974.

Lacking a more suitable place in the still incomplete White House, John and Abigail Adams used the East Room during their tenancy to hang their wet laundry. Thomas Jefferson partitioned it to create office space for his secretary Meriwether Lewis who had to hurriedly move when the ceiling collapsed. Restored after the War of 1812, the room was then used as an encampment for Union soldiers.

In the 1880s President Chester Arthur called on the famed New York designer Louis Tiffany to completely redecorate not only this room, which was soon covered in velvet swags and fringe and outfitted with an ornate tin ceiling, but all the others as well. The Victorian transformation was so heavy-handed that afterward the White House was rather snidely referred to as the "steamboat palace" until Teddy Roosevelt arrived. He swept away all the heavy adornments and, moving from the sublime to the ridiculous, turned this room into a sort of gymnasium. In the evenings, wrestling matches were held here for the president's entertainment, and during the day the Roosevelt children circled the perimeter in their roller skates.

The Green Room

Although it too has served many purposes, the **Green Room** was originally designed as a dining room. Thomas Jefferson is thought to have been first to choose the color scheme, covering the floor with a then stylish green-painted cloth in lieu of carpeting. It's full of early

LEFT: Dolley Madison.

DOLLEY MADISON

When the British burned the mansion during their retreat from Washington in the War of 1812, First Lady Dolley Madison managed to save the Gilbert Stuart portrait of George Washington that now hangs in the East Room, the only object known to be original to the White House. As she was being evacuated, she also somehow found time to pen a letter to her sister. "I insist on waiting until the large picture of General Washington is secured," she scribbled as the flames advanced, " and it requires to be unscrewed from the wall. This process was found too tedious for these perilous moments; I have ordered the frame to be broken, and the canvas taken out; it is done – and the precious portrait placed in the hands of two gentlemen of New York, for safe keeping. And now, dear sister, I must leave this house, or the retreating army will make me a prison in it."

Dolley Payne Todd Madison (1768–1849), a Quaker, was a 26-year-old widow when she married James Madison, 17 years her senior. Her social graces and political skills were put to good use when Madison became president in 1809. A contemporary wrote: "She looked like a Queen… It would be absolutely impossible for anyone to behave with more perfect propriety than she did."

Map on page 119

19th-century furniture, acquired during the Coolidge administration and fashioned by the New York Scottish-born cabinet maker Duncan Fyfe, including a pair of rare mahogany work tables ingeniously designed with hidden compartments. The present green silk wall coverings were chosen by Jacqueline Kennedy and installed by Pat Nixon. Several historic pieces of presidential silver are on display here, probably the most important being a Sheffield coffee urn made about 1785 that belong to John Adams and was considered among his most prized possessions.

The Blue Room

The **Blue Room** is a small, oval-shaped chamber where the President and First Lady often receive visitors. Its elliptical shape is owing to George Washington's preference for having his visitors standing in a semi-circle around him, although he never had the privilege of greeting them here since the White House was not completed until after his death. With the

identically shaped Yellow Room above and the Diplomatic Reception Room below, it forms the most elegant feature of original architect James Hoban's plan. Decorated in the Empire style and in a strikingly blue-and-gold color scheme, this is where the First Family usually has a Christmas tree. An 1800 portrait by Rembrandt Peale of Thomas Jefferson hangs in this room along with John Trumbull's portrait of John Adams.

The Red Room

The **Red Room** has always been the province of the first ladies, who entertain their guests here. During Dolley Madison's day, the room was the site of her famous Wednesday night receptions and musicals, which featured exquisite French food, fine wines, and entertainment. No invitation was required and Dolley, a Quaker by upbringing but a gregarious socialite by nature, encouraged anyone in Washington with any sort of talent to attend. Guests sang and told jokes;

BELOW: the Blue Room.

romances, encouraged by Dolley, sprang up. The room was yellow in Dolley's day. Before he set fire to the White House in 1814, British Admiral George Cockburn snatched one of the first lady's yellow seat cushions as a souvenir. Now decorated in red, the present color was inspired by a chair shown in the portrait of Dolley Madison that hangs here.

The **State Dining Room**, whose 1902 wood paneling was painted a pale green under President Truman's incumbency, is now the antique ivory color selected by Jacqueline Kennedy to complement the gold curtains and gold chandeliers. The room originally could hold no more than six, but during the 1902 renovations during Teddy Roosevelt's administration it was expanded to allow for seating 140. Roosevelt's roughrider tastes dictated that the room be decorated with his big-game trophies and that a moose head hang over the fireplace. Carved into the fireplace mantel below George P. A. Healy's portrait of Lincoln, which eventually replaced

the moose head, is an earnest inscription taken from a letter written by John Adams on his second night in the White House. Adams, a Federalist, held the interests of the Union over that of the individual states and had serious doubts about the ability of a largely uneducated and self-interested populace to always show forbearance in electing its presidents. Reflecting that concern, Adams wrote, "I Pray Heaven to Bestow the Best of Blessings on THIS HOUSE and All that shall hereafter Inhabit it. May none but Honest and Wise Men ever rule under this Roof."

Map on page 119

Private rooms

Out of bounds to the public are nine rooms on the second floor, the **Yellow Oval Room**, the **Treaty Room**, the **Queen's Bedroom**, the family bedrooms, the **President's Dining Room,** the family's **West Sitting Hall**, and the one that sparks the most interest, the **Lincoln Bedroom**. Between 1830 and 1902 this room served as either an office or a cabinet meeting room, but never as Lincoln's bedroom. Lincoln used it as an office, heaping his newspapers, mail and paperwork on the floor and tacking to the walls the maps detailing Civil War troop movements. This is where he signed the Emancipation Proclamation on New Year's Day 1863, which officially, if not practically, ended slavery.

When the second floor offices were moved to the West Wing during the Roosevelt renovation, this room became a private family room. In the Truman years, Lincoln's imposing rosewood bed, more than 8 ft long and 6 ft wide (2.4 by 1.8 meters) to accommodate the tall Lincoln, was moved here. Mrs Lincoln bought the bed in 1861, but Lincoln never slept in it.

The West Wing is also closed to the public. Here is where the President's **Oval Office** lies, his desk made of the timbers from *HMS Resolute*. Also in the West Wing are the **Cabinet Room**, the **Appointments Lobby** and the **Roosevelt Room**, where staff conferences take place and which was formerly known as the Fish Room during Franklin Roosevelt's time for the aquarium it contained. ❏

LEFT: the State Dining Room. **RIGHT:** George and Laura Bush in the Oval Office.

THE MALL EAST

People have protested here and partied here, but what most visitors come to see are the treasures of the Smithsonian, the world's largest museum complex

When someone mentions the Mall, chances are that they are referring to that great, grassy, pebble-pathed pedestrian strip stretching between 3rd and 14th streets, lined on both sides with world-class museums. Of course, this wasn't exactly what Pierre L'Enfant had in mind for the Mall. The city planner had envisioned something along the lines of a grand, commercial and residential central boulevard running along the main east-west axis from the Capitol to somewhere near the present site of the Washington Monument.

Until ground was broken in 1848 for the Washington Monument, and then for the Smithsonian castle the following year, the Mall sat largely undeveloped. Then in 1850 landscape architect Andrew Jackson Downing was commissioned to transform L'Enfant's plan for a grand avenue into a "public museum of living trees and shrubs." But before he could execute his plans he died in a steamboat explosion, thus further delaying plans for the Mall. The next half century saw the great, grassy strip used as a cow pasture before it deteriorated into an unsightly collection of ramshackle houses and business establishments, a railroad station on the present site of the National Gallery of Art its most prominent feature.

The 1901 plan

The colossal green expanse you see today, which is flanked by two narrow drives named for presidents Madison and Jefferson, bears the mark of the McMillan Commission, established in 1901 to translate L'Enfant's ideas into a new city plan. The railroad was moved to the present site of Union Station, building restrictions went into effect, and land was set aside for the future Lincoln Memorial. The Mall had been cleared, but little else was done to improve it for another half-century, until after World War II when museum building began in earnest.

The Mall has many moods, determined by the light, by the time of day, and by the seasons. When empty it is a reflective space where you can appreciate the scale and the grandeur of the monuments, and the precision of the planner. Best known as the front yard of the Smithsonian museums and the National Gallery, it also serves as a giant jogging track and a softball field for congressional leagues. Frequently the setting for concerts, festivals, gatherings, and all manner of happenings, it is also the traditional rallying point for protest marches and demonstrations, especially during the 1960s. Among the first protestors to gather here were women suffragists and Bonus

Map on page 133

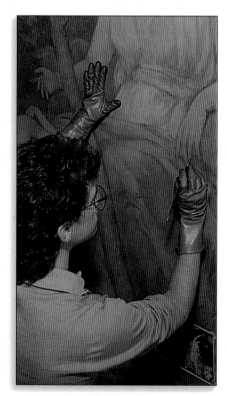

marchers, veterans of World War I demanding payment of their bonuses.

Among the major events staged here is the annual Smithsonian Folklife Festival, held from late June through the July 4th holiday, which transforms the Mall into a sizzling circus of smells, tastes, sights and sounds from the US and cultures farther afield. If you don't mind picnicking blanket-to-blanket with the crowds, this is the place to be on the Fourth of July, where the gala celebration features a free concert by the National Symphony Orchestra and, of course, a fireworks extravaganza.

Museums on the Mall

Silhouetted in late afternoon light, the sleek obelisk of the Washington Monument contrasts dramatically with the Norman-style towers of the **Smithsonian Institution Building**, aptly nicknamed **the Castle ❶**. James Renwick, Jr, architect of the Renwick Gallery and New York's St Patrick's Cathedral, designed this red sandstone structure, which was completed in 1855 – the Smithsonian Institution's first building. Today it holds administrative offices and the Woodrow Wilson International Center for Scholars, as well as an **information center**. The Smithsonian is open every day except December 25, from 10am to 5.30pm. The Castle opens at 9am and closes at 5.30 pm. Tel: 202-357-2700 or check the website at www.si.edu for information about special exhibits and tours.

To get your bearings, stop first at the Castle. It has maps and brochures, several interactive screens, a 20-minute orientation film, and several friendly docents to help you decide what to see. It also holds the crypt containing the body of Smithsonian founder James Smithson (*see page 143*). All museums and attractions on the Mall are free, except for the IMAX features and the planetarium in the Air and Space Museum, and the IMAX and Immersion Theaters in the Natural History Museum.

The elegant Florentine Renaissance palace nearby – small-scale compared to

LEFT: the Smithsonian's "Castle".

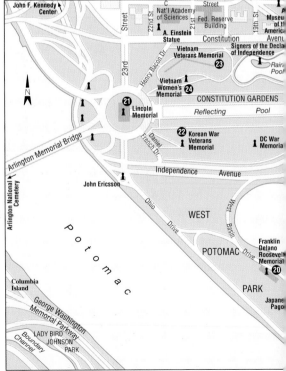

Map below

the other Mall-side monoliths – is the **Freer Gallery of Art ②**. The Smithsonian's first art museum, opened in 1923, is an intimate collection of mainly Asian art donated by a self-made Detroit industrialist and art connoisseur, Charles Lang Freer. Freer's gift included his collection of etchings, plus the only collaborative assemblage of works by his friend, the painter James McNeill Whistler. Freer retired from railroad-car manufacturing when he was 45, having already made his fortune, and by his death in 1919 had bequeathed 9,000 works to the Smithsonian. Its Asian collection has grown to over 27,000 pieces and includes ornamental objects from the Ming and Quing dynasties, plus Chinese jades and lacquerware dating from 3500BC, Korean ceramics, and fine works of Japanese calligraphy.

The American collection contains several works by Whistler, including the impressive **Harmony in Blue and Gold: The Peacock Room**, the artist's only existing interior design which had been com-

missioned by an English shipping tycoon to show off his collection of Chinese porcelains. Intricate and luminescent paintings of peacocks embellish every surface, from walls to ceilings. In 1904, Freer bought the room and its contents from a London dealer and shipped it to his home in Detroit before donating it to the Smithsonian. In 1988 the museum was renovated and includes an underground exhibit area that permits passage to the Sackler Gallery.

Asian and African art

Between the Freer and the Castle, a copper-domed kiosk marks the entrance to the S. Dillon Ripley Center, part of the quadrangle complex on Independence Avenue opened in 1987, which also includes the **Arthur M. Sackler Gallery**, the **National Museum of African Art**, and the **Enid A. Haupt Garden**.

The garden itself (open daily at 7am, closing times determined by the season) is a magical mosaic of decorative sub-gardens, whose themes clearly reflect those

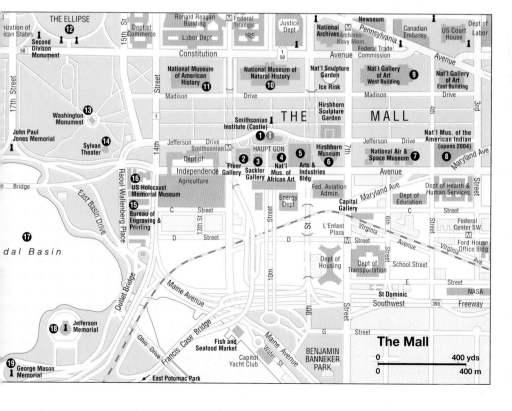

of the surrounding museums. In fact, the whole complex is a marvelous harmony of form: the Sackler's diamond motif complements the spires of the Arts and Industries Building, while the African's circles relate to the Freer's arches.

The Ripley Center, which contains classrooms and offices, is named for S. Dillon Ripley, secretary of the Smithsonian between 1964 and 1984. During that time New York psychiatrist and researcher Dr Arthur M. Sackler made a generous pledge of his Asian artifacts, but Ripley lacked a suitable space for them. Ripley was also struggling with the problem of what to do with the African collection. Begun privately in 1964, then given to the institution in 1979, the Smithsonian's collection of African art was rapidly outgrowing its space in a former home of abolitionist and educator Frederick Douglass on Captiol Hill. The solution to both problems turned out to be this innovative underground complex linking the two museums.

While the descent to the subterranean galleries of the Sackler and the African is akin to entering a mausoleum, the galleries are warm and the collections – each of which runs the gamut from ancient to contemporary – are often remarkable.

Aside from the art, you can admire the galleries' remarkable feat of engineering. Ninety-six percent of the quadrangle is below ground to a depth of 57 ft (17 meters), which required waterproofing the entire structure and designing a special roof to support the 4-acre (1.6-hectare) garden above – much of which is deliberately landscaped to conceal structural elements below. As if that weren't enough, the architect also had to figure out how to preserve a century-old linden tree.

Highlights from the **Sackler Gallery ❸** include ritual bronze and jade objects, more than 450 of them dating from about 3000BC. Also on display are ancient metalwork vessels and ornaments from Iran and Turkey. The **National Museum of African Art ❹** has more than 7,000 items, mostly of the traditional arts of the peoples

LEFT: Asian art in the Sackler Gallery of the Smithsonian.

SURVIVAL TIPS FOR MUSEUMS

The Mall is more than a mile long and is lined with millions of square feet of museum space, so the essential advice is: wear comfortable shoes. There's too much to see in one visit, so set your preferences by checking in first at the Castle, Smithsonian's "information central", or by visiting the website at www.si.edu.

Driving to the Mall is not recommended. Parking is timed and extremely limited, and with two nearby Metro stops, there's no need for a car. The Smithsonian station is on the Orange and Blue lines and will let you off near the center of the Mall. The L'Enfant Plaza station, near Independence Avenue, is a major transfer point on the Green, Yellow, Orange, and Blue Lines.

Anti-terrorist security has become much tighter. Guards will greet you at the entrance of every museum, and you can expect that your bags will be searched.

If you can, try to visit in the fall and winter when there are fewer crowds. If you must come during the summer, be aware that Washington's summers are brutally hot and dress accordingly.

As for restroom facilities, one of the many good features you'll discover when visiting the Mall is that comfort stations are plentiful, easily accessible, and clean.

Map on page 133

of sub-Sahara. Many of the objects – masks used in rites of passage, fertility figures, medicinal objects and tools of divination, all of which are made of bone, fiber and other organic materials – are religious in nature. Standouts of the collection include objects from the kingdom of Benin.

The nation's attic

The Philadelphia Centennial of 1876, which celebrated the Industrial Revolution, lives on inside the **Arts and Industries Building** ❺ – the cheerful Victorian structure flanking the quad's east side. When this 19th-century World's Fair closed, some 40 freight-car loads of leftover international exhibits were shipped to the Smithsonian, prompting the Institution to build the National Museum, as it was then called, to store all of the items – an episode which earned the Smithsonian its reputation as "the nation's attic."

Finished in 1881, the exposition-style building was the most modern museum in the country. Every item displayed in its

dizzyingly busy interior was either exhibited at the Expo or produced during that era, including furniture, jewelry, and even horse-drawn carriages. The museum shop is good for inexpensive Victorian-repro gifts and, in keeping with the mood, an old-fashioned carousel still turns and grinds on the Mall outside the museum's entrance just as it has since 1940. Children enjoy this museum, especially the Discovery Theater, which has live educational performances.

Modern art

The mood next door at the **Hirshhorn Museum and Sculpture Garden** ❻ is quite different. This contemporary museum-in-the-round has been aptly described by writer E. J. Applewhite as a concrete doughnut in a walled garden. In summer, there is an outdoor café among the bevy of enormous abstract sculptures.

It was the wish of Latvian immigrant and self-made millionaire Joseph Hirshhorn to donate his entire collection of 4,000 paint-

BELOW:
the Arts and Industries Building – "the nation's attic."

ings and 2,000 sculptures – notable both for its size and its variety of late 19th- and 20th-century art – to the country that had served him so well. The bequest, twice as large as the collection amassed in 50 years by New York's Museum of Modern Art, made the Hirshhorn an instant treasure, long before it even had walls. Several other countries tried to entice Hirshhorn to leave his collection to them, but in the mid-1960s Secretary Ripley with assistance from President Lyndon Johnson, who had a reputation for never taking no for an answer, convinced the Wall Street broker and uranium magnate that Washington, DC was the only place worth considering.

Paintings are arranged chronologically top to bottom beginning with the works of Edward Hopper, Georgia O'Keeffe, Mark Rothko, and Willem de Kooning among others. Late 20th-century artists such as Andy Warhol and Jasper Johns are featured on the bottom floor. The circular galleries along the circumference of the second and third floors display small sculptural works by such artists as Henri Matisse and Pablo Picasso.

Outside in the fountained plaza there are works by Alexander Calder and Claes Oldenburg. And it's worth visiting the landscaped, sunken sculpture garden, which has on display Auguste Rodin's famous *Burghers of Calais*.

Air and Space

The **National Air and Space Museum** ❼ (tel: 202-357-1400 to schedule a tour, or visit www.nasm.edu) next door is the world's most popular museum, drawing 10 million visitors a year. Although the experts who designed the three-block-long glass and marble building lacked museum experience, it only takes a glance around the main exhibit hall at the Mall entrance to realize that the concept, like the 90-ft (27-meter) tall hangar-like structure, not only works – it soars. Suspended from the ceiling are such air-age stars as the Wright brothers' *Flyer*, Lindbergh's *Spirit of St Louis*, and the *Apollo 11* space craft.

BELOW:
Juan Munoz
sculptures at
the Hirshhorn.

Map on page 133

And that's just the beginning. There are 23 galleries here, a theater, a planetarium, a museum store (well stocked with puzzles, games, kites, and "space food" for children), and a research library which you can access by prior appointment. The information desk offers a detailed guide.

The Smithsonian has always had an interest in flight. In 1861, Joseph Henry, the institution's first secretary, was involved with balloon experiments that led Lincoln to use them for military purposes during the Civil War. The institution's third secretary, Samuel Pierpont Langley, a renowned astronomer and inventor, was deeply involved in the theory of flight when he joined in 1887. In 1946 Congress established a National Air Museum, and in 1976 the Air and Space museum opened in time for the bicentennial celebrations.

Perhaps the most popular stop is the **Lockheed Martin IMAX Theater** (tel: 202-357-1686; admission fee), to see the special film presentations, especially the perennial favorite "To Fly." The screen is 70 ft wide and 50 ft high (21 by 15 meters) and the sensation is close to reality as you soar above the treetops and look out across the earth's landscape from the gondola of a hot air balloon.

The Albert Einstein Planetarium (tel: 202-357-1686; admission fee) is equipped with a Zeiss VI projector to create a regular schedule of starry shows beamed on an overhead dome. In the **Golden Age of Flight** gallery you'll see the history of aviation between the two world wars. In the **World War II Aviation** gallery you can see a Messerschmitt Bf 109, a Supermarine Spitfire, and a North American P-51D Mustang, among others.

Space Hall has the infamous V-2 rocket, a model of the Columbia Space Shuttle, and a replica of the Hubble Space Telescope. **The Pioneers of Flight** gallery features several planes that made historic flights, including the Wright EX Vin Fiz (so named for the sponsor's grape drink). Galbraith Perry Rodgers made the first coast-to-coast flight in this plane between

BELOW: flights of fancy at the National Air and Space Museum.

September 17 and November 5, 1911, covering 4,300 miles (6,900 km) at an average speed of 52 mph (84 km/h).

A new attraction (opening date: December 2003) is Air and Space's cavernous annex in Virginia near Dulles Airport. **The Steven F. Udvar-Hazy Center**, named for the benefactor who donated $66 million for the center's construction, will display more than 200 historic aircraft, from spy planes to World War I-era biplanes.

Right next door to the Air and Space Museum, on the Mall's last building site, is a welcome center for the new **National Museum of the American Indian ❽** (Mon–Sat, 10am–4pm). Scheduled to open in September 2004, this latest Mall attraction will house all manner of American Indian relics and artifacts.

The National Gallery

Cross the east end of the Mall, which is dominated by the imposing edifice of the Capitol, to Madison Drive and you will run into the West and East Buildings of the **National Gallery of Art ❾** (Mon–Sat, 10am–5pm; Sun 11am–6pm; closed Dec 25 and Jan 1; tel: 202-737-4215 or visit www.nga.gov).

Incongruous as they seem side by side – the one neoclassical, the other starkly contemporary – these structures complement one another architecturally as well as artistically. Although the National Gallery is run by the federal government, it relies exclusively on private and corporate contributions for acquisitions and is not part of the Smithsonian.

The seed for a national gallery was planted by the industrialist and Treasury secretary Andrew W. Mellon, who built his collection of 121 Old Masters, including 21 paintings purchased from Russia's Hermitage Museum, with the idea that he would eventually give the collection away. Following Mellon's lead, 1,300 donors have so far given their treasures to the National Gallery since its doors officially opened in 1941.

The esteemed John Russell Pope was

LEFT AND BELOW: milestones of flight at the Air and Space Museum.

Map
on page
133

commissioned as architect for the **West Building**. Its dome happens to look distinctly like that crowning the Jefferson Memorial, which Pope also designed. Being a traditionalist, Pope chose a classical design – firstly, because official Washington was classical, and secondly, because he believed that it conveyed the appropriate image for a serious art gallery. One of the world's largest marble structures, it was to be the last of the District's heroic, neoclassical monuments.

Start your visit in the art information room where you can find a listing of current exhibits, lectures, concerts, and special events. Here you'll also find the **Micro Gallery** with its touch-screen monitors to guide you through the museum and allow you to research artists and locate more than 1,7000 works in the collection.

From there, you can proceed to the rotunda with its marble columns and fountain centerpiece and then into the the sculpture halls. The **East Sculpture Hall** displays classically inspired marble works while the **West Sculpture Hall** houses works in bronze. Each Hall ends in a garden court, and in the **West Garden Court** on Sunday evenings from October through June you can attend a free concert performed by one of Washington's most distinguished string orchestras (tel: 202-842-6941 or visit www.nga.gov for a listing of concerts).

As for the galleries themselves, the National Gallery has distinguished itself as a repository of the paintings that plot the evolution of art in the West from the late Middle Ages through the early 20th century. Among the 90 or so galleries, 34 are dedicated to the Italian painters alone. The Dutch and Flemish masters, including Rembrandt and Vermeer, are also represented here, along with several galleries of French and British painters including Turner and Hogarth, and of course the American painters who include Winslow Homer and John Singleton Copley.

In 2002 the Gallery reopened its much anticipated and renovated ground floor

BELOW:
garden courts
at the National
Gallery of Art.

Sculpture Galleries with 22 oak-floored and Doric-columned rooms displaying 800 sculptures. The Degas display, which includes four versions of his *Little Dancer Aged Fourteen*, is perhaps the collection's highlight and is the largest group of his sculptures to be seen anywhere.

An underground concourse with a moving walkway, a museum shop, a café, and the Mall's only espresso bar, links the West with the **East Building**. When Mellon stipulated that the area to the east of the gallery be reserved for future expansion, the challenge of designing a structure for this odd, trapezoidal lot fell to the creative master architect, I.M. Pei.

He resolved the problem by slicing the trapezoid diagonally and creating two interlocking, symmetrical triangles. He further balanced the site by lining up the central axis of the West Building with the midpoint of the triangular base forming the main block of the East. You don't have to be a mathematician to appreciate this bold and luminous building-as-sculpture, mercurial as the light that plays across its blade-sharp edges and planes.

Pei's brilliant use of triangular geometry infuses the interior with a sense of movement. Open balconies and bridges sweep across the atrium, decried by critics as "wasted space" or Pei-ian self-indulgence. The undeniably dramatic museum, which exhibits all manner of 20th-century art, from Henry Moore's *Knife Edge Mirror Two Piece* sculpture, displayed outside, to the huge red, black, and blue mobile by Alexander Calder specially commissioned for the musuem and dominating the low-ceilinged lobby.

There are five floors of European and American masters of the 20th century, including Wassily Kandinsky, Pablo Picasso, and Joan Miró. The ground-floor galleries feature rotating exhibits, the mezzanine galleries hold the larger traveling exhibits, and the concourse galleries are home to the late 20th-century Americans including Robert Motherwell, Andy Warhol, and Roy Lichtenstein.

BELOW: intimations of Egypt at the East Building of the National Gallery of Art.

Map
on page
133

Natural History

As you walk west, be sure to pass through the **Sculpture Garden,** which faces the National Archives on Constitution Avenue and features two dozen works from the National Gallery's collection in a peaceful setting of winding paths and flowerbeds. The pool is used as an ice-skating rink in winter (Mon–Thurs 10am–11pm; Fri–Sat 10am–midnight; Sun 11am–9pm; admission fee) and offers a respite in summer from the heat and crowds.

The **National Museum of Natural History ⑩,** subtitled the **National Museum of Man,** is a hulk of a building. When the Hornblower and Marshall building opened in 1910, it was dubbed the "new" National Museum. Be forewarned that it is today one of the more popular museums on the Mall, a fact immediately apparent if you venture into the four-story-high rotunda on a typical summer day. As you wander through its three floors and labyrinthine halls, which exhibit everything under the sun from

dinosaur skeletons to the Hope diamond, the biggest blue diamond known to exist, bear in mind that all that you see represents only a fraction of the more than 124 million items stored here.

Stop at the information desk near the mall entrance for a map and for information about the IMAX features and the **Immersion Theater** (tel: 202-633-7400; admission charge), where you can manipulate the onscreen action with your own control panel.

As you enter the rotunda you'll notice the enormous elephant on display. Even for an African elephant, he's a big specimen, weighing in at 12 tons – about 5 tons more than the average. He was killed in 1955 and his skin was shipped to the Smithsonian where workers mounted it on a wooden frame. Every year, more than 6 million people stop for a look at him.

In the **Dinosaur Hall,** the giant reptile known as "Hatcher" is on display. Discovered by Wyoming bone collector John Bell Hatcher in 1891, this triceratops is

BELOW:
the rotunda of
the National
Museum of
Natural History.

shown frozen in battle with a Tyranno-saurus Rex. Encircling the dinosaurs are the **fossil galleries** where you can see evidence of life on display from the Big Bang 4.6 million years ago up to the present.

The second floor holds the **gem and mineral collections**, including the **Hope Diamond**, the largest in the world at 45.5 carats and celebrated for its color and clarity. Other gems include sapphires, rubies, emeralds, and a display of royal jewellry including the 950-diamond necklace that Napoleon presented to Empress Marie-Louise on their wedding day in 1810.

If you have children in tow, be sure not to miss the **Insect Zoo** where the cases are low to accommodate young visitors who can watch a giant mound of African termites at work and catch the popular tarantula feeding (Tues–Fri, 10.30am, 11.30am, 1.30pm; weekend times vary).

American History

The pink marble box next to the National History Museum is the **National Museum of American History** ⓫, designed by the Beaux-Arts firm of McKim, Mead and White, and eventually completed in 1964 by Steinman, Cain and White. Originally called the National Museum of History and Technology, it was created to house the Philadelphia Centennial leftovers that would not fit in the Arts and Industries Building. With its lively focus on *things* – more than 16 million objects – it's aptly called the "nation's attic." Perhaps garage would be a better term, given that exhibits include a Ford Model T automobile and a Conestoga wagon like the kind that traveled to the West Coast "frontier" during the California Gold Rush.

Other exhibits include a display of inaugural gowns worn by presidential wives, the first-ever typewriter, the flag which inspired Francis Scott Key to write the words to America's national anthem, Archie Bunker's armchair, Dorothy's red slippers from the 1939 film *The Wizard of Oz*, Mr Rodgers's zippered cardigan, and George Washington's false teeth. ❑

Map on page 133

LEFT: antique dolls in the National Museum of American History.

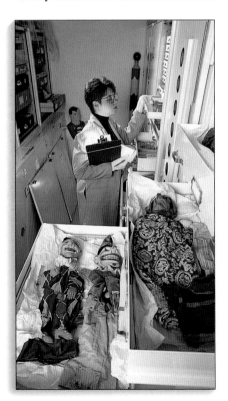

WHERE TO EAT

When it's time to stop for something to eat, you can always find a street vendor for an (overpriced) ice cream or hot dog. There's also a food kiosk located in the middle of the Mall.

But for a more pleasant sit-down treat, try one of the museum eateries. Not every museum offers food service. Here's a short list of those that do:

In the National Gallery: Try the Cascade Café in the concourse between the West and East buildings, which serves light fare and has an espresso bar and assorted sweets. A cascading waterfall behind glass adds to the atmosphere. The Garden Café, on the ground floor of the West Building is a quiet spot for sandwiches and coffee. The Pavillion Café has outdoor seating and views of the sculpture garden with pizza, sandwiches and salads, and the Terrace Café in the East Building's upper level serves coffee and light meals.

In the Museum of Natural History: The Atrium Café offers upscale cafeteria fare.

In the Museum of American History: Try the cafeteria on the lower level, or The Palm Court, an old-fashioned ice cream parlor complete with a "Star-Spangled Banana Split."

James Smithson

No one knows why, exactly, independently wealthy Englishman James Smithson left his fortune to the US government, although it's safe to say that resentment probably had something to do with it.

Born in 1765, Smithson, originally named James Lewis Macie, was born out of wedlock to Hugh Percy, the newly enobled Duke of Northumberland, and Elizabeth Keate Macie. As a student at Oxford, Smithson distinguished himself by publishing scholarly scientific papers and discovering a carbonate of the mineral zinc. In 1800, he inherited his mother's fortune, and in 1806 legally adopted his father's original surname of Smithson.

Illegitimacy precluded his rising through the ranks of English aristocracy. His plan was to leave his money to his nephew Henry James Hungerford, but only if Hungerford produced heirs. When Hungerford died childless, Smithson rewrote his will. He left his fortune of $500,000 to the United States, which he had never visited, "to found at Washington, under the name of the Smithsonian Institution, an establishment for the increase and diffusion of knowledge." Smithson, it seemed, was determined to make sure that, despite his beginnings, his name lived "in the memory of men, when the titles of the Northumberlands are extinct and forgotten." He died in 1829 and nine years later, after some legal wrangling during which his relations contested the will, 105 bags of gold coins were shipped across the Atlantic to the Philadelphia Mint.

For the next eight years, despite Smithson's clear stipulations, Congress argued over what to do with the money. Finally, in 1846, President James Polk signed the bill that established the Smithsonian Institution.

Congress mandated that the Smithsonian be governed by a board of regents and that the new institution be comprised of a library, a museum and an art gallery. As its first secretary, the board elected Joseph Henry, a physicist from Princeton.

Between 1846 and 1878 Henry devoted his energies to the "increase of knowledge," shaping the Smithsonian into a research institution. The second secretary, Spencer Fullerton Baird, changed course, moving away from research to supervise the new National Museum Building, now the Arts and Industries Building, and putting his efforts toward increasing the institution's collections.

As the new nation expanded westward, the Smithsonian proved to be an ideal repository for such items as American-Indian artifacts and recently discovered plants. A large donation came from the US Exploring Expedition of 1838–42. Led by Lt Charles Wilkes, the group sailed around the world under orders from the US government to map uncharted lands, and study and collect animals and plants. During their four-year voyage they covered 87,000 miles (140,000 km), mapping 1,500 miles (2,400 km) of uncharted Antarctic coastline. The thousands of rare items given to the Smithsonian included an armadillo from South America, a Hawaiian human-hair necklace, and collections of rare gems, rocks, minerals and tropical plants.

The Smithsonian's "pack rat" mentality lives on. It is comprised of 16 museums, 9 of which are on the Mall, and has in its holdings some 140 million objects. ❏

RIGHT: James Smithson, whose fortune founded the Smithsonian.

THE MALL WEST

Reserved for remembrance and reflection, the grassy expanse between the museums on the Mall and the Potomac has become a place of national tribute for those who have served the nation

Map on page 133

S outh of Constitution Avenue and west of 14th Street to the banks of the Potomac sprawls an oasis of urban parkland, tree-shaded and watered, that comprises the westward extension of the Mall. Unofficially, it also takes in the **Ellipse** ⑫, the 52-acre (21-hectare) oval field south of the White House, where the national Christmas tree stands each year. Set apart from the museum-lined corridor across the way, this is sacred ground – the heart of monumental Washington.

One of the most enjoyable ways to explore this area, apart from walking, is on a bicycle. Miles of paths weave through the trees, around the memorials, and along the riverfront. You may find, though, that you prefer to drive or to take a **Tourmobile** shuttle bus (tel: 202-554-7950) which loops through the entire Mall area allowing visitors to get on and off between sites. Distances are farther away than they originally seem once you strike out on foot.

The Washington Monument

The prevailing vertical in the horizontal city is the **Washington Monument** ⑬ (daily Jun–Aug 8am–11.45 pm, the rest of the year daily 9am–4.45pm; closed Dec 25. Tickets are required, but are free at the kiosk or tel: 202-426-6841 or 1-800-967-2283 or visit www.nps.gov/nacc).

This is the essential memorial to the nation's first president, the District's quintessential symbol and landmark – and the invariable object of ribaldry. A few hundred feet west of the monument, you will find the Jefferson Pier marker, designating the monument's *intended* site at the intersection of the city's east-west and north-south axes – in line with the White House.

The spot proved to be too marshy (Constitution Avenue was at that time part of the C&O Canal), and so the monument had to be constructed on higher and drier ground. However, this threw the true east-west axis off by one degree to the south.

In 1783 the Continental Congress passed a resolution to erect a statue in honor of George Washington, though what they had in mind was something equestrian. Lack of funds delayed the project and finally in 1833 a group of private citizens raised $28,000. The cornerstone for architect Robert Mills's simpler and more elegant monument was laid July 4, 1848, but construction was delayed again, in part because of the intervening Civil War but also because of additional funding troubles that had the appearance of malfeasance. "A great many citizens of this country will never feel entirely comfortable until they know exactly what has been done with the cart-load of money

collected in behalf of Gen. George Washington's monument," wrote a journalist in the *Washington Post* in 1877. "The public ear is elevated to the proper angle in anticipation of a real nice explanation. Will some gentleman arise?"

Work resumed when blocks of stones were donated by several nations, states, and other organizations, inscribed with mottos and good wishes. One large stone donated by Pope Pius IX provoked the anti-Catholic contingent involved in the building, who smashed the stone and threw the pieces into the Potomac.

Ask a National Park Service ranger about the color change in the monument's shaft and you may be told that it is a high-water mark. Or that it is the height to which the monument is folded at night so planes don't crash into it. Actually, the color change marks the switch from Maryland to Massachusetts marble after construction resumed. When the discrepancy was noticed, building stopped again until a more compatible quarry was found.

Work started up again in 1880 when Congress came up with the needed funds to finish the job, and eight years later, the Washington Monument was finally compled. At 555 ft (169 meters), it was the tallest structure in the world until the Eiffel Tower was completed five years later. An elevator will take you in 70 seconds to the top, where from the observation deck you can have a panoramic view of the city. The monument is crowded during the day, especially in summer, so it's best to arrive early, or go at night when the view of the city is even more spectacular.

Down the slope and off to the side of the monument stands the small outdoor stage of the **Sylvan Theater** ⓮. On sultry summer nights it's a great spot to spread a picnic, share a bottle of wine, and enjoy a concert. Beginning at 8pm on every summer evening except Saturday, you can attend free concerts by the US Army Band (tel: 703-696-3399), "The President's Own" Marine Band (tel: 202-433-4011), the US Navy Band (tel: 202-433-2525) , and the US Air Force Band (tel: 202-767-5658), which perform on a rotating schedule.

Money for nothing

While in the vicinity, you might consider a detour down to the "new" **Bureau of Engraving and Printing** ⓯ at Raoul Wallenberg Place (formerly 15th Street) and C Street (tours Mon–Fri, 9am–2pm; closed federal holidays; tickets required but free with valid photo ID, and available at the 15th St. kiosk which opens at 8am; tel: 202-874-2330 or 1-800-967-2283). Until World War I, the "old" bureau was housed next door in what is now called the Auditor's Building.

All of the country's paper currency – about $20 billion a year – is produced here, along with treasury notes, postage stamps, White House invitations, and other printed matter. On a tour, you can watch the stuff roll off the presses. This is a popular family attraction and lines can be long.

A sobering experience

The bureau's neighbor at 100 Raoul Wallenberg Place is the privately funded **United States Holocaust Memorial**

LEFT: lining u to see dollar bills being produced at the Bureau o Engraving an Printing.

Map on page 133

Museum **16** (daily 10am–5.30pm; closed Dec 25 and Yom Kippur; tel: 202-488-0444 for information about "timed tickets" which may be needed for permanent exhibits, or get in line by 8am to receive a ticket for that day. Advance tickets may be obtained from tickets.com or by phoning 1-800-400-9373, but they carry a service charge. Website: www.ushmm.org). The museum is a sobering tribute to the 6 million Jews and 5 million other victims of the World War II Holocaust.

As its designers intended, this five-story brick and limestone simulated concentration camp with its rough gray surfaces and exposed beams and "watchtowers," delivers an emotionally charged message about one of the world's darkest times. The atmosphere is grim and even when crowded, the museum is unnaturally quiet. The story is told through documentary films, photographs, artifacts and oral histories. To add to the intensity, visitors are given a photo identification card of an actual Holocaust victim, someone of the same age and gender. Many exhibits focus on Jewish life just before the Holocaust and explain the political and military events leading up to it, but the most powerful are the archival films documenting executions, medical experiments performed on prisoners, and suicide victims.

Collections of personal articles such as shoes, glasses, and eating utensiles help to quantify and make tangible the loss of human life and are interspersed with an actual railroad car used to carry prisoners to the camps and a scale model of a gas chamber. One display is devoted to Raoul Wallenberg, a Swedish diplomat who was stationed in Budapest and who led the effort to save the Hungarian Jews.

Particularly moving is the **Hall of Remembrance**, a hexagonal, skylit spiritual space for reflection and contemplation, where visitors can light memorial candles and where a perpetual flame burns for the dead.

There is nothing remotely lighthearted about visiting this museum, nothing that can ease the horror of the subject matter, so plan accordingly, especially if you're traveling with young children.

Cherry blossoms

After this, you may be ready for some fresh air and the serenity of the **Tidal Basin 17** nearby. Not a monument but undeniably monumental, this free-form "lake" was created in 1897 to trap the overflow from the estuarial Potomac River and drain it into the Washington Channel. A photographic cliché when the cherry trees are in bloom, it is nevertheless one of the city's beauty spots, great for picnicking, paddleboating, or perambulating.

The **cherry blossom trees**, Washington's most distinctive horticultural image, originally came from Japan. According to one local correspondent who has researched their background, since the trees were planted around 1912, there have been some local difficulties. When President Franklin D. Roosevelt ordered ground broken for the Jefferson Memorial in 1938, bulldozers went into the cherry groves and found some local matrons chained to the trunks in protest. During World War II, there were spasmodic ax

attacks on this symbol of Imperial Japan, and such vandalism was made a federal offense. Occasionally, a hapless tourist trying to break off some blossoms is giving a stern warning of imprisonment by one of the vigilant Park Police.

Bill Anderson is the chief scientist for the National Park Service, and it is his annual duty to stroll through the groves to establish which is the day to announce peak blossoming, which he reckons at 70 percent in bloom on 70 percent of the trees. Mr Anderson explains that this is a tricky calculation because the Japanese cunningly sent two varieties of cherry tree. Roughly two-thirds are Yoshino, and these bloom an average 10 days earlier than the other variety, the Kwanzan. March 16, in 1990, was the earliest day ever recorded for peak blossoming. Nevertheless, Washington's Cherry Blossom Festival kicks off the first weekend in April, whether there are blossoms or not (check the *Washington Post* Friday "Weekend" section for information on scheduled events).

If you want to join the annual pilgrimage, park your car elsewhere and walk in, or use public transportation. A nocturnal stroll under the glowing trees, set off by stunning views of the monuments, is a magical experience, but it's best to exercise caution if strolling around once the sun has gone down.

The Jefferson Memorial

The Basin's shoreline path will lead you to the steps of the domed and graceful **Jefferson Memorial** ⓳ (daily 8am–11.45pm; closed Dec 25; tel: 202-426-6841 or visit www.nps.gov/nacc) arguably one of Washington's prettiest monuments, especially at night.

Designed by John Russell Pope, architect of the National Gallery of Art and the National Archives, and landscaped by Frederick Law Olmsted, Jr, the monument-in-the-round recalls the Roman Pantheon, a nod to Jefferson's own fondness for that ancient structure and similar to the president's designs for his own home, Monticello.

Built on the soft site of a dredged marsh which characterized much of the Mall, the memorial, completed in 1943, needed a foundation of concrete-filled steel cylinders sunk 135 ft (41 meters) down to bedrock. The domed building is ringed by 54 Ionic columns.

Not everyone loved the monument. Critics decried its form as too "feminine," or as a "cage for Jefferson's statue," a 19-ft-high (5.8-meter) bronze sculpted by Rudolph Evans. The memorial marks the southern point of the Mall's north-south axis, with the White House forming the counterpoint.

To the south is **East Potomac Park**, a peninsula that dangles between the Potomac River and the Washington Channel, and which forms the southern half of a 700-acre (280-hectare) riverside park. The views are superb, and you can watch planes take off and touch down across the river at Virginia's Reagan National Airport. This is also a good place to enjoy the springtime cherry blossom trees, where there are fewer crowds than around the Tidal Basin.

You'll also come to the new **George**

LEFT: cherry blossoms overlooking the Jefferson Memorial.

Map on page 133

Mason Memorial ⑲ (tel: 703-550-9220 or visit www.gunstonhall.org), dedicated in 2002 to the memory of Virginian and patriot George Mason whose Virginia Declaration was the basis for the Bill of Rights. Mason refused to sign the original constitution, drawn up in 1787, because it did not guarantee freedom of the press and of religion, and freedom from unreasonable searches and the right to a fair and speedy trial.

Thanks to his efforts, the Bill of Rights was finally added in 1791 and form the basis of American democracy and political freedom. Mason's restored plantation home, Gunston Hall, is in Virginia, about 20 miles (32 km) south of DC, and is open for tours *(see page 248)*.

The peninsula's tip, **Hains Point**, holds a wonderful surprise: a giant figure breaking through the earth. The sculpture, *The Awakening* by J. Seward Johnson, Jr, was installed in 1980 as part of a citywide sculpture show. Climbing on the sculpture is encouraged by the artist.

The FDR Memorial

Stay a northwestern course into **West Potomac Park** and you'll reach the **Franklin Delano Roosevelt Memorial ⑳** (daily 8am–11.45pm; closed Dec 25; tel: 202-426-6841; www.nps.gov/nacc). The memorial was opened in 1997, although there already existed a monument to Roosevelt, a stone slab on Pennsylvania Avenue in front of the National Archives that was, according to wishes he expressed to his friend and Supreme Court Justice Felix Frankfurter, no bigger than his desk. This one, designed by Lawrence Halprin, is much grander.

Spread over 7½ acres (3 hectares), this memorial to the nation's 32nd president, who helped lift the country out of the Depression and guided it through World War II, combines sculpture and natural landscape. Soft red South Dakota granite walls constitute four open-air rooms, each signifying one of Roosevelt's four terms of office. Shade trees and waterfalls are interspersed, with quotes and carvings on

ELOW: the DR Memorial.

THEY (WHO) SEEK TO ESTABLISH SYSTEMS OF GOVERNMENT BASED ON THE REGIMENTATION OF ALL HUMAN BEINGS BY A HANDFUL OF INDIVIDUAL RULERS. CALL THIS A NEW ORDER. IT IS NOT NEW AND IT IS NOT ORDER.

the walls. A 9-ft. (2.7-meter) bronze statue shows Roosevelt seated with Fala, his faithful Scotch terrier, by his side.

The Lincoln Memorial

Continue further into the park and you'll reach the incomparable shrine to Abraham Lincoln: a neoclassical temple of gleaming, white marble cresting its acropolis. Familiar as its image, which graces the back of the penny and the $5 bill, Henry Bacon's design for a **Lincoln Memorial** ㉑ (open daily 8am–11.45pm; closed Dec 25, tel: 202-426-6841 or visit www.nps.gov/nacc) was selected over John Russell Pope's and was dedicated on Memorial Day in 1922. It was a sad irony that day that a key speaker, the black president of Tuskegee Institute, was ushered away from the speaker's platform and seated in the segregated black section of the audience across the road.

On Easter Sunday 1939 operatic singer Marian Anderson sang from the steps for an audience of 75,000, having been denied permission to perform at DAR Constitution Hall because she was black. The First Lady, Eleanor Roosevelt, arranged for the outdoor concert, then resigned her own DAR membership.

Similar to Athens' Parthenon, the building is fronted by a colonnade of 36 Doric columns, the exact number of American states in the Union at the time of Lincoln's death. The columns slope inward to avoid looking out of proportion. Above the columns is a frieze with the names of these 36 states; higher still are the names of the 48 states that were in the Union at the time the monument was dedicated.

Daniel Chester French designed the famous figure of Lincoln, the tallest president, seated inside the memorial. If this marble Lincoln were to stand up, he would measure 28 ft (8.5 meters). When it was discovered that Lincoln's naturally-lit face was nearly obscured, General Electric was called in to install special artificial lighting to create the shadows on his hair, brows, cheeks, and chin. Adding

BELOW: listening to Abe inside the Lincoln Memorial.

The man

who gave

meaning,

honor, and

purpose

to the Nation

speaks to us still.

Map on page 133

a personal touch to his masterpiece, French, who had a deaf son and who knew American sign language, so shaped Lincoln's hands so that the left holds the sign for "A," and the right "L."

From the memorial, you can savor Lincoln's magnificent view across the **Reflecting Pool**, the elegant, 2,000-ft (600-meter) waterway and promenade inspired by Versailles and the Taj Mahal, to the Washington Monument and beyond to the Capitol. Attractive at any time, the scene is most dramatic in the light of dawn or at dusk.

War memorials

To the south of the Reflecting Pool is the **Korean War Veterans Memorial** ㉒, (daily 8am–11.45pm; closed Dec 25, tel: 202-426-6841; www.nps.gov/nacc) dedicated to those who served in the conflict which, although it lasted from 1950 until 1953, was never officially declared a war. Regardless, 1½ million Americans served, with more than 54,000 being killed,

BELOW: the Korean War Memorial.

110,000 captured or wounded, and another 8,000 declared missing. The site features an American flag at the point of a triangle which thrusts into a circular pool. Nineteen statues of soldiers, sculpted by Frank Gaylord and a mural with many faces, the work of Louis Nelson, are also featured.

Sandwiched between the Reflecting Pool and Constitution Avenue is the green swath known as **Constitution Gardens**. After the last of the temporary wartime structures was demolished in 1966, President Richard Nixon proposed that the area be developed as a Disney-esque amusement park. The idea was soundly vetoed. Instead, you will find a tranquil setting of shady trees and bench-lined paths which meander alongside an artificial lake. A lovely little island in the lake contains a memorial to the signers of the Declaration of Independence.

The most subtle and emotionally powerful of the Mall's monuments, the **Vietnam Veterans Memorial** ㉓, (daily 8am–11.45pm; closed Dec 25; tel: 202-

426-6841 or visit www.nps.gov/nacc) stands just a short distance away through the trees to the west of the lake. It was a Vietnam vet who proposed the idea of a memorial, which Congress authorized in 1980. Jan Scruggs came up with the idea for the memorial after he saw Michael Cimino's 1978 movie *The Deer Hunter*, which addressed the struggles returning vets had faced. To raise funds, Scruggs founded the Vietnam Veterans Memorial Fund which he still operates from a downtown office, raising funds for upkeep and overseeing an educational program.

Selected from nearly 1,500 entries in an open design competition, the contemplative V-shaped wall of names – 58,229 casualties – in polished black granite was designed by Maya Lin, a 21-year-old Yale University architecture student and the daughter of Chinese immigrants, who explained: "The design was purposefully simple because embellishment upon something so moving [as the war] would only diminish its effect."

To assuage the critics of Lin's abstract "rift in the earth," a representational statue of three soldiers by sculpture Frederick Hart and a memorial flagpole were erected. The names are inscribed on the wall in chronological order in which the soldiers died. At the two entrance ramps leading down into the memorial, visitors can locate the names of the deceased alphabetically and find the exact location of each soldier's name on the wall. In the 20 years since the memorial was completed, 40 million Americans have come to the wall to find solace and closure, making it the country's most popular memorial.

Veterans' groups have set up makeshift shelters and tents near the memorial which are used as 24-hour vigil sites. In addition to selling mementoes and advertising their political grievances, the groups and shelters are there for the vets themselves, who come to reclaim the wall in the privacy of night. By day the wall belongs to the public – who file reverently by to gaze, to discover their own reflections, to touch a name, to leave flowers or a poem, or to make a wall rubbing.

Nearby is another Vietnam War memorial, this one to the 265,000 women who served either in the military or as volunteers with such organizations as the USO and the Red Cross. **The Vietnam Women's Memorial** ❷ was dedicated in 1993 and features a 7-ft (2.1-meter) high bronze statue of three women tending a wounded soldier – most of the women who served were nurses.

Memorials to come

With all the memorials that have been erected on the Mall lately in commemoration of those who served in America's recent wars, the Iwo Jima Memorial in Arlington, which pays tribute to those who served in World War II, no longer seems sufficient. In 2001 Congress approved a 7.4-acre (3-hectare) site between the Lincoln and Washington memorials for the future **National World War II Memorial**. A 4-acre (1.6-hectare) site on the northwest edge of the Tidal Basin will be the new home of the **Martin Luther King, Jr. National Memorial**. ❑

Map
on page
133

LEFT:
the Vietnam
Women's
Memorial.

A monumental city

At last count, Washington had 750 monuments, with more on the way, including one in the works to honor civil rights leader Martin Luther King, Jr. and another to commemorate those who served in World War II. Besides the memorials everyone visits – Washington, Jefferson, Lincoln and the Vietnam War Memorial – there are other lesser known, even obscure monuments.

Take, for instance, the monument to Taras Shevchenko, a 19th-century Ukrainian serf turned poet and revolutionary, whose enormous bronze statue on P Street between 22nd and 23rd was erected in 1964 as a Cold War-era snub to the Soviets, for whom Shevchenko's most famous work *The Bard* was a threat. The Lebanese poet Khalil Gibran, best known for *The Prophet*, is remembered with a statue at Normanstone Park near Observatory Circle. There's a memorial to Maine lobstermen on Water Street, one to the Boy Scouts on the Ellipse, and one to Joan of Arc on Meridian Hill.

A former French ambassador, Jean Adrien Antoine Jules Jusserand, has a memorial in Rock Creek Park. A birdwatching buddy of President Theodore Roosevelt's, the ambassador was also a fellow skinny-dipper in the Potomac, although – unlike T.R. – the more modest Jusserand, as the story goes, was always careful to wear his gloves so as not, technically, to be caught naked.

The craze for monuments began with Congress in 1815 when it decreed that the grave of every senator or representative buried in Congressional Cemetery be marked with a shrine. Sixty years and 200 monuments later, the practice was stopped as space ran out.

One of the more striking monuments is in Rock Creek Cemetery (off North Capitol Street near Rock Creek Church Road, not to be confused with Rock Creek Park on the other side of town). The bronze statue of a young woman by Augustus Saint-Gaudens was commissioned in 1890 by writer Henry Adams as a memorial to his wife Clover, a suicide. The figure is cloaked and seated on a rough stone and evokes such a sense of remorse that among Washingtonians, even though the memorial bears no inscription or title, she has always been known simply as Grief.

It's hard to miss the rather glorious Columbus Memorial Fountain by Lorado Taft that's been in front of Union Station since 1908, but inside, in the West Hall, is a lesser known memorial to William Frederick Allen, single-handedly responsible for making the trains run on time. In the 1880s, the country had 50 different railroad time schedules, none coordinated, all confusing, and sometimes the cause of tragic accidents when trains simultaneously found themselves on the same tracks. To alleviate the problem, Allen divided the country into four time zones, a system that since 1883 has kept more than just the railroads running like clockwork. ❑

RIGHT: Gandhi statue outside the Indian Embassy.

Map on page 160

CAPITOL HILL AREA

With two of the three branches of the federal government headquartered here, Capitol Hill's few short blocks form the critical mass of Washington's political scene

From the long green sweep of the Mall, the **US Capitol** ❶ floats like a white mirage above the city. Icon of the federal republic and symbol of Washington, the Capitol has been the center of the city's political life since 1800 when the first joint session of Congress was called to order. Today, nearly 20,000 Congressional staff members, hordes of lobbyists and camera-carrying and note-book-toting members of the media swarm like bees around the Capitol hive.

The building is open Mon–Sat, 9am–4.30pm; closed Sundays, Thanksgiving and Dec 25, but open federal holidays. Tours are by free timed tickets only from the West Front facing the Mall at the Capitol Guide Service kiosk near the Garfield traffic circle at Independence Avenue and First Street; tickets are distributed from 8.15am on a first-come first-served basis (tel: 202-225-6827 or visit www.aco.gov). Senate and House chambers are not included.

Capitol classic

The original land patent for what would become Capitol Hill was granted in 1663 to a George Thompson and eventually passed to the prominent Carroll family of Maryland. Known then as Jenkins Hill, this 500-acre (200-hectare) plot above the Potomac was renamed Federal Hill by city architect Pierre L'Enfant who described it in a letter to George Washington as "a pedestal waiting for a superstructure." The superstructure L'Enfant had in mind was the Capitol, which he envisioned as the center of a thriving neighborhood of elegant shops and homes, though absurdly high real estate prices thwarted development. Even George Washington lost his original investment on a pair of townhouses he had built near the Capitol. Only since the 20th century has this section of the city really come into its own, displaying the elegance L'Enfant intended.

He envisioned the Capitol as the geographical centre of the city, the heart of the four equally developed quadrants, but the city developed primarily westward instead. Although the Capitol itself is surrounded by beautiful grounds and massive marble buildings, this grandeur doesn't extend much beyond Lincoln Park. The Capitol Hill neighborhood includes many of Washington's most important sites and a small historic district of townhouses located east of North and South Capitol streets and behind the Capitol. Here, you'll also find clusters of restaurants and bars – and more of the city's lower-income housing. Caution is advised, especially at night, since these neighborhoods between Lincoln Park and the Anacostia River have

PRECEDING PAGES: the Folger Library, Capitol Hill. **LEFT:** the Great Welcoming Hall of the Library of Congress. **RIGHT:** the Capitol took 60 years to complete.

contributed significantly to DC's high rate of violent crime.

A new **Capitol Visitors Center** will make it much easier to accommodate the crowds, which can exceed 18,000 curious tourists in a day. The three-story underground Center, with exhibit space, concessions and restrooms and convenient access to the Capitol, is scheduled to open in mid-2005.

Design by committee

Dr William Thornton, an amateur architect and physician from the Virgin Islands, designed the original US Capitol after, as he put it, he "got some books and worked a few days." Nevertheless, Washington, who had chosen Thornton's plan from among the 16 submitted in a design competition for the Capitol, praised the plan for its "grandeur" and "simplicity." Construction commenced when Washington, using a silver trowel and a marble-headed gavel in the traditional Masonic style, laid the cornerstone in 1793. Thornton's casual attitude, however, soon cast doubt on his ability to carry through with the project and he was replaced by the runner-up in the competition whose egotism soon got the better of him. James Hoban, who won

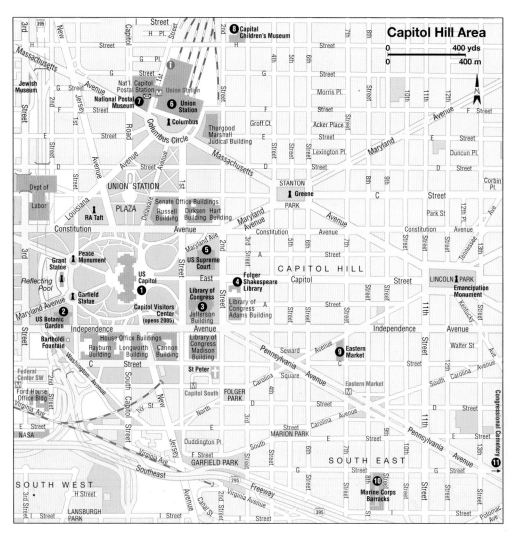

Map on facing page

the competition to build the president's house, was called in to save the day, and by November 1800 enough of the work was complete to enable Congress to meet in the north wing, the original Senate chamber, for the first time.

Jefferson then called in his friend Benjamin Henry Latrobe, a professional and by this time the Capitol's fourth architect, to supervise construction of the south wing, which was the House's original chamber. Except for the U-shaped gap between the two wings, the Capitol was more or less complete – or so everyone thought. On August 24, 1814, British Admiral George Cockburn torched "this harbor of Yankee democracy," leaving it a ruin. Latrobe once again set to work to restore the Capitol according to Thornton's plan while Congress carried on its work across the street on the site that the Supreme Court now occupies.

BELOW: The Capitol in 1850.

Temperamental and outspoken, Latrobe eventually fell victim to politics and his own indiscretions and was forced to resign his post. He was replaced by Charles Bulfinch, a New England architect noted not only for his brilliance but also for his tact and even-handedness in dealing with difficult congressmen who saw the Capitol not as a place to conduct the nation's business but as a monument to their own egos. "Architects expect criticism," the self-effacing Bulfinch wrote, "and must learn to bear it patiently."

He managed to link the two wings that Thornton and Latrobe had built before abandoning the project and was responsible for adding a 55-ft (17-meter) high dome, the Capitol's first.

Bulfinch also finished the **old House Chamber**, which was in use until 1857 when it was vacated for larger quarters and eventually became **Statuary Hall** where each state has on display a statue commemorating favorite sons and daughters. (The story goes that the Missouri delegation would have included Mark Twain but decided against it after the famous author wisecracked, "Suppose you were

an idiot. And suppose you were a member of Congress. But I repeat myself.")

By 1850, Congress had clearly outgrown the Thornton-Latrobe-Bulfinch building and ordered an expansion, which included a bigger dome. This dome, the one you see today, is the work of Philadelphia architect Thomas U. Walter and consists of two trussed cast-iron shells, one superimposed upon the other, and painted to resemble marble.

Oddly enough, the Civil War, which should have slowed construction, didn't, even though Union soldiers (who referred to it as the "Big Tent") were bivouacked under the dome. Lincoln insisted that construction on the Capitol continue on the theory that "if people see the Capitol going on… it is a sign we intend the Union shall go on."

In 1863, Thomas Crawford's 19-ft (5.8-meter) bronze **Statue of Freedom** was lifted into place atop the dome. The statue features a robed woman, her right hand on the sheath of her sword, her left holding a laurel wreath and a shield with 13 stripes, one for each of the original colonies.

Spectacular view

Inside, the dome forms the Capitol's **Great Rotunda**, 180 ft (55 meters) high, where painter Constantino Brumidi's *Apotheosis of Washington* depicts George Washington and other colonial statesmen mingling with a bevy of loosely robed allegorical figures. Brumidi, born in Rome in 1805, worked at the Vatican before arriving in the US in 1852. "My one ambition," he wrote, "is that I may live long enough to make beautiful the Capitol of the one country on earth in which there is liberty."

Unfortunately, he didn't live long enough to see his wish fulfilled. At the age of 72, as he was working on the ceiling, his chair slipped from the scaffolding and he was left dangling 60 ft (18 meters) above the floor until he could be rescued. This proved too much of a strain, and a few months later he died. The grisaille frieze around the base of the dome, begun

BELOW: touring the Capitol.

Map on page 160

by Brumidi, runs a total length of 300 ft (90 meters) and depicts 400 years of American history. It was finally completed in 1953 by American artist Allyn Cox. A little closer to eye level, you'll notice the Rotunda walls, decorated with paintings by John Trumbull, an aide to General Washington during the Revolutionary War.

You can also visit **Statuary Hall**, whose semi-circular and half-domed shape amplifies the slightest sound, an unintended consequence of the original design and a feature that many of the members of the House of Representatives who met here originally objected to for obvious political reasons. In the **Hall of Columns** you'll find the overflow of works from Statuary Hall. Running parallel to that is the **Hall of Capitols**, decorated with murals by Allyn Cox which depict historical and everyday moments in American life.

The handsome **Old Senate Chamber** features a Rembrandt Peale portrait of George Washington. The **Old Supreme Court Chamber** was used by the court

RIGHT: the Capitol's Rotunda is 180ft (55 meters) high.

between 1810 and 1860 and contains many original furnishings.

Beneath the Rotunda is the **Crypt**, intended originally to enshrine the body of George Washington. His survivors insisted instead that he be laid to rest at Mt Vernon. The crypt also features a landmark embedded in the floor marking the zero point from which the city's quadrants were drawn. There is also on display a bust of Lincoln by Mt Rushmore sculptor Gutzon Borglum who chose to leave off the president's left ear, this to symbolize Lincoln's unfinished life.

On either side of the Capitol are the buildings that house the offices and committee rooms where the everyday work of Congress takes place. On the House side of the Capitol, running along Independence Avenue, are the Cannon, Longworth and Rayburn buildings. The Senate office buildings, the Russell, Dirksen, and Hart buildings, are on the north side of the Capitol along Constitution Avenue.

On the Capitol's **West Terrace**, free

GRANT'S STATUE

Look for the statue of General Ulysses S. Grant astride his horse. It's outside at the base of the Capitol. As equestrian statues go, it's the second largest in the world (only Victor Emmanuel's, the first king of modern Italy, is bigger), and the work of the little known sculptor Henry Shrady, a law school drop-out and failure in the match stick business whose father was one of Grant's physicians. Sited directly in line with the Lincoln Memorial, it was intended to anchor the east end of the Mall. Unfortunately, Grant, big as he is (40 ft/12 meters high), is no competition for the even bigger Capitol building.

In 1885 Grant's funeral drew a million mourners who turned out to pay homage to the general who, as commander in chief of the Union's forces, quite literally saved the country. He served two terms in the White House and despite the scandals that marred his presidency, his popularity was still holding steady when in 1922, after 21 years of work, Shrady's statue was finally ready to be unveiled. That April day, federal offices were closed and cheering crowds lined the streets as units from the army, navy, and marines ceremoniously marched from the White House to the statue. Yet, these days, with the Capitol looming behind it, the General is seldom noticed.

summer concerts are offered by the armed services bands on many evenings beginning at 8pm. (Check the listings in the *Washington Post* or phone the US Army Band (tel: 703-696-3399), "The President's Own" Marine Band (tel: 202-433-4011), the US Navy Band (tel: 202-433-2525) , and the US Air Force Band (tel: 202-767-5658) for performance schedules.)

There is no more quintessential Washington entertainment than to sit on the Capitol steps, tap your feet to a Sousa march, and watch the sun set behind the Washington Monument.

Take time to stroll the **Capitol grounds** – a former mud flat filled with alders and transformed into a shady parkland by the landscape architect Frederick Law Olmsted, Jr, in 1874. Now encompassing 200 acres (80 hectares), the grounds include 5,000 trees, some planted in the late 1800s, and in spring a magnificent display of red and yellow tulips and jonquils.

For more of the lush life, visit the **US Botanic Garden ❷** (First Street and

Maryland Avenue; entrance on Maryland Avenue; open daily 10am–5pm; free admittance; tel: 202-225-8333) on the southwest side of the Capitol. The gardens were originally established to hold the specimen plants brought back from South America and the Pacific by the congressionally mandated Wilkes Expedition in 1842. A conservatory was erected in 1933 as the collection expanded and has recently been renovated to include a huge glass atrium packed with tropical plants and flanked by two smaller pavilions.

The gardens make a great stop on cold, rainy days and during special seasonal shows. Across the street, in a charming pocket park (open daily, dawn to dusk), is the city's graceful bronze **Bartholdi Fountain**, designed by Frederic August Bartholdi (of Statue of Liberty fame) and installed here in 1877.

The Library of Congress

The favorite room of many a Washingtonian is the Main Reading Room at the **Library of Congress ❸**, located inside the stunning **Thomas Jefferson Building** just behind the Capitol (open Mon–Sat, 10am–5.30pm; closed Sunday and federal holidays; free guided tours of the Jefferson Building by ticket Mon–Sat, 10.30am, 11.30am, 2.30pm, 3.30pm from the Visitors Center at the First Street entrance; tel: 202-707-5000 or 202-707-9779 or visit www.loc.gov).

It takes dedicated readers to concentrate on their research in this room with its 160-ft (49-meter) high domed ceiling, clusters of richly veined marble columns, and its allegorical murals that chronicle the progress of civilization and human understanding.

The original library was housed in the Capitol and intended only for the legislators. When the library was destroyed in the fire of 1814, Thomas Jefferson offered his own 6,500 volumes to Congress, which they bought for $24,000, roughly half the price he would have received at auction. Unfortunately, most of that collection was lost in another fire, this one in 1851. Congress once again allotted money for the rebuilding of the Library, a Beaux-

LEFT: the Bartholdi Fountain.

Map on page 160

Arts gem and probably the city's most beautiful public building, opened in 1897. As a way to increase its holdings, the Library gradually opened its doors to the public and now has more than 150 million books, rare manuscripts, maps, musical scores, recordings and musical instruments, films, photographs, and other materials. Among its treasures are a Gutenberg Bible, a rough draft of Jefferson's Declaration of Independence, and Abraham Lincoln's Gettysburg Address.

Besides the suburban warehouses it operates to house all this material, the Library complex also includes two other buildings, the **John Adams Building**, behind the main library, and the **James Madison Memorial Building**, across the street, which are staff offices.

At the foot of the Jefferson Building, by the entrance, you'll notice the **Neptune Fountain** by Roland Hinton Perry. Inside is a truly impressive two-story white marble **Great Hall** decorated with murals and sculptures and gold leaf. A grand staircase leads to the second-floor gallery where changing exhibits showcase some of the Library's vast collection.

The Library of Congress also offers concerts by noted musicians, lectures by historians and important public figures, readings by some of the country's best fiction writers and poets (it appoints the nation's Poet Laureate), and other events, all of them free. Check at the Visitors Center or phone the main information line for a schedule.

Folger Shakespeare Library

The austere Art Deco exterior of the nearby **Folger Shakespeare Library** ❹ (201 East Capitol Street, SE; Mon–Sat, 10am–4pm; closed federal holidays; tel: 202-544-4600, or visit www.folder.edu) belies the cozy, Elizabethan-style interior of this great library, museum, and center of literary and performing arts. Previously the site of 14 overpriced townhouses erected in 1871 that never sold and then fell into disrepair, the Folger was built by Standard Oil president

Henry C. Folger who bought the land in 1928 with the idea of building a library to hold his expansive collection.

The Folger, which opened in 1932, houses the world's largest collection of Shakespeare's printed works, including a set of precious First Folios from 1623. The Elizabethan and Renaissance worlds are brought to life in the **Great Hall** where many of the Folger's treasures are displayed. It's easy to imagine this hall full of mead-drinking, boar-eating revelers feasting at sturdy banquet tables.

The intimate Elizabethan-style **Folger Theater** presents performances of medieval and Renaissance music, and other educational programs. The handsome **Reading Room** is open to the public during events such as the Poetry Series, the PEN/Faulkner Fiction Readings, and Shakespeare's birthday celebration. Herbs and flowers grown in Shakespeare's time are planted in the secluded Elizabethan garden, a place to contemplate "nature's infinite book of secrecy."

The Supreme Court

"Oyez! Oyez! Oyez!" With this dramatic cry the **US Supreme Court** ❺ is brought to order (open Mon–Fri, 9am–4.30pm; closed major holidays; sessions open on a first-come first-served basis; lectures and guided tours are offered when the Court is not in session; tel: 202-479-3211 or visit www.supremecourt.gov). This imposing marble edifice at 2nd and East Capitol streets wasn't opened until 1935 and, with just nine members, houses the smallest branch of the federal government.

The Court was established by the US Constitution for the purpose of balancing the federal and legislative branches of government and to act as court of last appeal. It met for the first time in 1790 in New York, and for its first 145 years had no permanent home. Taverns, hotels, private homes, and the Capitol itself served as a gathering place for the justices. Finally, in 1928, William Howard Taft, the only president to also serve as a justice on the court, convinced Congress to allocate

LEFT: researching in the Folger Library. **BELOW:** the Supreme Court.

Map on page 160

funds for the building, which was designed by architect Cass Gilbert. A broad staircase flanked by the *Contemplation of Justice* and the *Authority of Law*, both sculpted by James Fraser, leads up to this mammoth marble building. Carved into the pediment are the words "Equal Justice Under the Law."

In interpreting the Constitution, the Court has effectively shaped the course of American democracy by ruling in such pivotal cases as Brown v. the Board of Education of Topeka, which in 1954 ended school segregation. In the 1966 case of Miranda v. Arizona, the Court ruled that arresting police officers must inform suspects of their legal right to counsel. In 1973, the Court ruled in favor of making abortion legal in the case of Roe v. Wade.

Other historic cases the Court has heard include Marbury v. Madison, which in 1790 established the Court's ability to declare certain acts of Congress unconstitutional, and the 1857 Dred Scott decision in which the Court declared that Congress had no right to limit the spread of slavery and that slaves were chattel and not citizens and therefore not entitled to protection under the law. More recently, the outcome of the 2000 presidential election was controversially decided by the Court when it ruled in favor of the constitutionality of the electoral procedures.

From the first Monday in October through April, the court hears oral arguments. Of the 7,000-plus requests it receives each year, it actually only hears between 75 and 100. Most case arguments are scheduled for just one hour, each attorney having half an hour to present the case. The justices then withdraw to deliberate. In May and June it delivers its opinions.

RIGHT: the doors of the Supreme Court building.

Union Station

North of the Supreme Court on Massachusetts Avenue is one of the city's great successes of architectural preservation: **Union Station ❻** (50 Massachusetts Avenue, NE; tel: 202-371-9441). This glo-

ABOUT THE JUSTICES

During the Supreme Court's first two centuries, only 107 persons, nominated by the President and confirmed by the Senate, have served as justices. Of these, two are women (Sandra Day O'Connor and Ruth Bader Ginsburg), and two African-American (the late Thurgood Marshall and Clarence Thomas). All but 14 have been Protestants, most have been educated at Harvard, though one (Stanley F. Reed who served from 1938 to 1957) had no law degree at all. Justices have come from 31 of the 50 states, most often from New York, which has sent 16 justices. In 1967, President Lyndon Johnson nominated the first black justice, Thurgood Marshall, a noted civil rights lawyer who won quick Senate confirmation. The same wasn't true for the second black justice, Clarence Thomas, a George Bush, Sr. conservative nominated in 1991 who was humiliated when former colleague Anita Hill testified to his sexual harassment of her. President Ronald Reagan nominated the first woman justice, Sandra Day O'Connor, in 1981. An Arizona judge, she won unanimous Senate approval, although when she began her career in 1952 she couldn't find a job. In 1993, President Bill Clinton nominated Ruth Bader Ginsburg, the Court's second woman and the first Democratic appointee in over 35 years.

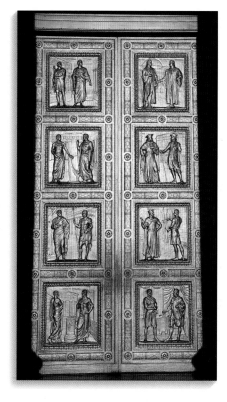

rious Beaux-Arts train station was built in 1908 by architect Daniel Burnham and when it opened was the largest train station in the world at 750 ft by 344 ft (229 by 105 meters). With the decline of train travel, it had fallen into serious disrepair by the 1970s. In 1981, private and public funds provided $160 million for a rescue job. In 1988 Union Station's former grandeur and the romance of the rail were restored. Today, 40,000 visitors a day pass through Union Station's doors.

The entrance features a grand triple-arched portico modeled after the Arch of Constantine. In the Main Hall – inspired by the Roman Baths of Diocletian – Amtrak passengers, Metrorail riders (there's a Metro stop here), lunching Capitol Hill staffers, shoppers and visitors surge and flow in perpetual motion. Several cafés and first-class restaurants, including **America** (tel: 202-682-9555), plus a 9-screen movie theater, and dozens of shops and fast-food stands keep the station lively day and night.

East of the station, around North Capitol and F streets, you'll find an Irish flavor and a sprinkling of politicians at **The Dubliner** (520 North Capitol Street, NW; tel: 202-737-3773). Alternatives are the **Hawk and Dove** (329 Pennsylvania Avenue, SE, tel: 202-543-3300), a long-time favorite among Capitol Hill staffers, and the upscale **La Colline** (400 North Capitol Street, NW; tel: 202-737-0400), the powerbrokers' choice.

Next to Union Station is another stately Beaux-Arts building, the old City Post Office building, now home to the **National Postal Museum ❼** (Massachusetts Avenue and First Street, NW; open daily 10am–5.30pm; closed Dec 25; free admittance; guided tours available daily at 11am and 1pm; tel: 202-357-2991 or visit www.si.edu). The museum, operated by the Smithsonian Institution, was opened in 1993 to house an interesting but otherwise worthless collection of Confederate postage stamps and has since grown to become the world's largest philatelic col-

BELOW: inaurugal balls are held in Union Station's marbled hall.

Map on page 160

lection. The museum charts America's mail service from its Pony Express days, and features three vintage mail planes suspended from the 90-ft (27-meter) glass atrium ceiling, a walk-through railroad car, and six galleries with changing exhibits.

A museum for children

Behind Union Station at 800 3rd Street is the **Capital Children's Museum 8** (open late May–early Sept daily 10am–5pm; remainder of the year Tues– Sun, 10am–6pm; closed Thanksgiving, Dec 25, New Year's Day; admission fee; tel: 202-675-4120 or visit www.ccm.org). The exhibits here are designed to let children learn through experiences as diverse as play-riding a bus or taxi or learning about the Mexican culture by making tortillas and hot chocolate. The museum's greatest accidental PR event is staged near closing time when parents begin coaxing their children away from the museum's captivating gizmos and gadgets.

Kids can learn Morse code and how to type on a Braille typewriter, print a poster on a Ben Franklin printing press, and star in a cartoon. Experience is the theme at this museum whose motto is a Chinese proverb: "I see and I forget. I hear and I remember. I do and I understand."

Eastern Market

Once, Washington had several fresh food emporiums, but **Eastern Market 9** at 7th and C streets, SE is the only survivor and the unofficial center of the neighborhoods surrounding the Capitol. This block-long building dates from 1873 and houses vendors selling fresh produce, meat, poultry, seafood, and cheeses every day but Sunday. Prices aren't exactly cheap, but the old-market atmosphere and the quality of the goods are worth every penny.

Get in line at **Market Lunch** (open Tues–Sat, 7.30am–3pm; Sun 11am–3.30; tel: 202-547-8444), famous for its hearty breakfasts, crab cakes, and oyster sandwiches. On Saturdays and Sundays, vendors set up shop outside and sell baked goods, flowers, antiques, clothing, jewelry, and a variety of junk.

On the south side of Pennsylvania is the

Marine Corps Barracks 10 (tel: 202-433-6060) at 8th and I streets, home of the "Eighth and Eye Marines." Make reservations well in advance for the Friday evening Marine Corp parade held here in summer. With unmatched precision and patriotism, the Band, the Drum, and the Bugle Corps as well as the silent drill team assemble within the quadrangle for a ceremony that makes you proud to be American, even if you're not. The sight of a solitary, spotlighted marine playing "Taps" from the parapet of the main tower sends chills up some spines and brings tears to some eyes.

Ten blocks east of the barracks is **Congressional Cemetery 11**, at 1801 E Street. Opened in 1807, these burial grounds include the graves of senators, diplomats, prominent members of Congress as well as Capitol architect William Thornton, "March King" John Philip Sousa, Civil War photographer Matthew Brady, FBI director J. Edgar Hoover, and Choctaw chief Pusha-ma-ta-ha. ❑

RIGHT:
Eastern
Market.

Map on page 174

DOWNTOWN

Washington's revitalized center city is a mix of historic homes and museums, theaters, parks, trendy restaurants, classic hotels and office buildings where the work of the federal government is carried out

Downtown Washington is as much a concept as a locale. Geographically, it stretches roughly between Foggy Bottom and Chinatown, runs from the White House north toward P Street, and includes much of Pennsylvania Avenue.

Revitalized after long neglect, downtown burns with power, glitters with wealth, and oozes with culture. It's power-lunching at restaurants such as **The Palm** (1225 19th Street, NW; tel: 202-293-9091) and **The Caucus Room** (401 9th Street, NW; tel: 202-393-1300), and dealing with panhandlers on the street corners. It's K Street lawyers, Pennsylvania Avenue bureaucrats, and a new young breed of moneyed professionals taking up residence in the new Pennsylvania Quarter along the 7th Street corridor, fresh with pricey condos, shops, and the Shakespeare Theatre *(see page 179)*. It's limo stretched curbside and yuppies at their happy-hour meccas. Side by side and overlapped, it's all downtown.

A walk in the park

The best place to start is **Lafayette Square ❶**, directly across Pennsylvania Avenue from the White House. Thanks to that populist Thomas Jefferson, this once-private presidential park is enjoyed by chess players, office workers on their lunch breaks, tourists, and, alas, the homeless, whom you'll notice camped out with their shopping carts and placards. Most are harmless, though they may seem intimidating. If you want to help, see the sidebar on page 48.

Contrary to popular belief, Lafayette Square is not named for the French major-general whose statue you'll see tucked into the southeast corner and who served in Washington's army during the Revolutionary War, but rather for the day itself when in 1824 the Marquis visited Washington and the adoring crowds overflowed into the park just to get a glimpse.

The main sculptural feature of the square is actually the statue of General Andrew Jackson – the first equestrian statue, by the way, erected in America. As a tribute to the General, Congress resolved that the statue include metal from the brass guns which Jackson captured when he defeated the British in the War of 1812 at the Battle of Pensacola, in Florida, and then finally in New Orleans in 1815.

Among the landmarks ringing the park and facing the **Hay-Adams Hotel** (One Lafayette Square; tel: 202-638-6600) on the corner of H and 16th streets is the pretty yellow-and-white **St John's Church ❷** (Mon–Sat, 9am–3pm; Sun 8am–3pm; tel: 202-347-8766) with its adjacent Federal-style **Parish House**.

PRECEDING PAGES: the Great Hall of the National Building Museum. **LEFT:** the Willard Hotel. **RIGHT:** exuberant staff at The Palm.

Completed in 1816, the socially-correct Episcopal church has been visited by every US president since James Madison, earning it the accolade "Church of the Presidents." Pew 54 is reserved as the "President's Pew," should he decide to drop by.

The brick townhouse across H Street and Jackson Place and facing the Square is the 1818 **Decatur House ❸** (748 Jackson Place, NW; Tues, Wed, Fri, Sat, 10am–5pm; Thur, 10am–8pm; closed Mon, Jan. 1, Thanksgiving, Dec 25; free admission; tel: 202-842-0920). This was the first private residence on the Square. Stephen Decatur, a wildly popular 19th-century naval hero, had the house built near the White House, where he was frequently a guest, reputedly because he had designs on the executive mansion himself. Unfortunately, he only lived here for 14 months before he was killed in a duel.

After Decatur's death, several notable statesmen and diplomats lived in the house before Edward Fitzgerald Beale bought it. An adventurer, Beale was the messenger who brought news from California twice, first in 1847 to announce the state's accession into the Union, and then, after a 47-day mad dash across the country on horseback, to announce the discovery of gold there. The first floor of the house displays furnishings and possessions from the Decatur era, and the second floor is resplendent in Victoriana.

If it's time for a meal, **Café 15** in the Sofitel Lafayette Square Hotel (806 15th Street, NW; open for breakfast, lunch, and dinner; tel: 202-730-8800) is one of Washington's impressive new trailblazing restaurants, though it's a bit pricey.

Head back toward the White House and turn right and you'll reach the **Blair House ❹** at 1651 Pennsylvania Avenue. It's not open to the public, but is where visiting dignitaries and other presidential guests stay when they're in town on business. The house was originally the property of Francis Preston Blair, a prominent 19th-century publisher. It was here that the command of the Union Army at the

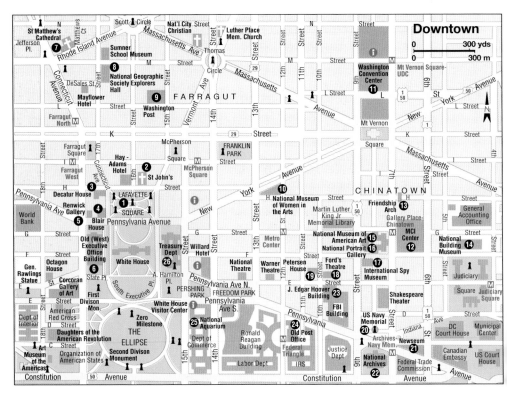

Map on facing page

start of the Civil War was offered to General Robert E. Lee, who refused the post.

The government purchased the house in 1942, and during the Truman renovation of the White House, the president and his family lived here. In 1950, two Puerto Rican nationalists attempted to assassinate the president while he was inside. In the ensuing shoot-out on the street in front of the house, one guard was killed along with one of the intruders. The other was sent to prison where he remained until, by presidential decree, he was released in 1979 and returned to Puerto Rico.

Anchoring the block on the corner of Pennsylvania and 17th Street is the palatial **Renwick Gallery** ❺ (open daily 10am–5:30pm; closed Dec 25; tel: 202-357-2700 or visit www.si.edu), an extension of the Smithsonian's National Museum of American Art, devoted to crafts and decorative arts from the traditional to the contemporary and beyond. Built in 1859, it was the District's first art gallery. When William Corcoran commissioned James Renwick to design an avant-garde museum for his art collection, he unwittingly set an architectural trend responsible for the likes of the overdone French Second-Empire-style Old Executive Office Building across the street.

A grand red-carpeted staircase leads to the Renwick's second floor **Grand Salon**, all very red-velvet Victorian and stacked with paintings à la Louvre that include Rembrandt Peale and George Catlin, best known for his portraits of Native Americans. Also notable is the six-sided **Octagon Room**, designed to display Hiram Powers' notorious nude, *The Greek Slave*. Victorian etiquette dictated that men and women view this sculpture separately. The statue is now shamelessly on display at the **Corcoran Gallery of Art** *(see page 190)* just down the street. The rest of the museum is devoted to American crafts, everything from basketry and clay, to wood and fiber, with five galleries on the first floor reserved for changing exhibits.

Across the street from the Renwick is the **Old Executive Office Building** ❻ a Second Empire style "wedding cake" designed by Alfred Mullett and built in 1888. It was originally the home of the State, War, and Navy departments until each gradually moved out to bigger quarters. In 1947, the Executive Office of the President took it over as a much-needed annex to ease its own crowded office space, but little was done in the way of upkeep. Largely neglected, the building underwent a wholesale restoration in the 1980s, and now stands as the country's finest example of this architectural style. There are more than 500 rooms here, occupied today by the Office of Management and Budget, the National Security Council, and the Office of the Vice President.

Downtown favorites

Walk north through **Farragut Square** and up Connecticut Avenue and you'll come to the sumptuous **Mayflower Hotel** (1127 Connecticut Avenue, NW; tel: 202-347-3000), a Washington institution completed just in time to serve as the site of Calvin Coolidge's inaugural ball in 1925. FDR stayed here while his rooms were being

readied at the White House, FBI director J. Edgar Hoover dined here nearly every night for 20 years, and cowboy actor and singer Gene Autry once rode his horse Champion right through the middle of the banquet room. More recently, famed Clinton paramour Monica Lewinsky bunked here when she was in town testifying before Special Prosecutor Kenneth Starr.

Just off Connecticut on Rhode Island Avenue is the deceptively plain edifice of **St Matthew's Cathedral ❼** (1725 Rhode Island Avenue, NW; tel: 202-347-3215), which President John F. Kennedy attended. The interior, a profusion of mosaic and marble, displays an altarside marker where Kennedy's casket rested during his funeral Mass in 1963.

Monopolizing the M Street block between 16th and 17th streets is the imperious, three-building HQ of the **National Geographic Society ❽** (Mon–Sat, 9am–5pm, Sundays and holidays, 10am–5pm; closed Dec 25; free admission; tel: 202-857-7588; www.nationalgeographic.com)

Oddly, the Grosvenors, the conservative first family of the society, hired Edward Durell Stone, the avant-garde architect of New York's Museum of Modern Art and the Kennedy Center, to design the modernist 17th Street building. The society, characteristically, documented the construction with a time-lapse camera.

Inside, **Explorers Hall** tackles exactly what you'd expect of National Geographic, everything from weather and biology, to anthropology and outer space, and regularly changes its exhibits. Free films, usually well-done National Geographic productions, are shown on Tuesdays at noon, and the gift shop has a large stock of videos and books. Perfect for kids, but just as engaging for adults.

If you're hungry, stop at the cozy and reasonably priced **15 Ria** at 1515 Rhode Island Avenue (tel: 202-742-0015), hence the name, at the Washington Terrace Hotel, then head over to **The Washington Post ❾** (1150 15th Street, NW; free guided tours available Mondays with advance reservations; tel: 202 334-7969). Second only to *The New York Times* in journalistic influence, the *Post*, in print since 1877. outshined them all with its legendary Watergate coverage in the early 1970s which led eventually to President Richard Nixon's resignation in 1974. News junkies especially will appreciate the tour of the newsroom and pressroom.

Walk through **McPherson Square** and stop at **Chapters Literary Bookstore** (1512 K Street, NW; tel: 202-347-5495), pleasant, well-stocked and the site of a regular schedule of literary events, on your way to the **National Museum of Women in the Arts ❿** (1250 New York Avenue, NW; open Mon–Sat, 10am–5pm; Sun noon–5pm; closed Jan 1, Thanksgiving, Dec 25; admission fee; tel: 202-783-5000 or visit www.nmwa.org). Housed in a former Masonic Temple, a National Historic Landmark, the museum opened in 1987 after an $8 million renovation.

There are more than 2,700 pieces of art here, all the work of women, everything from *Portrait of a Noblewoman* by Lavinia Fontana, a 16th-century Italian considered the first professional woman

LEFT: on display in the National Museum of Women in the Arts.

Map on page 174

artist, and Elisabeth Vigee-Lebrun, court painter to Marie Antoinette, to renowned Impressionist Mary Cassatt and 20th-century painters Helen Frankenthaler and Lee Krasner.

Nearby is the **Washington Convention Center** ⓫ (801 Mt Vernon Place, NW; tel: 202-789-1600), which opened in 2003. Its dramatic 100-ft (30-meter) high curved glass entrance gives onto more than 2 million sq. ft. (186,000 sq. meters) of space and it is expected to draw 3 million visitors a year to conventions and trade shows.

In the old days, Washington, like big cities across America, had a thriving downtown anchored by more than one department store to draw crowds. Those days, of course, are gone, but one lone survivor, the **Hecht Company** (corner of 12th and G streets, NW; tel: 202-628-6661), spruced up and catering mainly to visitors who stay at the new hotels nearby, offers shoppers an array of choice soft goods. The centerpiece of this revitalized neighborhood is the **MCI Center** ⓬ (601 F Street, NW; tel: 202-628-3200), a 20,000-seat sports arena complete with shops and restaurants and which serves as the home of the Washington Wizards basketball team and the Capitals hockey team.

Chinatown

At the Gallery Place Metro station at 7th and H streets, NW, you'll notice the striking red pagoda-styled **Friendship Arch** ⓭ marking the beginning of Washington's **Chinatown**. It's definitely not San Francisco or New York and only spans the few blocks between 5th and 9th streets, but the restaurants are authentic, and the shops offer lots of imported products, including a wide selection of teas.

From Chinatown, you're not far from the **National Building Museum** ⓮ (401 F Street, NW, Mon-Sat, 10am–5pm, Sun noon–5pm; closed Jan 1, Thanksgiving, Dec 25; tel: 202-272-2448; www.nbm.org), which occupies an entire city block and commemorates American architecture in a series of changing exhibits. In the 19th

RIGHT: gateway to Chinatown.

century, the building was the headquarters of the Pension Bureau, a federal agency charged with distributing pension funds to war veterans and their families. The building was declared a National Landmark, in part because of its terracotta frieze, the work of sculptor Casper Buberi, that extends around the entire outside of the structure and which depicts Civil War scenes. Inside, 75-ft (23-meter) high Corinthian columns mark the enormous space, the setting for the inaugural balls of several presidents.

The Old Patent Office Building on 8th Street at F and G streets, NW is home to two of the Smithsonian's best off-the-Mall museums, the **Smithsonian National Museum of American Art ⓯** and the **National Portrait Gallery ⓰**. Both are getting much-need facelifts, and both, unfortunately, are closed until at least 2006, though their collections are traveling around the country, and you can take a virtual tour by visiting www.si.edu. The building, which housed the Patent Office, the agency that

issued patents to inventors, dates from 1867, and at the time it was opened was the city's largest. In 1958, Congress officially gave the building to the Smithsonian. The American Art Museum began as a mix of private and public collections put on display in the office in 1841. The collection is dedicated strictly to American art and has in its holdings more than 30,000 objects.

The adjacent Portrait Gallery, a sort of national family photo album, took up residence in the Patent Office in 1968 and counts among its 15,000 portraits, sculptures, photos, and drawings of American men and women of significance one of the celebrated Landsowne portraits of George Washington.

Where the spies are

Across the street is the **International Spy Museum ⓱** (800 F Street, NW; open daily 10am-8pm; closed Jan 1, Thanksgiving, Dec 25; admission fee; tel: 202-393-7798; www.spymuseum.org). As soon as they walk through the door, visitors are put under surveillance and assume a "cover" identity to get a sense of what it's like to live as spies do, constantly on red-alert. The emphasis here is on the Cold War with exhibits on the Rosenbergs and Gary Powers, the U-2 pilot shot down over Soviet airspace in 1960, though spies from others eras are represented, including everyone from Mata Hari to the Navajo Codetalkers of World War II. James Bond, Maxwell Smart, and Austin Powers also get billing. Turncoats Aldrich Ames and former FBI agent Robert Hanssen make their mark here amid the cases of spy gadgetry, including a tube of lipstick that doubles as a miniature pistol. Note that lines can be up to three hours long, according to the museum itself. To avoid the crowds, visit after 4pm, and stop for dinner at **Zola** (tel: 202-654-0999), the museum's classy eatery.

Where Lincoln was shot

In the middle of 10th Street between E and F Streets, NW is **Ford's Theater ⓲** (open daily for tours 9am–5pm; closed Dec 25 and during rehearsals and performances; tel: 202-426-6924 or visit

LEFT: it's not a well-kept secret.

Map
on page
174

www.nps.gov/foth). It was here that Abraham Lincoln was shot on April 14, 1865. The theater was then just two years old, lavishly outfitted in the Victorian style, and considered one of the finest theaters in the country. Five days after the end of the Civil War, President and Mrs Lincoln were enjoying a performance of the English comedy *Our American Cousin* when John Wilkes Booth, a popular actor and impassioned Southern sympathizer, crept into the presidential balcony box and fired a bullet into Lincoln's head, just behind the left ear. Booth then leapt for the stage, snagged his foot in the draped flag and broke his left leg. But he still managed to escape on horseback.

While Booth galloped toward Maryland to make his crossing at Port Tobacco into safe territory in Virginia, the unconscious Lincoln was carried across the street to a boarding house owned by tailor William Petersen. Lying in a back room diagonally across a bed because he was too tall to fit in it, Lincoln succumbed to his wound at 7.22am the following morning. On April 26, a detachment of soldiers, in search of Booth, surrounded a Virginia tobacco barn where the assassin had holed up. Refusing to surrender, Booth shot himself.

After the president's death, the theater was closed by the government, which ultimately bought it, then re-opened it in 1932 as a museum. Restored to its original condition, it reopened as a theater in 1968. Check the *Washington Post* for a schedule of plays, usually comedies and seasonal classics such as *A Christmas Carol*, a Ford's Theater annual tradition. The presidential box is as it was on the day Lincoln was shot, and many artifacts from the assassination are displayed in the museum in the basement, given a thorough renovation during its 2003 closure.

The small and humble **Petersen House** ⑲ across the street at 516 10th Street, NW, has been closed for renovation and will offer tours when it reopens.

One of the best theaters in the city is the new **Shakespeare Theatre** (450 7th

Street, NW; tel: 202-547-1122 or visit www.shakespearetheatre.org) whose resident company gives new life to the bard's classic tales. Several good restaurants line the block, including **Teaism** (400 8th Street, NW; tel: 202-638-6010), less expensive and more casual than the others but with a good choice of teas and light Asian fare. **Olsson's Books and Music** (418 7th Street, NW; tel: 202-638-7610), is a well-stocked chain with knowledgeable staff and several locations across the area, including a branch at Metro Center (1200 F Street, NW; tel: 202-347-3686)

The nation's main street

As part of city architect Pierre L'Enfant's link between Capitol Hill and the White House, Pennsylvania Avenue was designed to be the center of its commercial activity, lined with farmers' markets, shops, government offices and residences. The avenue thrived well into the 20th century until the growth of the suburbs after World War II drew business and housing away from the city. By the 1960s, the avenue, the traditional inaugural parade route of presidents, was an embarrassment, lined with ramshackle buildings and dreary government offices. Not any more. Improvement has been gradual, but Pennsylvania Avenue is restored, and once again worthy of being "the nation's main street."

Continue your downtown circuit by heading up Pennsylvania Avenue toward the White House. Those with an interest in things nautical will enjoy the **US Navy Memorial** ❷ (7th Street and Pennsylvania Avenue; open Tues–Sat, 9:30–5pm; free admission; tel: 202-737-2300 or 1-800-723-3557 or visit www.lonesailor.org). It charts the Navy's story since the War of Independence. Those who have served can add their name to the memorial's register, call up photos of their ship on a computer, and connect with their shipmates. On summer evenings, the Navy and Marines stage free concerts on the plaza.

Where Pennsylvania and Constitution avenues come together, you'll notice the

BELOW: the US Navy Memorial

Map on page 174

very modern and gleaming glass wedge that serves as the **Canadian Embassy** (501 Pennsylvania Avenue, NW; tel: 202-682-1740), and across from it at 6th Street and Pennsylvania Avenue what will eventually be the **Newseum** ㉑, a museum for serious news aficionados.

For five years, the Newseum was a popular attraction in the headquarters building of the Gannett newspaper chain in nearby Rosslyn, Virginia. When it outgrew its space, newspaper executives also decided that a move into the city proper would add to its prestige. All glass with a 90-ft (27-meter) high atrium, the building will come equipped with a large LED media screen attached to its exterior, a sort of giant television facing the nation's main street, to project a constant flow of up-to-the-minute breaking news. It will also feature a 60-ft (18-meter) high slab of stone engraved with the words of the First Amendment, which guarantees freedom of the press, lest anyone on Capitol Hill, deliberately within clear sight of the new building, have ideas otherwise. Inside, the museum promises to be an "interactive museum of news" with videos, photo displays, a 500-seat auditorium, the requisite museum store, and a "news café."

The National Archives

As the federal government's repository of records essential to the country, the **National Archives** ㉒ (Pennsylvania Avenue between 7th and 9th streets, NW; tel: 202-501-5000) has everything from the original US Constitution and census records to materials related to the Kennedy assassination and more. Following a renovation in 2003, the Archives are highlighting artist Barry Faulkner's restored *Declaration of Independence*. Additional features include a new display for the Constitution and other documents that make them easier to view, a theater showing documentary films, and special galleries devoted to timely topics and showcasing some of the Archive's holdings.

Next is the **J. Edgar Hoover Building** ㉓ (tours unavailable until 2004; tel: 202-324-3000 or visit www.fbi.gov), home of the Federal Bureau of Investigation.

Opened in 1975 and named for the bureau's most famous leader, who served for nearly half a century, this stark and cold-looking building, erected in the off-putting architectural style known as New Brutalism, is meant to give an impression of no-nonsense, though that doesn't quite match the bureau's tarnished image of the past few years. The tour, a hit with children, has traditionally focused on gangsters and has included a short firearms demonstration. Plans are in the works for a new tour, which promises to bring things a little more up-to-date.

Spared the wrecker's ball during the city's revitalization campaign, the imposing castle-looking granite building across the street is the **Old Post Office** ㉔ (Pennsylvania Avenue and 12th Street, NW; tel: 202-606-8691; open Easter through early Sept, 8am–10.45pm; the remainder of the year 9am–4.45pm). The building dates from 1899 and has been converted into space for offices and trendy shops. Take the elevator 270 ft (82 meters) to the top

of the tower for a commanding view of the city, especially if you've missed the timed tickets for the Washington Monument. The food court below offers lots of good, inexpensive choices.

Next door is **Federal Triangle**, a modern office complex including the **Ronald Reagan Building**, and across Pennsylvania are two of the city's oldest theaters, the **National** (1321 Pennsylvania Avenue, NW; tel: 202-628-6161), on this site since 1864, and the **Warner** (1299 Pennsylvania Avenue, NW; tel: 202-783-4000).

Stop to rest in **Freedom Plaza**, all flags, fountains and pavement into which has been embedded L'Enfant's original plan for the city, then head over to the Department of Commerce where tucked away in the basement is the **National Aquarium**  (14th Street and Constitution Avenue; daily 9am-5pm; closed Dec 25; admission fee; tel: 202-482-2825 or visit www.nationalaquarium.com). Though not the biggest or the splashiest, it is the country's oldest aquarium. Every day at 2pm, the keepers feed will feed either the sharks, the piranha, or the alligators, which you're welcome to watch, and give impromptu talks. Children like it here, and it's a good place to stop when the Washington heat becomes unbearable.

Nearby **Pershing Park**, a pleasant city green space, offers shade on a hot day. The pond attracts ducks, and in winter you can ice skate. (Rink open Mon–Thurs, 10am–9pm; Fri–Sat, 10am–11pm; Sun 10am–7pm; admittance fee; tel: 202-737-6938).

A renowned hotel

At 14th Street and Pennsylvania Avenue is the **Willard Hotel** (tel: 202-628-9100), built in 1901 as a replacement for the original Willard built on this site in 1801. Lincoln stayed here while he was awaiting the White House to be readied for him. Charles Dickens was a guest here, as was Edward VII when he was Prince of Wales. Julia Ward Howe was a guest of the hotel during the Civil War when she composed *The Battle Hymn of the Republic*. The Willard also gave us the term "lobbyist" since it was in the lobby of this grand hotel during the 19th century that White House office seekers and others pressing their cases in Congress routinely gathered.

The dining room is a lovely treat, but for something a little less expensive but just as tasty (it all comes from the same kitchen) try the coffee shop instead, especially for Sunday brunch. Those seeking a good American hamburger should try the **Old Ebbitt Grill** (675 15th Street; tel: 202-347-4800), another Washington institution dating back more than a century.

Lastly, at 15th Street on the east side of the White House, you'll come to a gray and rather impenetrable looking building that may seem vaguely familiar. This is the **US Treasury Department Building**  and it's pictured on the back of the $10 bill. Designed in the Greek Revival style by Robert Mills, who also designed the Washington Monument, it dates from 1833. The Treasury Department itself was established in 1789 to oversee the nation's finances. Several government bureaus, including the Mint, the Internal Revenue Service and the Secret Service, fall under its purview. ❑

LEFT: the food court at the Old Post Office.

Map on page 174

Art Treasures

Washington is not just about politics and marble monuments. There's art here, too, and plenty of it, as one would expect in a capital city. The National Gallery, in particular, is packed floor joists to rafters with some of the world's best.

The city's first art gallery, the Corcoran (now the Renwick), which opened in 1859, and the smaller private museums such as the Kreeger Museum *(see page 183)* and the Phillips Collection, claim prizes of their own. The good news is that there's so much to see here that Washington's been able to slough off its image as a cultural backwater to become one of the world's most important centers of art. The bad news is that there's so much to see here that you can't possibly cover it all, certainly not in one visit.

Of course, seeing it all isn't the point. The point is to choose what pleases you and come away from the experience restored, inspired and reminded that "art," as operatic singer Beverly Sills once said, "is the signature of civilization." Here's just a smidgen of what you'll find.

The National Gallery *(see page 138)* has the only painting by Leonardo da Vinci outside of Europe, the haunting portrait of *Ginevra di Benci* (c. 1480), whose expression he seems to have perfectly captured after she was, so the story goes, deserted by her Venetian lover. It is his earliest known portrait. Several of the Dutch masters are here, too, including Jan Vermeer. Not much is known about the man from Delft beyond the fact that he fathered 11 children, died bankrupt, and produced only a few paintings, among them the lively *Girl with the Red Hat* (1665–67). Enter the Gallery's modern East Wing and you'll see Alexander Calder's massive *Mobile* (1978) suspended from the ceiling and gently paddling the air.

Visit the Corcoran Gallery *(see page 190)* to see Frederick Church's *Niagara* (1857). A major figure in the Hudson River School, which celebrated American landscape, Church's grand painting was a tribute to the Falls, at one time regarded as a place for spiritual renewal. *The Luncheon of the Boating Party* (1881) by Pierre-Auguste Renoir is in the Phillips Collection *(see page 208)*. The comforts and pleasures of good friends and good wine on a pleasant afternoon, this is one of the best loved in the gallery and a favorite of founder Duncan Phillips.

"I am the dog," sculptor Albert Giacometti said of his famous elongated and existential creature *The Dog* (1951), which you can see in the Hirshhorn Sculpture Garden *(see page 135)*. If you stop at the National Museum of Women in the Arts *(see page 176)*, you'll find the Impressionist Mary Cassatt's *Mother Louise Nursing Her Child* (1899), a loving reminder of the special link between mother and child. ❑

● *For a listing of DC's smaller galleries, see the Culture section of Travel Tips.*

RIGHT: checking out John Quidor's *Reproof* at the Museum of American Art.

FOGGY BOTTOM AND THE WEST END

Despite its unassuming name, this neighborhood offers several quality museums and galleries and includes the city's cultural crown jewel, the Kennedy Center

Map on page 188

Croaking frogs, marshy lowlands, and a smoke-belching brewery – such were the salient features of Foggy Bottom in the mid-1700s. By the early 20th century, much of this soggy land along the Potomac River was filled in to accommodate the expanding city. Centered around **Washington Circle**, Foggy Bottom's slow and awkward growth created not a few disputes between developers and preservationists. The successes and failures of these groups have created a neighborhood seemingly at odds with itself.

Curious mix

Foggy Bottom and the adjacent West End are a mix of 19th-century town houses, the modern structures of George Washington University, the oddly-placed Kennedy Center (replacing the former brewery), the quirky Watergate complex, magnificent marble edifices, government buildings, new hotels, and a ganglion of poorly marked highways that speed traffic in and out of Virginia.

This section of DC – bounded by the Potomac River, Constitution Avenue, 17th Street, and N Street – is best on foot by day, and car by night, especially if you plan to take in the night life. Tourism really isn't big business in Foggy Bottom as it is on the nearby Mall. Entrances are often oblique, reservations for tours may be required weeks in advance, admission might be granted only by showing a photo identification. But the extra effort pays off in the form of smaller (if any) crowds and the feeling you're discovering some unusual attractions just off the beaten path.

To see Washington like a diplomat, begin at 23rd and C streets for one of the city's lesser-known but worthwhile attractions, a guided tour of the **State Department's Diplomatic Reception Rooms ❶**

(2201 C Street, NW; tel: 202-647-3241 to arrange in advance for a guided tour, available Mon–Fri, 9.30am, 10.30am, and 2.45pm). These five drawing rooms and dining rooms are where the Secretary of State entertains distinguished foreign guests several nights a week. The exquisite 18th- and 19th-century-style rooms contain a collection of donated antiques, including the mahogany desk upon which Thomas Jefferson drafted the Declaration of Independence, one of John Jay's punch bowls and a portrait by John Singleton Copley. The oddest item in the collection is the *Landing of the Pilgrims*, a work that depicts a British man-of-war flying the

PRECEDING PAGES: marching orders from the big brass. **LEFT:** a resident of the Watergate complex. **RIGHT:** Einstein's memorial at the National Academy of Sciences.

US flag, redcoats as pilgrims, and a dangerously rocky Massachusetts coast lined with welcoming Indians. More sublime scenes of the Mall and the Potomac River beyond are framed by the south rooms' graceful Palladian windows.

Secluded in a grove of elm and holly trees at the corner of 22nd and Constitution is the **Einstein Memorial** at the National Academy of Sciences. The memorial consists of a 7,000-lb (3,200-kg) bronze statue of Albert Einstein seated casually before a circular sky map. The granite map is embedded with 2,700 metal studs representing the planets, sun, moon and stars visible to the naked eye. So endearing is this avuncular Einstein that most people can't resist climbing up onto his lap to pose for a photograph.

The seldom-visited **National Academy of Sciences** ❷ (2201 C Street, NW; tel: 202-334-2000) building features a Foucault's Pendulum in the ornate Great Hall, two-story-high window panels illustrating the history of scientific progress, and an auditorium where free concerts are frequently held.

At the **Federal Reserve** ❸ (20th and C streets; Mon–Fri, 11am, 1pm, and 3pm with 24-hour pre-registration required; tel: 202-452-3778 www.federalreserve.gov), which sets the country's monetary policies, you'll find a changing art exhibit in the building's central atrium. Here you might see anything from 19th-century formal portraiture to an on-the-spot graffiti artist armed with a spray can. The Fed's board meetings are sometimes open to the public, so if you just can't resist sitting in on a discussion of inflation, prime interest, or bank discount rates, check the website for schedules.

Americana

The **Department of the Interior** ❹ building at 18th and C streets (Mon–Fri, 8.30am–4.30pm with valid photo ID; closed weekends and federal holidays; tel: 202-208-4743) houses one of the more eclectic museums. The department is charged with overseeing America's public lands and natural resources, including everything from its coal mines to its

national parks, and its American Indian reservations and US territories, and as a result it has in its possession a little bit of everything. Dioramas depict an Indian trading post and scenes from the 19th-century Oklahoma land rush, and there are collections of artifacts from the Oceanic peoples of the Marshall Islands and American Samoa. The **Indian Craft Shop** (Mon–Friday and the third Sat of every month, 8.30am–4.30pm) across the hall sells museum-quality turquoise and silver jewelry, baskets, sculpture, weavings and other fine works produced by the artisans of 36 Native American tribes.

In the exact geographical center of DC, at 17th Street and Constitution Avenue, is the **Organization of American States** ❺ (Tues–Sun, 10am–5pm; tel: 202-458-3000). The OAS, a largely ceremonial coalition of 35 North and South American countries, also occupies one of the city's most ornate Beaux-Arts buildings. The interior courtyard, with its pre-Columbian-style fountain and jungle of tropical trees, pro-

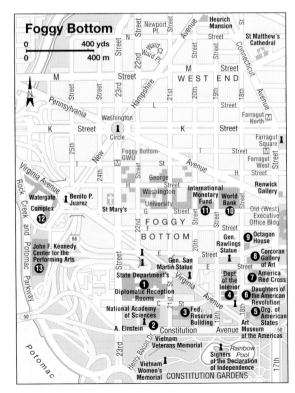

Map on page 188

vides respite from the summer heat and a south-of-the-border getaway in the winter. Ascend the grand staircase to the **Hall of the Americas**, a magnificent room with barrel-vaulted ceiling and Tiffany chandeliers. Behind the building is the slightly neglected **Aztec Garden** and, on 18th Street, the OAS-operated **Art Museum of the Americas** (Tues–Sun,10am–5pm; closed major holidays; guided tours Tues–Friday with reservations; tel: 202-458-6016 or visit www.museum.oas.org) – a treasure house devoted to contemporary Caribbean and Latin American art and also offering a series of films and lectures.

Continental Hall at 1776 D Street (Mon–Fri, 8.30am–4pm, Sun 1pm–5pm; closed major holidays; tel: 202-879-3241 or visit www.dar.org) serves as the headquarters and museum of the **Daughters of the American Revolution ❻**. The organization, founded in 1890 by a group of women descendants of Revolutionary War patriots, has more than 170,000 members. The **DAR museum** contains 34 period rooms furnished to evoke the state-by-state and historical differences in interior design in the United States. You'll see a California adobe-style parlor from 1850; a New Hampshire attic full of children's dolls and toys of the 18th and 19th centuries; and a nautical-style New Jersey room with a chandelier made from the recast anchor and chains from a British frigate.

The DAR's **Library of Genealogy and Local History** is the second largest in the country. Visit this four-story library if only to gaze up at its skylight ceiling that sheds light over a cascade of balconies and onto the tomes perused by diligent researchers below. This room served as the DAR's convention hall until 1920 when the annual convention outgrew this space. In 1929, **Constitution Hall**, designed by architect John Russell Pope, opened next door on 18th Street to accommodate the DAR assembly. DC's second largest auditorium also hosts a variety of public lectures and concerts.

Just north of the DAR at 17th and D

BELOW: the Department of the Interior's museum.

streets is the **American Red Cross** (Visitors Center Mon–Friday, 8.30am–4pm; tours Tues and Fri, 9am, with appointment; tel: 202-639-3038). Here you'll find a trio of stained-glass windows by Louis Comfort Tiffany in the upstairs board room. These beautifully iridescent, opalescent windows feature St Filomena, famed for her healing powers, plus gallant knights of the Red Cross and fortitudinous Una from Spenser's *Faerie Queene*.

The Corcoran Gallery

One block north on 17th Street is the city's oldest and largest private art museum, the **Corcoran Gallery of Art** (Wed–Mon, 10am–5pm, Thurs until 9pm; closed Jan 1, Thanksgiving, Dec 25; admission fee; tel: 202-639-1700; www.corcoran.org). The present gallery was opened in 1888 to house the expanding collection of William Wilson Corcoran, a Washington philanthropist and banker, which had outgrown its original red-brick quarters near the White House. That original gallery, now the Renwick Gallery, was begun in 1859. Construction stopped at the outbreak of the Civil War when Corcoran, a Southern sympathizer, found himself unwelcome in Washington and headed off for Europe.

The museum, finally opened in 1874, was hailed as one of the country's first major art galleries, only surpassed 14 years later by the "new" and bigger Beaux-Arts gem that is the gallery's permanent home. As you enter the Corcoran, look up at the frieze where 11 names, chosen by the architect, are carved: PHIDIAS, GIOTTO, DURER, MICHAELANGELO, RAPHAEL, VELASQUEZ, REMBRANDT, RUBENS, REYNOLDS, ALLSTON, and INGRES.

The trustees discovered too late that the much-lesser talent, Washington Allston, was an uncle of the architect's mother. Correcting in print what they can't in stone, gallery publications have replaced Allston with the more deserving Da Vinci.

Inside is a comprehensive collection of 19th- and 20th-century American art as well as European carpets, tapestries, mar-

LEFT: Tiffany window in the Red Cross building. **BELOW:** Octagon House.

Map on page 188

ble sculptures, and European paintings including those by well-known artists such as Rembrandt, Rubens, Renoir, and Degas. Perhaps the best known work is a nude sculpture by artist Hiram Powers, *The Greek Slave*, which scandalized Victorian museum-goers when first unveiled.

In 1925 Senator William Clark of Montana bequeathed to the Corcoran his extensive European collection, including tapestries and stained glass. He also donated an entire French salon, the *Salon Dore*, considered one of the finest examples of late French-Rococo interior design.

Just west of the Corcoran is the **Octagon House ⑨** (Tues–Sun, 10am–4pm; closed Jan 1, Thanksgiving, Dec 25, admission fee; tel: 202-638-3105 or visit www.archfoundation.org) designed in 1798 by Dr William Thornton, the first architect of the Capitol, for Colonel John Tayloe, a wealthy Virginian and friend of George Washington's. Although its name suggests eight sides, this Federal-style structure, which is now the headquarters

of the American Institute of Architects, actually has only six.

The city's second largest land holder (after the federal government) is **George Washington University**, occupying 20 square blocks from Washington Circle south to G Street. The school settled here in 1912 and expanded into the surrounding town homes, converting them into classrooms, dormitories, and offices. The university's well-known law library at 718 20th Street is an elegant brick and concrete structure and an improvement over GW's prosaic campus style.

If you see an ocean liner run aground in a row of townhouses, you're at **2000 Pennsylvania Avenue**, a thoroughly modern office building fronted by the **Lion's Row town houses**. The building now houses offices, Tower Records, several restaurants, and a number of other mall-type shops. Also in the neighborhood are the little-known private **international galleries** run by the arts societies of the **World Bank ⑩** (1818 H Street; daily 9am–5pm;

BELOW: inside the Corcoran Gallery of Art.

tel: 202-477-1234), which features sculpture, jewelry, and paintings by bank staffers and the **International Monetary Fund** ① (720 19th Street, NW; open Mon–Friday, 10am–4.30pm; tel: 202-623-6869) which displays member countries' currency and showcases international artists.

On the north side of Washington Circle, between Georgetown and Downtown, is DC's newly developed **West End**. The main strip along M Street is dominated by glitzy hotels and new office buildings. There are too few trees and pedestrians.

Watergate

Moving south toward the Potomac you'll come upon the round and layered **Watergate complex** ⑫ and the adjacent rectangular Kennedy Center. Distinctive if unattractive, they've been described together as a "wedding cake and the box it came in." Now infamous as the site of the break-in at the offices of the Democratic National Committee that ultimately resulted in the resignation of Republican

president Richard Nixon in 1974, the ultra-modern curvaceous complex of Watergate apartments, offices, shops and restaurants is now popular for other reasons. The Watergate's Jeffrey's Restaurant (2650 Virginia Avenue, NW; tel: 202-965-2300) is housed here, as is the Watergate Pastry Shop (2534 Virginia Avenue, NW; tel: 202-342-1777) where you can feast on scandalously rich desserts.

The Kennedy Center

The John F. Kennedy Center for the Performing Arts ⑬ (tel: 1-800-444-1324 or 202-467-4600; www.kennedy-center.org for information, schedule and box office) pays a more positive tribute to a former president. It serves as a living memorial to Kennedy's belief that the US should be remembered not for its "victories or defeats in battle or in politics," but for its "contribution to the human spirit." Legislation establishing the center was actually signed into law by President Dwight D. Eisenhower in 1958 and was designated

BELOW: the Watergate complex.

Map
on page
188

a memorial to President Kennedy following his assassination in 1963. Angular and flat-roofed, the building was designed by Edward Durrell Stone, and opened for its first performance in 1971.

Isolated by careless urban planning and a tangle of roadways, this imposing monument contains six theaters presenting opera, ballet, musical theater, chamber music, jazz concerts, silent films, recitals, workshops, and children's theater. The Concert Hall, its largest performance space with more than 2,400 seats, is home to the National Symphony Orchestra.

There are free performances at 6pm every evening at the Millennium Stage (actually two stages with one at either end of the Grand Foyer. Afterwards, you can take the elevator to the Roof Terrace for a splendid view of the city and a bite to eat in one of the three restaurants (tel: 202-416-8555 for reservations and hours). The Roof Terrace Restaurant serves dinner on performance evenings and brunch on Sundays, the Encore Café is good for salads and sandwiches, and the Hors D'Oeuverie serves drinks and light fare.

You can take a free guided tour (Mon–Fri, 10am–5pm; Sat–Sun, 10am–1pm; tel: 202-416-8524) to see the inside of the theaters, the glamorous **Hall of Nations**, which displays the flags of all the nations with which the US has diplomatic relations, and the **Hall of States**, where the flags of the states are displayed in the order that they entered the Union. The tour includes the **Performing Arts Library** and the **Grand Foyer,** one of the world's largest rooms at 630 ft (192 meters) long, 60 ft (18 meters) high, and 40 ft (12 meters) wide. It contains the Robert Berks sculpture of President Kennedy and 18 one-ton Orrefors crystal chandeliers.

Note that parking is limited and expensive. The easiest option is to take Metro to the Foggy Bottom/George Washington University station and catch the free shuttle bus, which operates every 15 minutes between 9.45am to midnight, Mon–Sat, and noon–8pm on Sun and holidays. ❏

BELOW:
JFK's bust in
the Center.
RIGHT:
the Kennedy
Center's Hall
of Nations.

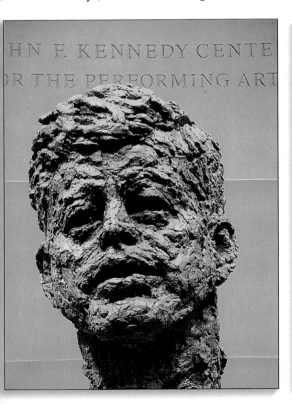

Map
on page
198

GEORGETOWN

Once a prosperous port city, Georgetown bustles with shops and restaurants and its shady streets are lined with the finely restored Federal-style townhouses of Washington's elite

Georgetown is the pedestrian heart of Washington. Shaded streets, brick and cobblestone sidewalks, row houses, and low-rise commercial buildings create an atmosphere that is quaint and on a distinctly human scale. Georgetown has the commercial bustle of a 19th-century port city, the youthful energy of a college town, the international style of a diplomatic community and the wealthy mien maintained by its resident population of senators, lawyers, socialites, and members of the media.

It is inaccessible by Metro, thanks to the residents who argued when the tunnels were being dug that a subway stop would only crowd the streets with the hoi polloi, so you'll be forced to either hoof it the 20 minutes or so it takes from Foggy Bottom, the nearest Metro stop, or spring for a cab. Georgetown's traffic gridlocks are legendary and its limited parking a source of frustration. It is also a place of rowdy just-legal imbibers cruising M Street and Wisconsin Avenue, and of Halloween madness and mayhem after a Redskins victory. Georgetown hosts a not-trivial number of muggings and burglaries that make some quiet side streets unsafe after dark.

Port authority

Originally part of Maryland, Georgetown was settled in 1703 on the banks of the Potomac River. Wharves, warehouses, and factories – many structures still extant – lined the riverbank of this flourishing port that shipped tobacco and flour world wide. The square-mile town was probably named after King George II (two owners of the original land grants were also named George and historians still squabble over who among the Georges the area was really named for). It was soon populated by plantation owners from Maryland and Virginia, merchants from New England, and a great number of slaves and laborers. In the 1780s, Georgetown's gracious homes, inns and taverns made it an ideal staging area for the planning of the nation's permanent capital city.

George Washington, Thomas Jefferson, John Adams, and a slew of architects, governors and foreign envoys frequented Georgetown, many establishing permanent homes. In the late 1780s Bishop John Carroll founded Georgetown Seminary – an institution that added to Georgetown's distinction. The seminary grew into **Georgetown University ❶**, the oldest Catholic college in the country. Two of George Washington's nephews were enrolled here, as was a grandnephew of Andrew Jackson. The gothic spires of the university's **Healy Hall** dominate the skyline of

PRECEDING PAGES: reflections on Georgetown. **LEFT:** Georgetown University. **RIGHT:** Georgetown Park Mall.

Georgetown from its perch at 37th and O streets. The building was named for Father Patrick J. Healy, a black man who served as the university's president between 1874 and 1882. The university's School of Foreign Service has turned out some of the nation's top diplomats, and its law school is among the best in the country.

Georgetown thrived as a cosmopolitan city, industrial center, and shipping canal terminus until the advent of the railroad and of steam navigation, which required deeper waters than the town's port could provide. The growth of the new capital after the Civil War sealed Georgetown's fate, especially after its territory was swallowed by Washington in 1871. By the end of the 19th century, Georgetown fell into neglect, but never lost its sense of a separate identity. In 1950, the area was declared a National Historic District, paving the way for restoration of many of the original Federal-style homes. Many of Washington's most powerful live here, conducting much of the business of government and politics at their exclusive private parties.

Georgetown is ringed like a walled city by Rock Creek Park, the Potomac River, Georgetown University and Whitehaven and Dumbarton Oaks parks. Though most of the trade here is conducted in shops and restaurants along M Street and Wisconsin Avenue, Georgetown's waterfront retains some of its earlier bustle.

Touring and tea

The 185-mile-long (300-km) **Chesapeake & Ohio Canal ❷** *(see panel, page 223)* begins in Georgetown and is easily accessible from several points here. Commuters, strollers, dog walkers and weekend athletes use the serene and shaded canal and its towpath. **Mule-drawn boat rides** from the Foundry Mall take you back to a more leisurely age.

On the Potomac River at 31st Street is **Washington Harbour ❸**, a modern monstrosity housing not-so-notable shops, eateries and offices. But it's worth a visit if only for its computer-choreographed fountain, sculptures and riverside promenade.

At 30th and M streets you'll find the **Old Stone House ❹** (3051 M Street,

NW; open Wed–Sun, 10am–4pm; closed major holidays; tel: 202-426-6851), one of the city's oldest and quaintest buildings. Built in 1765, this six-room house was used as a carpentry shop and home by its original owners. The architecture and furnishings reflect the modest lifestyle of the pre-Revolutionary days. Behind the house is a small and wonderfully wild garden where fruit trees and densely planted borders of flowers bloom with abandon spring through fall. This is a perfect retreat for lunchtime picnickers and weary pavement pounders.

At the west end of M Street is the oft-promoted **Georgetown Park ❺** mall. This $100-million Victorian-style mall is a success story of architectural preservation, but its 85 international boutiques and specialty shops can't hold a candle to the more interesting (and less pricey) stores along M and Wisconsin streets. These streets are lined with opportunities to buy everything from expensive Italian suits, cheap shoes, surplus military wear,

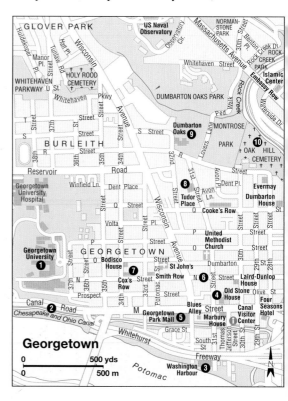

Georgetown

Map
on page
198

antiques, household goods, coffees and spices, and American crafts. For the slightly unusual in fashions and accessories, try **Commander Salamander's** (1420 Wisconsin Avenue, NW; tel: 202-337-2265), a Georgetown institution with a hip, knowledgeable and stylish staff.

The best cure for shopper's syndrome (exhaustion) is afternoon tea in the pretty **Garden Terrace** of the **Four Seasons Hotel** at 2800 Pennsylvania Avenue (tel: 202-342-0444). For the real thing served ceremonially in the Asian style, try **Ching Ching Cha** (1063 Wisconsin Avenue, NW; tel: 202-333-8288), a quiet little teahouse snuggled inauspiciously against a women's lingerie shop where you'll have your choice of the finest black, oolong, and green teas accompanied by a full menu of salmon or tofu Bento boxes.

Famous residents

Georgetown's greatest charm is its architecture – street after street of elegant town homes, best seen on foot. In particular,

look for the fine Federal-style historic homes on **N Street ❻**. The **Laird-Dunlop House** at **number 3014**, originally the home of a wealthy tobacco merchant, was owned by Abraham Lincoln's son, Robert Todd. The house at **number 3017,** built in the 1790s by a descendant of an original Georgetown landowner, was where Jacqueline Kennedy lived after her husband's assassination. Before he was president, Jacqueline and John Kennedy lived at **number 3307** in the stately **Marbury House.** The houses from numbers 3327 to 3339 are known as **Cox's Row ❼**, named for Colonel John Cox, Georgetown's first mayor, who served for 22 years and lived at **number 3339.**

Evermay is a spectacular private residence you can admire from the sidewalk at 1623 28th Street, along with **Cooke's Row**, four whimsical Victorian town homes on the 3000 block of Q Street.

One of the grandest residences in the neighborhood is **Tudor Place ❽** at 1644 31st Street (by guided tour only with

advance reservations; Feb–Dec, Tues–Fri 10am, 11.30am, 1pm, 2.30 pm, and Sat on the hour 10am–3pm; closed major holidays; donation required; Christmas tours available in December; garden open year round Mon–Sat, 10am–4pm, no reservation needed; tel: 202-965-0400 or visit www.tudorplace.org).

From 1805 to 1983, this neoclassical mansion was occupied by generations of the same family. Martha Custis Peter, a granddaughter of Martha Washington, purchased a city-block's worth of property with $8,000 left to her by the Washingtons. Martha and her husband Thomas hired Dr William Thornton, architect of the original US Capitol and the Octagon House, to design their home. As a friend of Thomas Jefferson, Thornton created a gracious home with some clever Monticello-esque features. The rooms are eclectically decorated with formal neoclassical portraits, Civil War-era daguerreotypes, moderately masterful paintings by "artistic" family members, and modern snapshots.

Dumbarton Oaks

Just north of Tudor Place is **Dumbarton Oaks** ❾ (Tues–Sun, 2pm–5pm; closed major holidays; tel: 202-339-6401 or visit www.doaks.org) a museum-house and garden that is beautiful in every season. The 1801 home and property were bought in 1920 by Mr and Mrs Robert Woods Bliss as a "country retreat in the city." Mr Bliss, a former ambassador to Argentina, was a collector of Byzantine and pre-Columbian art. His wife Mildred, and noted landscape gardener Beatrix Farrand, designed 10 acres (4 hectares) of formal gardens.

In 1940, the Blisses donated their property, library, and collections to Harvard University, which subsequently opened it to the public. Several famous musicians performed for the Blisses and their guests in the lavish **Music Room**, including Igor Stravinsky whose *Dumbarton Oaks Concerto* was commissioned by the couple for their 30th wedding anniversary. In 1944, amid the room's Flemish tapestries and frescoes and under its hand-decorated ceiling, representatives from the US, the UK, China, and the Soviet Union gathered for talks that led to the formation of the United Nations.

The museum's **Byzantine collection** is comprised of some 1,500 artifacts, including textiles, mosaics, crosses and other liturgical items, plus 12,000 Byzantine coins, making it an exceptionally complete collection. The **pre-Columbian collection** is housed in eight exquisite glass pavilions designed by Philip Johnson. The glass invites the leafy green outdoors in to envelope the brilliant gold jewelry, jade statues, and stone masks on display.

What was envisioned for Dumbarton Oaks' neglected grounds was nothing short of brilliant. The old barnyards, cow paths, and steep slopes were replaced by terraced gardens, boxwood hedges, 10 pools, nine fountains, an orangery, a **Roman-style amphitheater**, a pebble garden, and three seasons of blooming flowers. A full-time crew of a dozen gardeners work year-round to maintain the gardens, whose beauty and number of visitors peak with the blooming of bulbs, forsythia, and cherry trees in the spring.

LEFT: Dumbarton Oaks gardens.

Map on page 198

Nearby **Oak Hill Cemetery**  provides the perfect classroom for an education in urban landscaping, local history, and the architecture of the afterlife. In continuous use since 1849, this comfortable place includes the graves of John Howard Payne (author of *Home, Sweet Home*), statesmen Edwin M. Stanton and Dean Acheson, and socialite Peggy O'Neill. The gatehouse offers a brochure with a map locating the graves of other Washingtonians.

A stroll down the brick paths and mossy steps leads you past a **Gothic-style chapel**, a miniature **Temple of Vesta**, marble obelisks, **pensive angels**, and forlorn women carved in Phidian robes.

Georgetown's nightlife

By night, Georgetown is where you'll find the greatest variety and densest concentration of restaurants and bars, from well-established French restaurants and top-rated hotel dining rooms to trendy nouveau Italian spots, upper-crust pizza joints, cafés, bakeries and pubs. You can choose from Moroccan, Mexican, Indian, Thai, Vietnamese, Indonesian, Japanese, Ethiopian and American cuisines.

For suppers, **Bistro Francais** (3128 M Street, NW; tel: 202-338-3830) is Georgetown's pick, open until 3am during the week, and 4am on Friday and Saturday. **Fettoosh** (3277 M Street, NW; tel: 202-342-1199) offers fabulous Lebanese cuisine, **1789 Restaurant** (1226 36th Street, NW; tel: 202-965-1789) is an elegant Georgetown favorite, and **Nathan's** (3150 M Street, NW; tel: 202-338-2000), a bar and white-tablecloth restaurant, is a long-time favorite with the locals. For something chi chi and romantic, try **Citronelle** at 3000 M Street in the **Latham Hotel** (tel: 202-625-2150), pricey but worth it.

While the Georgetown bar scene tends to be fairly run-of-the mill (J. Paul's, Clyde's), one nightclub is worth seeking out. **Blues Alley** (in the alley at Wisconsin Avenue below M Street; tel: 202-337-4141; www.bluesalley.com) is DC's oldest and most prominent jazz club where stars such as Dizzy Gillespie, Wynton Marsalis and Nancy Wilson have performed.

RIGHT: strolling by the Potomac at Georgetown.

Reservations are essential for this intimate club. The better tables are given to patrons who show up early for dinner, which features Louisiana Creole cuisine.

Geared to the international set, the local nightclubs offer entertainment such as salsa and merengue dance bands, live Persian music, and dinner-theater performances of the type performed by the political satire revue "Capitol Steps," a sort of Gilbert-and-Sullivan tongue-in-cheek troupe made up of Capitol Hill staffers with on-target but often politically irreverent points-of-view (check the listings in Friday's *Washington Post* for their performance venues, dates, and ticket information). If you're dancing until the wee hours, a couple of Georgetown hangouts serve continental breakfast Saturday and Sunday mornings.

If you still have energy to burn, you can wander over to **3600 Prospect Street** to gaze down the 75-step staircase where the priest in the film *The Exorcist* (played by Max von Sydow) met his ill-timed fate. ❑

Map on page 206

DUPONT CIRCLE TO ADAMS MORGAN

Dupont Circle is Washington's artiest neighborhood, packed with galleries, eateries, small museums, and a bevy of Beaux-Arts mansions, while Adams Morgan is best known for its ethnic mix

Dupont Circle ❶ – where Connecticut, Massachusetts, and New Hampshire avenues intersect – is the heart of the artsy neighborhood that goes by the same name. Since the 1960s the circle has been a rallying point for demonstrations and a stage for ad hoc concerts and impromptu happenings. When the sun shines, the circle is a veritable theater-in-the-round, whose colorful cast of characters includes chess players, lunchtime picnickers, lovers and potential lovers, the generally wacky and weird – and the down and out.

The circle, originally known as Pacific Circle, was part of the neighborhood colorfully named The Slashes after Slash Run, a stream used as the dump for the aromatic offal of several nearby slaughterhouses. Thanks to the massive public works projects of the 1870s, Slash Run was finally diverted into a sewer system and buried, and Connecticut Avenue, until then a muddy lane, was widened and paved, making the area suddenly attractive to wealthy real estate developers.

Architecturally, Dupont Circle is a treasure. The town houses lining its shady side streets are infinitely, and often whimsically, outfitted with English-style gardens, keyhole porticoes, stained- and leaded-glass windows and – if you look up – slate roofs, turrets and copper bays.

In 1882 the circle was renamed for Admiral Samuel F. Dupont, a member of the famous Delaware chemical family, who gave the Union its first naval victory of the Civil War when he captured Port Royal, South Carolina. Put in charge of the new ironclad fleet, Dupont then went off to take Charleston from the Confederates, but failed miserably, was relieved of his command, and died in disgrace in 1865. His widow Sophie worked to revive her husband's honor, which eventually led to Congress's approving the funds for the **fountain** you see in the circle's center and which commemorates the admiral. The fountain was dedicated in 1921 and designed by the team of Daniel Chester French and Henry Bacon, collaborators on the Lincoln Memorial, and features upper and lower basins connected by nautically inspired marble figures representing the sea, the stars, and the wind.

Architectural attractions

An antidote to Washington's monumental and decidedly conservative tendencies, Dupont Circle is comfortably human-scaled

PRECEDING PAGES: muscling in on Adams Morgan Day. **LEFT:** wall art in the style of Toulouse-Lautrec. **RIGHT:** Dupont Circle.

and playfully trendy, with a European touch. Two dozen **art galleries**, which hold a joint **open house** the first Friday of each month are scattered throughout (check the *Washington Post* or visit www.artgalleriesdc.com for a current list of openings and events). You'll also find myriad outdoor cafés, bistros, and bars such as the **Childe Harold** (1610 20th St., NW; tel: 202-483-6702), a smoky, down-under neighborhood institution frequented by writers, and **Kramerbooks and Afterwords Café** (1517 Connecticut Avenue, NW; Sun–Thurs, 7.30am–1am; Fri–Sat 24 hours; tel: 202-387-1400), Dupont Circle's Grand Central for browsing, cruising, or shmoozing. You'll also find funky shops, chic boutiques, and a good second-hand bookstore, **Second Story Books** (2000 P Street, NW; tel: 202-659-8884). This is also the hub of DC's gay community.

The Beaux-Arts style

The Dupont Circle area offers wonderful examples of the classically inspired Beaux-Arts style of architecture that characterized so many of the homes of America's 19th-century industrialists. Before the advent of income tax, when those with money were less concerned with having to shelter it, wealthy Americans put their money into their homes. In Washington, the elite looked to architects Waddy Butler Wood, Nathan C. Wyeth, George Oakley Totten, Jr. and others who studied in Paris, where the movement began, to erect their ornately decorated and elegantly symmetrical residences. Though arguably monuments to capitalism's excesses, the houses, with their carvings and porticos and mansard roofs, stand as fine examples of American artistry and craftsmanship.

Two palatial landmarks front Dupont Circle, both typical of the neighborhood's fashionable heyday. The ornate Patterson House at number 15 – now the **Washington Club**, a private women's social club – temporarily housed the Calvin Coolidges during White House renovations in 1927. When Charles Lindbergh returned from Paris, he received his presidential welcome here. The **Sulgrave Club**, another private social club housed in the triangu-

lar manse at Massachusetts and P Street, was named after George Washington's English ancestral home.

If you follow New Hampshire Avenue to 18th Street, you'll find an unusual pocket park cornering Church Street. With its backdrop of 19th-century ruins from the original Gothic-style **St Thomas' Episcopal Church** (1772 Church Street, NW; tel: 202-332-0607), it conjures romantic images of the English countryside. Toward the middle of this narrow, gas-lighted Victorian street is the **Church Street Theater** (1742 Church Street, NW; tel: 202-265-3748), a "rental house" to local and non-local companies, which stages plays, dance and readings from works of literature.

Another impressive Beaux-Arts landmark commands the corner of 18th and Massachusetts. Now the headquarters of the **National Trust for Historic Preservation**, the original McCormick Apartments, built in 1917, were once *the* prestige address in town. Each of its six apartments

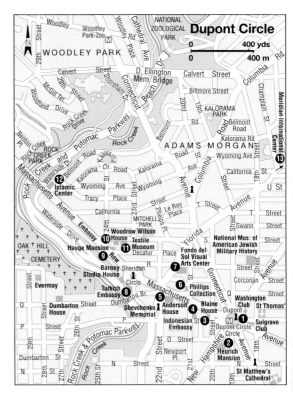

Map on page 206

– one per floor – featured such amenities as silver- and gold-plated doorknobs, wine closets, and silver vaults. Andrew Mellon, one of many notable residents, amassed the art collection here which spawned the National Gallery of Art. The Trust bought the building in 1977 from its think-tank neighbor on Massachusetts Avenue, the **Brookings Institution**.

Places to eat

Also in the neighborhood, N Street between 17th and 18th offers a few choice options for drinking or dining. The Victorian **Tabard Inn**, (1739 N Street NW; tel: 202 331-8528) with its homey ambience, is particularly cozy in winter when the fire is going. There is the Asian-inspired **Topaz Restaurant** in the **Topaz Hotel** (1733 N Street NW; tel: 202-393-3000). Across the street, the portals of the **Iron Gate Inn** (1734 N St NW; tel: 202 737-1370) open on a romantic high-walled garden, leading back to a secluded courtyard and the carriage-house restaurant. There are lots of other choices a few blocks up 17th Street.

When the German brewer and entrepreneur Christian Heurich came to Washington in 1871, he founded the successful Heurich Brewing Company in Foggy Bottom on the present site of the Kennedy Center and subsequently built a 31-room, Romanesque-Revival castle on the corner of New Hampshire and 20th. Today the lavish **Heurich Mansion** ❷ (1307 New Hampshire Avenue NW; Mon–Sat, 10am–4pm; free admission; tel: 202-785-2068), fully restored down to the wallpaper, stands as a monument to Victoriana. In the late 1980s Heurich's grandson reincarnated the family business as the Olde Heurich Brewing Company. While the new designer version of the family lager is now bottled in Utica, New York, DC claims Heurich as its own – and only – brew.

At 2000 Massachusetts Avenue is the **Blaine House** ❸, Dupont Circle's oldest surviving mansion. It was built in 1881 by Maine Senator James G. Blaine, co-founder of the Republican Party, two-time

BELOW: artist at work, Dupont Circle.

secretary of state, and unsuccessful Republican presidential candidate (he lost to Grover Cleveland). In 1901, the inventor George Westinghouse bought the house and lived there until his death in 1914. In the 1920s Blaine House was the site of the Japanese legation, and today it houses several professional offices.

Diplomat's alley

Fittingly, Massachusetts Avenue west of Dupont Circle is known as **Embassy Row**, where Washington's diplomatic community is concentrated. Colorful flags and coats-of-arms of more than 175 nations decorate the embassies and chanceries, both on and off the avenue. While most of the embassies are open only for official business, several offer concerts and other cultural events *(see page 213)* to which the public is invited. If you call ahead you may be invited inside for a tour of the opulent **Indonesian Embassy** ❹ at 2020 Massachusetts Avenue (tel: 202-775-5200), the former home of Thomas Walsh,

an Irish immigrant who struck it rich in the gold mines of Colorado. Walsh's daughter Evalyn – the last private owner of the Hope Diamond, which is now on display in the National Museum of Natural History – sold the home to the Indonesian government in 1951.

When the diplomat Larz Anderson and his wife planned their 50-room palace at 2118 Massachusetts Avenue, it was with the idea that it would become the future headquarters for the Society of the Cincinnati, an elite fraternity founded by Revolutionary War officers and restricted to their male descendants. The facade of the **Anderson House** ❺ (2118 Massachusetts Avenue NW; open Tues–Sat, 1pm–4pm; free admittance; tel: 202-785-2040) only hints at the impossibly opulent interior of this turn-of-the-century house-museum, filled with the couple's astounding international collection of art and furnishings, and the ghosts of privilege and power. The first floor houses Revolutionary War artifacts and a library of reference works on the war and local history, and the second remains as it was originally furnished with 18th-century paintings, Belgian tapestries, and enormous chandeliers.

The prestigious and private **Cosmos Club** across the street occupies the former mansion of railroad tycoon Richard Townsend. In contrast to Dupont Circle's typical lavish style is the austere **Friends Meeting House** (2111 Florida Avenue, NW; tel: 202-483-3310), built in 1930 for Quaker President Herbert Hoover, a few blocks away. The plain interior has a traditional "facing bench," designed for "weighty" Friends to pronounce their messages. At 1523 22nd Street, the **Brickskeller** (tel:202-293-1885) has more than 500 brands of beer from around the world.

An intimate museum

Of all the museums in Washington, the **Phillips Collection** ❻ at 21st and Q streets, (open Tues–Sat, 10am–5pm; Sun, noon–7pm; Thurs until 8.30 pm; closed Mon, Jan 1, July 4, Thanksgiving, Dec 25; admission fee; tel: 202- 387-2151 or visit www.phillipscollection.org) across from the Ritz-Carlton Hotel, is certainly the homiest

LEFT: chefs' market at Dupont Circle.

Map
on page
206

and most intimate. Indeed, the country's oldest museum of modern art started out in 1921 as a two-room gallery in Duncan and Marjorie Phillips's brownstone. Duncan was the grandson of a Pittsburgh steel magnate who put his fortune into his eclectic art collection, some 2,500 works that include everything from Degas and Renoir (*Luncheon of the Boating Party*, the museum's most renowned treasure, is here) to Picasso, O'Keeffe, Klee, and Rothko. There is a café next to the museum shop downstairs, and free concert are held in the grand music room. Call for details about the museum's popular program of "Artful Evenings," held on Thursday nights and featuring lectures and discussion groups, concerts, and receptions.

In a different vein, a block up on R Street is the **Fondo del Sol Visual Arts Center ❼** (2112 R St., NW; open Wed–Sat, 12.30pm–5.30pm; $5 donation; tel: 202-483-2777) three floors of exhibits, art and video cultural history of the Americas. A community-oriented multimedia

museum with a Hispanic focus, it was founded by a group of pan-American artists and writers to highlight the region's multicultural heritage. The gallery primarily showcases work by contemporary artists and craftspeople, but it also has an eclectic collection of folk and pre-Columbian art.

Kalorama

To the west, Massachusetts Avenue makes an elegant turn at **Sheridan Circle**, the embassy-ringed and tightly secured nucleus of the **Kalorama** neighborhood. Formerly a diplomat's country estate, the name means "beautiful view" in Greek. A few blocks to the north, the quiet streets of this exclusive neighborhood, epitomized by **Kalorama Circle**, are a feast of elegant Tudors, Normans, and Georgians dating from – and epitomizing – the 1920s.

In front of the **Romanian Embassy** (1607 23rd Street, NW), a curbside memorial marks where a car bomb killed the Allende-appointed Chilean ambassador

Orlando Letelier and, unintentionally, his passenger Ronni Moffitt, on September 21, 1976; the job was linked to Chilean secret police and army officials in the Pinochet regime.

Moving clockwise around the circle to 1606 23rd Street, NW, you'll come to the **Turkish Embassy ❽**, one of Washington's grandest mansions and the former home of Edward Everett who made his millions by, prosaically enough, inventing the bottle cap. Mrs Everett, an opera singer and popular Washington hostess, gave performances here and was joined by world-renown divas of the day.

On the circle at number 2306 Massachusetts Avenue, NW, is the **Alice Pike Barney Studio House**, a special but little-known landmark. At the turn of the 20th century, Barney, a flamboyant artist, playwright, and producer decided to liven up Washington's dull cultural life, and so she commissioned Waddy Butler Wood to design an all-purpose home and studio, where for years she ran the only Paris-style

salon in town. In 1960 Barney's daughters gave the well-used Studio House to the National Museum of American Art, which serves as custodian but does not offer tours to the public.

At 2349 Massachusetts Avenue is the stately **Hauge Mansion ❾**, now the **Cameroon Embassy.** Designed by George Oakley Totten, Jr., one of Washington's team of Beaux-Arts architects, it was first the home of Christian Hauge, Norway's first minister to the US. Hauge's American widow Louise hosted some of Washington's grandest parties in this home with its tower and candlesnuffer roof inspired by a 16th-century French château. In the 1930s the Czechoslovakian government established its foreign mission here, then in 1972 sold the home to Cameroon.

Woodrow Wilson House

The brick Georgian Revival townhouse at 2340 S Street, NW, the work of Waddy Butler Wood, is the **Woodrow Wilson House ❿** (open Tues–Sun, 10am–4pm by guided tour only; admittance fee; closed major holidays; tel: 202-387-4062 or visit www.woodrowwilsonhouse.org). It was the home of Woodrow and Edith Wilson after they left the White House in 1921.

The 28th president, who led the nation during World War I and in 1918 proposed the formation of the League of Nations to foster world peace, bought the house as a surprise for his second wife, presenting her with a piece of earth from the garden and the front door key, in Scottish tradition. Mrs Wilson lived here until she died in 1961.

Next door is the **Textile Museum ⓫** (2320 S Street, NW; open Mon–Sat, 10am–5pm, Sun 1pm–5pm; closed major holidays; $5 donation; tel: 202-667-0441 or visit www.textilemuseum.org), a fabulous private museum dedicated exclusively to the display and study of international textile arts, both historic and contemporary. Its collection of Oriental carpets is unmatched in the world, thanks to the museum's founder, George Hewitt Myers, who became hooked after buying his first rug for his Yale dorm room. He opened the museum in 1925 next door to his home and kept right on collecting. Today the

LEFT: Woodrow Wilson House.

Map on page 206

museum has 1,400 carpets and 13,000 textiles, and the its muted galleries feature permanent and special exhibits. The gift shop has a superb collection of books on textile arts, along with crafts, jewelry, and yarn. The small research library contains more than 13,000 volumes. Curators will personally advise about your own textiles, but it's best to call ahead.

You'll pass **Mitchell Park** if you continue up the hill. It was named for Mrs E. N. Mitchell who, in 1918, bequeathed the land to the city in exchange for the perpetual care of her pet poodle's grave. Dear Bosque's grave, an unmarked and unimpressive cement block encircled by a blue chain, is in the playground.

Across from Mitchell Park is a charming architectural interlude, a sweet spot, a secret garden called **Decatur Terrace**. The formal stairs and fountain, which the neighbors generously refer to as the "Spanish Steps," link 22nd and S streets.

Along Massachusetts Avenue, near Rock Creek Park, is the **Islamic Center** ⓬ (tel: 202-332-8343) at Massachusetts and Belmont Road. This is the spiritual and cultural mecca for Washington's estimated 65,000 Muslims. Designed by the Italian architect Mario Rossi, a convert to Islam, it's also Embassy Row's most exotic structure featuring the minaret, which rises to 160 ft (49 meters). The mosque itself, situated to face Mecca as all mosques do, is embellished with 7,000 deep blue Turkish tiles, a two-ton chandelier from Egypt, Persian carpets and Iraqi stained glass.

Ethnic interest

In the mid-1950s a neighborhood organization was formed to promote cooperation among racially segregated residents. The group took its name by combining the names of two area elementary schools: the all-white Adams and the all-black Morgan. The Adams Morgan organization left its name as a legacy to and symbol of this uniquely diverse community just north of Dupont Circle. Every September in honor of **Adams Morgan Day**, the neighborhood throws a giant street party, with music stages, ethnic food booths, crafts, and major crowds.

Along with the artists and bohemians, who moved in in the 1960s, came an influx of Hispanic refugees and immigrants, establishing Adams Morgan as DC's Latin quarter and Greenwich Village, rolled into one. The entire community comes out to celebrate itself, joined by thousands, during the **Hispanic Festival**, which is traditionally held the last weekend in July and features pan-Latin food, the sizzling sounds of salsa and marimba bands, wonderful African crafts, and a colorful parade of nations.

The heart of Adams Morgan is at the "T" intersection of **18th Street and Columbia Road**. The closest Metro stop is Woodley Park. Follow Calvert Street across the **"Duke" Ellington Memorial Bridge** – named, of course, for the preeminent jazzman and native son – and you're here.

The plaza by the Suntrust Bank is the village green of Adams Morgan. By day on Saturdays, when the neighborhood is a delight to explore, it becomes a **farmer's market**. On summer evenings you can

RIGHT: South American festival dancer.

expect to see anything from Andean pan-pipers to West African stilt-walkers, a kids' bucket-drum combo to a Delta-blues har-monica player, young rappers to Michael Jackson types.

Inevitable development – disparaged as "Georgetownization" – has transformed the once higgledy-piggledy stretch of 18th Street into a funky-hip quarter, where exotic aromas – especially Ethiopian and Eritrean – waft from the myriad ethnic restaurants, and the young and the restless come to see and be seen on crowded week-end nights. The flavor here is decidedly Left Bank à la Washington, an interna-tional smorgasbord of sidewalk cafés, trendy bars and clubs, bookstores and spe-cialty shops, antique and rummage stores, and galleries.

Alas, Adams Morgan is in danger of becoming a victim of its own success. Clubs, bars and restaurants spring up with regularity, only to close down or be taken over by someone else a few months later. At the moment, popular hot spots include

Habana Village (1834 Columbia Rd., NW; tel: 202-462-6310), where you can dance (or learn to dance, the club offers lessons) the salsa, the tango, and the samba, and the **Felix Lounge** (1834 Columbia Rd., NW; tel: 202-483-3549), a sort of little UN that mixes a French/Asian menu with a regular staple of local rock, Latin jazz, and country music performers.

Idle Time Books (2410 18th Street, NW; tel: 202-232-4774) is well-stocked with second-hand treasures, and local coffeehouses such as **Tryst** (2459 18th Street, NW; tel: 202-232-5500) and **Franklin's** (2000 18th Street, NW; tel: 202-319-1800) are hang-outs offering light and inventive menus.

Hispanic quarter

East of 18th Street, Columbia Road changes abruptly. Salsa blares from boom-boxes, Latino shops, and vendors' booths, and Spanish is the language of the street.

North of here Adams Morgan flows into the residential neighborhood of **Mount Pleasant**. The heart of Washington's poor and dispossessed Hispanic community is also undergoing block-by-block gentrifi-cation, though progress is slow. In the spring of 1991 a clash between black police officers and Hispanics here escalated into two days of riots that spilled into Adams Morgan. Seedy and with little to offer vis-itors, it's a section of the city best avoided.

A quiet refuge away from the fray is provided by **Meridian International Center ⑬** (1624–1630 Crescent Place, NW; Wed–Sun, 2pm–5pm; closed Mon, Tue, and major and federal holidays; admission free; tel: 202-939-5568), which crowns the ridge between Crescent Place and Belmont Street, off 16th Street. Set on the magnificently landscaped hillcrest are two historic mansions designed by the ever-versatile and prolific John Russell Pope. The one is a very French Louis XVI-style château, whose lovely pebbled gar-den is lined with pollarded Spanish linden trees, while the other recalls an English country manor. Both now belong to this international educational and cultural foundation, whose galleries and concerts are open to the public. ❏

LEFT: Adams Morgan annua street fair.

Map on page 204

Diplomatic Pleasures

One of Washington's hidden treasures is the on-going season of embassy-sponsored cultural events open to the public. As one cultural attaché quoted in the *Washington Post* put it, "Culture is the bearer of the most important information about a country. This is how we can really make ourselves understood."

Until the Great Depression forced many of Washington's wealthiest to sell their Massachusetts Avenue mansions, much of the city's diplomatic community was concentrated in the Meridian Hill section. In 1931, Great Britain and Japan built new embassy compounds along the avenue and were soon followed by several diplomatic missions, which bought the formerly private residences, thereby dubbing the area Embassy Row.

The embassies offer everything from jazz concerts and violin recitals to Warhol retrospectives (his parents were Slovakian immigrants), wine tastings, benefit dinners, plays, films, operas, birthday celebrations, rugby matches, mime performances, art exhibits and more. You can enjoy a glass of Burgundy or a Perugina chocolate, experience Finnish silk or Australian glass, taste an authentic Viennese pastry or a Polish kielbasa.

The embassies also stage culture events at venues such as public parks and art galleries and in cooperation with various other organizations such as the Russian Cultural Institute (1825 Phelps Place, NW; tel: 202-265-3840; www.russianembassy.org) and Germany's Goethe Institute (814 7th Street, NW; tel: 202-289-1200).

Besides being enjoyable, the embassy-sponsored events are a good way to meet new and interesting people, and to learn a little something else about the world. For schedules of events and details, check the following websites: www.EmbassyEvents.com; www.EmbassySeries.org; www.embassy.org; or contact the Inter-American Development Bank's Cultural Center at 1300 New York Avenue, NW, tel: 202-623-3558; www.iadb.org.

Some of the most active embassies are: **Australia**, 1601 Massachusetts Avenue, NW;

tel: 202-797-3176; www.austemb.org
Austria, 3524 International Court, NW; tel: 202-895-6776; www.austria.org
Canada, 501 Pennsylvania Avenue, NW; tel: 202-682-7712; www.canadianembassy.org
Ecuador, 2535 15th Street, NW; tel: 202-234-7200; www.ecuador.org
Egypt, 3521 International Court, NW; 202-895-5463; www.embassyofegyptwashingtondc.org
France, La Maison Française, 4101 Reservoir Road, NW; tel: 202-944-6091; www.la-maison-francaise.org
Italy, Instituto Italiano di Cultura, 2025 M Street, NW, Suite 610; tel: 202-223-9800; www.italcultusa.org
Japan, The Japan Information and Culture Center, 1155 21st Street, NW; tel: 202-238-6949; www.usemb-japan.go.jp/jicc/calendar
Korea, The Korean Cultural Service, 2370 Massachusetts Avenue, NW; tel: 202-797-6343; www.koreaemb.org
Switzerland, 2900 Cathedral Avenue, NW; tel: 202-745-7900; swissemb.org
United Kingdom, www.britainusa.com/arts/events (contact by web only).   ❏

Map on page 218

THE UPPER NORTHWEST

This pleasant stretch of country in the middle of the city features Rock Creek Park's shady acres and all creatures great and small at the National Zoo

"**B**ut in the country" was how heat-and-humidity-weary Washingtonians described Upper Northwest until well after the Civil War. Here, in this rural retreat above Rock Creek, woodland was dense and temperatures were 10 to 15 degrees cooler than downtown. On hot summer evenings, Presidents Van Buren, Tyler, Buchanan, and Cleveland came here by carriage from downtown to their gracious country "cottages."

Tempting triangle

With the completion of the first trolley bridges over Rock Creek at the turn of the 20th century, the city expanded northward. Cleveland Park and Woodley Park became fashionable year-round communities and luxury hotels and grand apartment buildings soon graced the broad Connecticut, Wisconsin, and Massachusetts avenues. Today, Upper Northwest – a triangle of land formed by 16th Street, Massachusetts Avenue, and DC's western border with Maryland – retains much of its original rural appeal. Despite commercial centers rapidly rising and expanding around Metro stations, unspoiled swaths of woodland, sylvan parks, and tree-lined streets give gracious sanctuary to the natural world.

Threading through the entire neighborhood is the city's largest park and prime natural attraction, **Rock Creek Park ❶** (open daylight hours; free admission; tel: 202-895-6000 or visit www.nps.gov/rocr). The park's 1,754 acres (710 hectares) were purchased by Congress in 1890 for its "pleasant valleys and deep ravines, primeval forests and open fields... its repose and tranquility, its light and shade..." About 4 miles (6.5 km) long and up to a mile wide, the park includes extensive trails and paths for hiking, a marked bike route, picnic areas, recreation fields, **tennis courts** (16th and Kennedy Streets, NW; tel: 202-722-5949), an 18-hole **golf course** (16th and Rittenhouse streets, NW; tel: 202-822-7332), a horse center (Military and Glover roads, NW; tel: 202-362-0117) and more.

Historical sites in the park include **Peirce Mill**, a 19th-century gristmill waiting to be restored, and the nearby 1880s **log cabin of Joaquin Miller**, "Poet of the Sierras."

An open-air theater within the park is **Carter Barron Amphitheatre ❷** (16th Street and Colorado Avenue, NW; box office open noon–9pm on day of concert; free tickets distributed on concert day; fee for other events; tel: 202-426-0486 for concert information; www.nps.gov/rocr). It features a summer festival of pop, rock,

PRECEDING PAGES: Hillwood House. **LEFT:** a porch in Cleveland Park. **RIGHT:** Rock Creek Park.

and jazz. Of particular appeal to children is the **Rock Creek Park Nature Center** ❸ (5200 Glover Road, NW; Wed–Sun, 9am–5pm; closed Jan 1, July 4, Thanksgiving, Dec 25; tel: 202-895-6070). Here, National Park Rangers lead guided nature walks and help children learn about the wildlife in the park. The hands-on Discovery Room is a favorite. So is the **Rock Creek Park Planetarium** (in the Nature Center; free shows at 1pm Sat and Sun for children 4 to 7 years and at 4 pm for children 7 and older).

The **tennis center** is the site of top-level tournaments, to which Washingtonians flock, especially in August. Despite the heat, it hosts the important Legg-Mason Tennis Classic.

For an old-fashioned motor tour, follow the 10-mile-long (16-km) **Beach Drive,** which winds its narrow way through the entire park. West of Rock Creek Park, the roadless **Glover-Archbold Park** ❹ offers picnic areas and a 3-mile (5-km) nature trail. **Battery-Kemble Park** ❺, once a Civil War outpost, is an escape from the sounds of the city. Narrow trails through this jungly park join the **C&O Canal towpath** (see page 223).

The main Upper Northwest thorough-

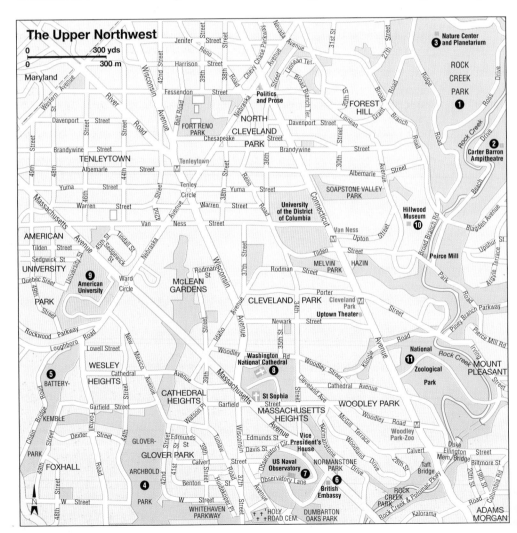

Map on facing page

fares – Massachusetts, Wisconsin, and Connecticut avenues – slice through the neighborhood on a northwest course into the Maryland suburbs. Just after you cross Rock Creek on Massachusetts from downtown is the **British Embassy** ❻ (3100 Massachusetts Avenue, NW; tel: 202-588-6500) on the left.

There is a larger-than-life statue of Winston Churchill giving the "V for Victory" salute (some joke that he's hailing a cab). The base of the statue contains a time capsule to be opened in 2063, the centenary of the conferring of Churchill's honorary US citizenship by President John F. Kennedy.

The British chancery and residence, an adaptation of an English country house, was designed by the illustrious British architect Sir Edwin Lutyens in 1931. The embassy's celebration of the Queen's birthday – which takes the form of a garden party with strawberries and Devonshire cream, plus a sprinkling of notables – is a highlight of Washington's spring social season.

BELOW: the Vice-President's House.

Observation decks

Massachusetts Avenue then curves in a gentle arc around **Observatory Circle** to distance rumbling automobiles from the sensitive instruments housed in the **US Naval Observatory** ❼ (free tours by reservation only on alternating Mondays, 8.30pm–10pm; tel: 202-762-1467 or visit www.usno.navy.mil). The Observatory charts the position and motion of the celestial bodies, measures the earth's rotation, determines precise time and maintains the nation's master clock. Weather permitting, visitors may look through the 26-inch refractor telescope used to discover the Martian moons in 1877.

Also on the Observatory grounds is the Victorian-era **house of the Vice President of the United States** originally built for the superintendent of the Naval Observatory. This turreted, informal-looking building is a far cry from the White House, for it is little known, little publicized, and seen even less. The house was co-opted from the Navy in the 1970s, as it

was felt that an official home located on the grounds of a military post would be easier – and less expensive – to protect than a private residence elsewhere in the city. Needless to say, the VP's home is not open to the public.

Dominating the city's skyline from the crown of Mount Saint Alban at Massachusetts and Wisconsin avenues is the towering **Cathedral Church of St Peter and Paul**, also known as the Washington Cathedral or **National Cathedral ❽** (May–Sept, Mon–Fri, 10am–9pm; Oct–Apr, Mon–Fri, 10am–5pm, Sat all year 10am–4.30pm, Sun all year 8am–4.30pm; tel: 202-364-6616). President Theodore Roosevelt laid the corner stone of this magnificent Gothic-style cathedral in 1907. Eighty-three years and 300 million tons of Indiana limestone later, one of the world's largest ecclesiastical structures was officially completed.

The cathedral serves as the seat of the Episcopal Diocese of Washington, but it welcomes people of all faiths to its services, concerts, and annual festivals. This is truly a national cathedral – a fact made gloriously evident through the stunning, quite modern stained-glass windows that tell stories about American history and American heroes. The **Space Window** in the nave commemorates the scientists and astronauts of Apollo 11 and includes a piece of moon rock retrieved on that mission. Other windows depict the home life of Martha and George Washington at Mount Vernon, Lewis and Clark's explorations of the American Northwest, the struggle for religious freedom in Maryland during the 17th century, and the events of World War II. Thomas Jefferson, Robert E. Lee, Abraham Lincoln as well as his mother and stepmother are glorified in other scenes.

Allow plenty of time not only to study these radiant windows, but to gaze at the nave's soaring vaulted ceiling, to explore the several small chapels, and to gaze at the kaleidoscopic **West Rose Window** as it glows in the setting sun. The best views

BELOW: the National Cathedral was built from 300 million tons of Indiana limestone.

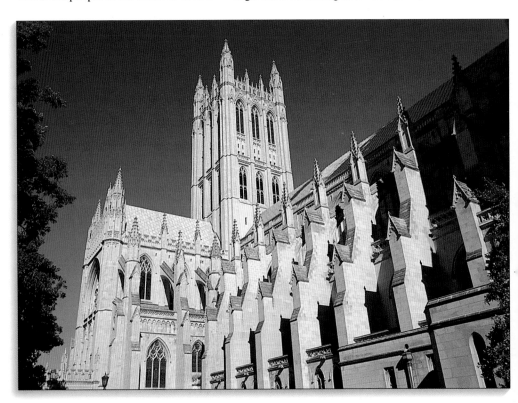

Map on page 218

of Washington and of the cathedral's exterior carvings are from the 70 windows in the **Pilgrim's Observation Gallery**, the highest vantage point in the city. The crypt level includes the London Brass Rubbing Center, the Rare Book Library, and the gift shop. A stroll on the peaceful cathedral grounds will lead you through a 12th-century Norman arch into the medieval-style **Bishop's Garden**, to the Herb Cottage shop, and greenhouse.

The largest Greek Orthodox church in the United States graces the block just south of the National Cathedral. **Saint Sophia** (36th Street and Massachusetts Avenue, NW; tel: 202-333-4710) is famous for its stunning and intricate mosaics that decorate the edifice and interior dome of the church. The best time to visit the church, and to hear the a cappella choir singing from the circular balcony, is during Sunday morning services.

Farther north on Massachusetts at **Ward Circle**, is **American University** ❾ (4400 Massachusetts Avenue, NW, tel: 202-885-1000), an independent university chartered by Congress in 1893. The public is invited to attend on-campus lectures, concerts, movies and art exhibitions.

From Ward Circle, take Nebraska Avenue northeast to Connecticut Avenue, then head south a few blocks to **Politics and Prose** (5015 Connecticut Avenue, NW; tel: 202-364-1919; www.politics-prose.com), a bookstore specializing in the books and interests of local authors. Readings, book-signing parties, and wine-and-cheese receptions are held regularly here.

Continuing south leads you into **Cleveland Park**, a residential neighborhood recently designated an historic district. Named after President Grover Cleveland, who spent the summers in the area, Cleveland Park is dominated by grand, late 19th-century homes, mostly in Queen Anne or Georgian revival styles. Stroll down Newark Street to Highland Place, then onto Macomb Street to see a panoply of porches, turrets, Palladian windows, gabled roofs, balconies, and white picket fences.

At Connecticut and Ordway is the **Uptown Theater** (tel: 202-966-5400, or check the listings in the *Washington Post*), an art-deco gem that features epic-size movies on a big screen. Traffic gets tangled in front of the theater when private premier shows lure stars and fans. There's a convenient strip of shops across the street with neighborhood restaurants including **Vace** (3315 Connecticut Avenue, NW; tel: 202-363-1999), an authentic delicatessen featuring handmade pasta.

Grand mansions

A bit farther north and off Connecticut Avenue is **Hillwood** ❿ (4155 Linnean Avenue, NW; $10 adults, $8 seniors, $5 students; by appointment only; admission fee; tel: 202-686-5807 or 1-877-HILL-WOOD; www.hillwoodmuseum.org), the estate of the late Marjorie Merriweather Post, Washington socialite and heiress of the Post cereal fortune. Hillwood is a rather quirky place, comprising a 40-room Georgian mansion, a museum of decorative arts and wonderful gardens. The inside of the house can be visited only by

RIGHT: Hillwood.

joining the two-hour guided tour to view 18th-century French decorative objects and masterpieces of Russian Imperial Art, many pieces collected when Mrs Post and her fourth husband served as the first American envoys to Moscow after the Russian Revolution.

If you prefer to go it alone, you can explore Hillwood's other buildings and 25 acres (10 hectares) of grounds. Here you'll find a *dacha* housing Russian folk art; an Adirondack-style cabin featuring American Indian artifacts, the greenhouse protecting thousands of orchids, and the pleasantly airy café building offering light fare and an afternoon tea with scones. The grounds also have formal Japanese and French gardens and a sweeping view over Rock Creek all the way to the Washington Monument. In the pet cemetery, Mrs Post's two dogs – Café au Lait and Crème de Cocoa – are buried amid dogtooth violets, weeping dogwoods and forget-me-nots.

Woodley Park, just south of Cleveland Park, was also a summer retreat for several US presidents. The neighborhood takes its name from Woodley Manor (now Maret School, at 3000 Cathedral Avenue), where presidents Van Buren, Tyler, Buchanan, and Cleveland headed in summer. Woodley Park's most famous residents these days, however, are **Marilyn Monroe** (painted on a huge mural at Calvert and Connecticut) and two giant pandas.

Map on page 218

At the zoo

The pandas live at the **National Zoological Park** ⓫ (3001 Connecticut Avenue, NW; free admission; open daily May 1–Sept 15, grounds 6am–8pm and buildings 10am–6pm; open daily Sept 16– Apr 30, grounds 6am–6pm and buildings 10am–4.30pm; tel: 202-673-4717 or visit www.si.edu/natzoo).

Established in 1889 by the Smithsonian Institution to protect an acquired herd of buffalo, the zoo now accommodates some 4,000 animals living in semi-natural environments on 163 wooded acres (66 hectares). The zoo's main trails have been beautifully re-landscaped according to the original plans of architect Frederick Law Olmsted, Jr. These paths link together a dozen looping side paths that lead to outdoor and indoor caged habitats supporting rare blue-eyed white tigers, lowland gorillas and other unusual animals.

Two recent additions to the zoo include the animal **Think Tank**, which helps to understand the way scientists investigate thinking, and the **Pollinarium**, inside the greenhouse. Popular attractions for children include the **Panda House**, home of Tian Tian and Mei Xiang, a breeding pair of panda bears on loan from the People's Republic of China, and the Reptile Discovery Center which houses the world's largest lizard, the Komodo dragon.

For the "peaceable kingdom" experience of the zoo, visit the grounds in the morning before the buildings open or in the evening hours after the buildings close. This way, you can enjoy the meandering pathways before crowds of homo sapiens and baby carriages create mile-long conga lines. Many of the animals are fed and are most active in the morning before 10am. ❏

LEFT: Marilyn looks down on Connecticut and Calvert.

The C & O Canal

Slicing a narrow liquid strip through the northwest sections of Washington is a piece of Victorian-era true grit. The C&O Canal, which begins in affluent Georgetown and lopes its way alongside the Potomac River for 185 miles (300 km) until it reaches the Allegheny Mountain town of Cumberland, Maryland, is a relic from another time. An ambitious transportation scheme that went awry, it survives as a playground for outdoor-loving Washingtonians.

The Georgetown section of the canal is well-preserved and upbeat. During warm weather months, restored canal boats are drawn by mules plodding their way along the towpaths, and guides in period costume entertain their audience with tales of the rise and demise of the canal system. (Hour-long rides depart from 1057 Thomas Jefferson Street and are offered April through mid-October; tel: 202-653-5190; www.nps.gov/choh).

The journey is slow, but not too slow. On board, the smells of dankness and fresh mint intermingle, and the mules vie for towpath space with joggers. Soon after, the vista opens up with views of the Potomac and, behind the river, high-rise suburban buildings. Here and there on the banks is a large cardboard box, which looks suspiciously as if it has been used as a temporary home.

About 10 miles (16 km) farther up the canal lies the wealthy commuter community of Great Falls. In this much more rural setting, passengers may catch a glimpse of a white-tailed deer or a raccoon. An abandoned goldmine, worked from the 1860s through the 1940s, is located nearby, and hastily dug graveyards in the hills pay tribute to those who labored to build this extensive waterway but succumbed to diseases.

The best times to view wildlife are either early or late in the day. The more exotic inhabitants include turkey vultures, the great horned owl, beaver and, rumor has it, bears.

Less than 10 miles (16 km) beyond Great Falls, the canal abruptly runs dry. It remains so with the exception of a brief rewatered section near its terminus 150 miles (240 km) away in Cumberland. As originally conceived, the canal was to have stretched well beyond Cumberland all the way to the Ohio River, hence its name: the Chesapeake and Ohio Canal. There, the plan went, canal boats would load up with bulky raw materials and carry them eastward to the Chesapeake Bay and beyond. But it never happened.

The C&O Canal had its problems from the start. On July 4, 1828, President John Quincy Adams turned the first shovelful of dirt only to encounter roots and then rocks. Seventy-four lift locks later, in 1850, the canal reached Cumberland, but there it halted, having cost slightly more than $11 million.

Labor shortages and unrest and frequent flooding from the nearby Potomac plagued construction. Materials were difficult to come by, there were never enough funds, and there were constant legal battles with the upstart Baltimore and Ohio Railroad (the B&O).

The dawning of the railroads sealed the fate of the C&O Canal as a viable means of carrying goods. The B&O was inaugurated the very same Independence Day as the canal, rendering this languid mode of transportation obsolete before it had even fully begun. ❑

RIGHT: the canal is now a leisure spot.

Map on page 228

LOGAN CIRCLE AND THE NORTHEAST

This fringe district outside the bounds of "tourist" Washington has several choice religious sites, two notable off-beat theaters, and, at the National Arboretum, a popular recreational expanse

Picture Logan Circle – that Victorian-ringed rotary intersected by Rhode Island and Vermont avenues – as the focal point for the vast, amorphous area north of downtown and Capitol Hill, and east of Dupont Circle, that extends through the northeast quadrant. Bounded on the west by 16th Street, it is raggedly defined on the south by Massachusetts Avenue, North Capitol Street, Florida Avenue, and Benning Road to the Anacostia River. The Maryland border contains the rest.

Once fashionable

The circle itself appeared originally on city architect Pierre L'Enfant's plan not as a circle, but as a triangle. The area around it sat neglected until the 1870s when what had been primarily a field for grazing cattle was transformed into a stylish Victorian neighborhood and called Iowa Circle. In the 1930s, the circle was renamed for Illinois Senator John A. Logan, energetic in both his support of Union veterans after the Civil War and his lifelong loathing of the South for what he regarded as its treachery. In the center of **Logan Circle** ❶, where a statue by American sculpture Franklin Simmons commemorates him, you can see the impassioned Union-general-turned-senator astride his horse.

Gentrified and exclusively white, Logan Circle was one of the city's poshest neighborhoods in the 1930s and 1940s, remarkable for its concentration of three- and four-story Victorian mansions and town-houses. As the residents moved up and out, many to still larger mansions around Massachusetts Avenue, black Washingtonians made the neighborhood theirs and for a time Logan Circle was a center of black intellectual and social life. By the end of World War II, the neighborhood fell into decline and remains one of the city's more troubled areas, known for drug dealers, small-time pimps, and prostitutes.

It was near here that DC's inimitable former mayor Marion Barry was arrested for crack cocaine possession at the Vista International Hotel on **Thomas Circle**. Of late, tighter controls and impressive restoration efforts seem to have reversed the trend and Logan Circle is clearly being yuppified. Still, it's wise to confine your wanderings around here to daylight hours.

Housed in a pristine Victorian house one refurbished block off the circle at 1318 Vermont Avenue is the **Bethune Museum-Archives for Black Women's History** ❷ (tel: 202-673-2402; open Mon–Sat 10am–

4pm; free admittance; guided tours given every hour on the hour). It is named for the pioneering civil rights activist and educator Mary McLeod Bethune, who founded the National Council of Negro Women here in her one-time residence. Don't expect to see a "preserved" home or even a single personal article. This no-frills museum tells the history of black women, but its exhibits are surprisingly lean.

A choice of stages

West of the circle, scattered along the seedy stretch of 14th Street now trendily tagged the "Uptown Arts District" – a reference to the area's city-supported cultural revival – is DC's funkiest and most enterprising alternative theaters, along with a growing batch of unpretentious restaurants and cafés, most notable the Spanish-inspired **Mar de Plata** (1410 14th Street, NW; tel: 202-234-2679).

On the corner of 14th and P streets, the **Studio Theatre ❸** (tel: 202-332-3300; www.studiotheatre.org) occupies an old car warehouse – complete with an industrial-sized elevator capable of transferring fully assembled sets from studio to stage – which has been outfitted as a state-of-the-art theater. Studio productions range from classic to contemporary, and feature artists from around the country.

Farther up the street, the **Source Theatre Company ❹** (1835 14th Street, NW; tel: 202-462-1073; www.sourcetheatre.com) has a stark, black interior where the audience is never more than six rows away from the actors. The Source spotlights work by new playwrights, and you can sometimes catch a late-night comedy show here.

A regular summer event, the month-long **Washington Theatre Festival** showcases new plays by local – and variable – talent, staged here and at venues around town.

The ambience changes considerably when you get to 16th Street, two blocks to the east. Perfectly and impressively aligned with the White House, the wide, terraced boulevard was designed as *the* approach to the District by car. It also runs

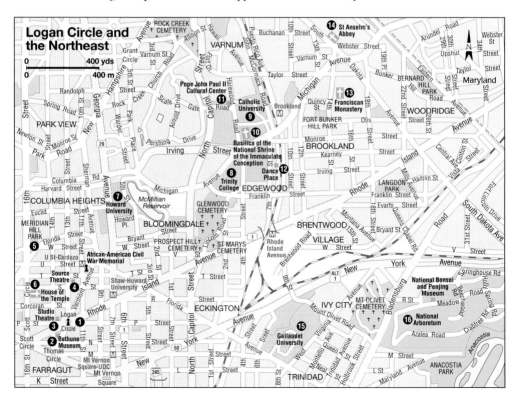

Map on facing page

alongside the city's central meridian, surveyed in 1816 as a possible alternative to the prime meridian at Greenwich – an unviable idea, as it turned out. The line's presence is recalled around **Meridian Hill**.

Once considered as the site for a new presidential residence, Congress instead purchased 12 hillside acres between Florida Avenue and Euclid Street for a public park. Not just a green space, **Meridian Hill Park ❺**, also known as Malcolm X Park, is reminiscent of the formal gardens of 17th-century France and Italy. Sadly, the park is ringed with crumbling and high-crime housing projects, making it a bad idea to visit, even during daylight hours.

Sacred places

Among the many houses of worship along 16th Street, which range from Baptist to Buddhist, Universalist to Unification, Swedenborgian to **Scottish Rite**. The latter – not a church, but nevertheless sacred – is between R and S streets and is without a doubt one of the most remarkable.

The Masons commissioned the capable John Russell Pope to design their **House of the Temple ❻** (1733 16th Street, NW; tel: 202-232-3579; free admittance; guided tours offered Mon–Fri, 8am–2pm) in the image of the Tomb of Mausolus, one of the Seven Wonders of the ancient world. Every inch of this colossal, sphinx-flanked temple is masonically symbolic, down to its very proportions and the sequences of its front steps, which represent the sacred numbers of Pythagoras.

As you veer northeast, you will run into a hive of activity on Georgia Avenue around U Street, where the 150-acre (60-hectare), city-worn campus of **Howard University ❼** (2400 6th Street, NW; tel: 202-806-6100) begins. This prestigious and predominantly black institution, established in 1867 by civil rights champion Oliver O. Howard, has produced such high-caliber alumni as Andrew Young and Thurgood Marshall. If you can find it, the small **art gallery** in the **College of Fine Art** has a wonderful collection of African art and exhibits the work of major black American artists, as well as those by students, faculty, and alumni.

Brookland

Once you cross North Capitol Street, you're in the city's northeast quadrant in the up-and-coming neighborhood of **Brookland**. The Catholic Church has a decided presence along this stretch of Michigan Avenue. In fact, it has a monopoly on the neighborhood's attractions.

At **Trinity College ❽** (125 Michigan Avenue NE; tel: 202-884-9000), a Catholic women's college founded in 1897, the prize-winning Byzantine **Chapel of Notre Dame** has a 67-ft (20-meter) dome, marvelously filled with a La Farge mosaic depicting a scene from Dante's *Divina Comedia*. You may have to hunt for someone to unlock the door, however.

Just beyond sprawls the gray-stone campus of **Catholic University ❾** (620 Michigan Avenue, NE; tel: 202-319-5000). Founded in 1887, it is the only university established by American Roman Catholic bishops. The majority of its board members belong to the clergy. The university's **Hartke Theatre** (box office tel: 202-319-

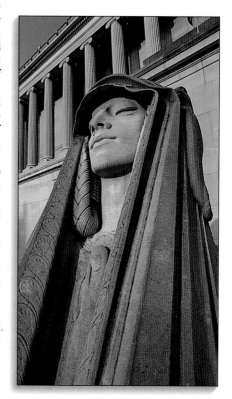

RIGHT: House of the Temple.

4000 during the academic year) is known not only for the quality and range of its productions, but for star alums like Jean Kerr, Jon Voight, and Susan Sarandon.

What commands attention here, with its mosaicked and gold-crowned dome, is the magnificent **Basilica of the National Shrine of the Immaculate Conception** ⑩ (4th Street and Michigan Avenue, NE, on the campus of Catholic University; tel: 202-526-8300; Nov 1–Mar 31, daily 7am–6pm; Apr 1–Oct 31, daily 7am–7pm; guided tours Mon–Sat, 9am–11am and 1pm–3pm; Sun, 1.30pm to 4pm; www. nationalshrine.com). The hemisphere's largest Roman Catholic church, and the seventh-largest church in the world, this Byzantine-Romanesque monument consumes 3 acres (1.2 hectares) donated by Catholic University – a gift sanctioned by Pope Pius X, who sent $400 along with his blessings. Dedicated to the Virgin Mary, the Marian shrine was constructed in fits and starts between 1920 and 1959 with funds from American parishes.

The basilica's ground floor holds the original dark and clammy **Crypt Church**; the sarcophagus of Bishop Thomas J. Shahan – the only person buried here; a hall of donors; and the tiara of Pope Paul VI. It all pales, though, compared to the glittering **Great Upper Church**, whose mosaic-covered walls contain an entire Italian quarry. Not to be missed are the three oratories and 57 unique chapels.

On the edge of Catholic University is the **Pope John Paul II Cultural Center** ⑪ (3900 Harewood Road, NE; tel: 202-635-5400; open Tues–Sat, 10am–5pm, Sun noon–5pm; admission fee; www. jp2cc.org). This three-story modern facility explores the Catholic faith in the modern world and highlights the life of Karol Joseph Wojtyla, the first non-Italian pope in 455 years. Several art treasures borrowed from the Vatican are on display and an interactive center covers a wide variety of topics from original sin to cloning.

On Monroe Street, near the Metro station, is **The Colonel Brooks Tavern** (901

BELOW: Basilica of the National Shrine of the Immaculate Conception.

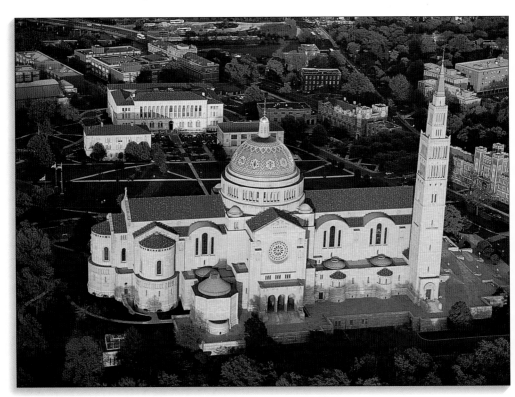

Map on page 228

Monroe Street, SE; tel: 202-529-4002), *the* neighborhood watering hole and good-time restaurant, which also serves up Dixieland jazz some nights of the week. If you're up for greasy southern soul food and ambience, try **Murry & Paul's** at 3513 12th Street (tel: 202-529-4078), Brookland's old-fashioned Main Street.

At least you won't go hungry if you venture out here for a performance at the **Dance Place** ⑫ at 3225 8th Street (tel: 202-269-1600; visit www.danceplace.org for a schedule of events), the enterprising force behind DC's contemporary, avant-garde, and ethnic dance scene. Undaunted by its move from more central Adams Morgan, Dance Alley (between 18th Street and Columbia Road) still stages the best in local, national, and international troupes each weekend, with special performances for Black History month in February and an annual African Dance Festival in June.

Places of pilgrimage

A pilgrimage to Brookland is incomplete without a visit to the **Franciscan Monastery** ⑬ at 1400 Quincy Street (tel: 202-526-6800; open Mon–Sat, 9am–4pm, Sun 1pm–4pm; guided tours on the hour except at noon; www.gardenvisit.com). The Franciscans built this hilltop American headquarters and named it Mount Saint Sepulchre. A sort of monastic theme park modeled after Istanbul's Hagia Sophia, it contains replicas of holy shrines, grottoes, and Roman catacombs. Byzantine in style with Renaissance touches, the blindingly opulent monastery church, completed in 1899, contains stained-glass windows from Bavaria and a 5-ton bronze altar canopy. In season the meticulous cloister garden is redolent with the scent of roses.

Much more modest in comparison is **St Anselm's Abbey** ⑭, the well-hidden Benedictine monastery at 4501 South Dakota Avenue and 14th Street (tel: 202-269-2300; www.stanselms.org). The monks annually host a spring flower show and sale on the grounds of the monastery.

South of Brookland at 800 Florida Avenue, **Gallaudet University** ⑮ (tel: 202-651-5000), the world's only college for the deaf, holds its own against the sur-rounding mean streets. Still, you can safely tour within the walled grounds of the Gothic-style campus, which were designed by Frederick Law Olmsted, Jr, the architect of New York's Central Park.

Not far to the east, by the banks of the Anacostia River, lies the **National Arboretum** ⑯ (tel: 202-245-2726; open daily 8am–5pm, closed Dec 25; visit www.usna.usda.gov). It is so little known that you can feel as though you have the entire rolling refuge of 444 acres (180 hectares) of gardens and woods plus 3 miles (5 km) of walking paths all to yourself – except in spring when the 70,000 azaleas, the arboretum's most prolific planting, are in bloom and it's mobbed.

For an olfactory treat, be sure to smell the historic roses and specialty herbs opposite the bonsai exhibit, and then wander over to the surreal-looking stand of 22 Corinthian columns salvaged from the old portico of the US Capitol building. There are no concessions, so consider packing a picnic if you plan to spend time here. ❑

RIGHT: the National Arboretum.

Map
on page
234

SOUTHWEST WATERFRONT AND THE SOUTHEAST

The neighborhoods that stretch beyond official Washington's gleaming marble epicenter are some of its toughest, but they contain several important historic sites and a beautiful garden

The heart of DC's Southwest and Southeast neighborhoods is the waterfront marina. At the **Maine Avenue Fish Market** ❶ along the Washington Channel, dozens of vendors hawk fresh seafood from the Chesapeake Bay and the lower Potomac and Delaware rivers. Boat-side stands are piled with everything from bluefish and rockfish, to oysters and soft-shell crabs.

The market is open daily, year round, but go on a summer weekend when vendors are shouting, families are hauling away bushels of still-snapping blue crabs, and cars are squeezing in and out of the parking lot. Even if you're not planning a feast, stop by to savor the scene and a plate of freshly shucked clams or oysters. This is one area of town where you're likely to find a cab driver who'll get you to your destination on time. If he's stopped locally before picking you up, chances are the cabbie will have a crate of dripping seafood tucked away in the trunk of the car. It's in *both* your interests for him to step on the gas pedal pronto.

Mixed fortunes

Not all of this part of DC is this lively, however. When the Federal City was laid out in 1791–92, the Southeast and Southwest quadrants north of the Anacostia River were slated as mixed residential and commercial areas. The mismanaged City Canal, which once flowed along Constitution Avenue, was intended to bring some of Georgetown's trade here. But the stinking canal and a new railroad depot sent the wealthier residents packing and created a squalid neighborhood of substandard buildings and high crime.

Beginning in the 1930s, redevelopment plans were introduced, then either abandoned or only partially realized. As a result, much of this area is dominated by a jumble of "innovative" architect-designed buildings of little interest to most visitors.

The rest of Southeast DC, separated from the city by the **Anacostia River**, developed slowly and independently from the nation's capital. The name Anacostia derives from the local tribe of Indians who originally settled the region before traders and tobacco farmers, and then freed blacks in the early 19th century claimed the area. Thanks to the middle-class flight into the suburbs and the 1968 riots, which destroyed many of the of the area's homes and businesses, Anacostia went into decline where, unfortunately, it remains.

LEFT: tipping the scales at Maine Avenue Fish Market. **RIGHT:** Anacostia.

Waterfront attractions

Both the **Washington Channel** ❷ and the riverfront are lined with private yacht clubs and boat yards. From the pier at 6th and Water streets you can take a scenic cruise on the Potomac River on the 145-ft *Spirit of Washington* ❸ (tel: 202-554-8000 or www.spiritofwashington.com for schedules and ticket information). This line offers day and nighttime cruises along the river, plus daytime trips to George Washington's home, Mount Vernon.

The waterfront is also popular for dining. Several cavernous restaurants along Water Street specialize in seafood and panoramic views of the Washington Channel, marina, and downtown monuments, and offer a variety of entertainments such as happy hours, comedy acts, dancing, and live jazz. Try especially **Le Rivage** (1000 Water St., SW; tel: 202-488-8111), which offers standard but well done French fare at reasonable prices. Other choices include **Pier 7** (650 Water St., SW; tel: 202-554-2500), **Zanzibar** (700 Water St., SW; tel: 202-554-9100), **Hogate's** (800 Water St., SW; tel: 202-484-6300), a Washington institution, and **Phillips Flagship** (900 Water St., SW; tel: 202-488-8515), famous for crabcakes and popular with tour buses.

One of the best places in DC for first-rate dramas by contemporary playwrights such as August Wilson is nearby **Arena Stage** ❹ at 6th and M streets (tel: 202-488-3300; www.arenastage.org for schedule and ticket sales). It also stages versions of Marx Brothers comedies and new takes on classic Broadway musicals. From its humble beginnings in the old vat room of the former Heurich Brewery (where the Kennedy Center now stands), the company performs in a complex that includes the 800-seat **Arena Theater** (in the round), the 500-seat **Kreeger Theater**, and the 180-seat cabaret-style **Old Vat Room**. Arena was the first theater outside of New York to be awarded a Tony, this for the overall quality of its productions.

Southwest of the Arena, occupying the mile-long peninsula near the confluence

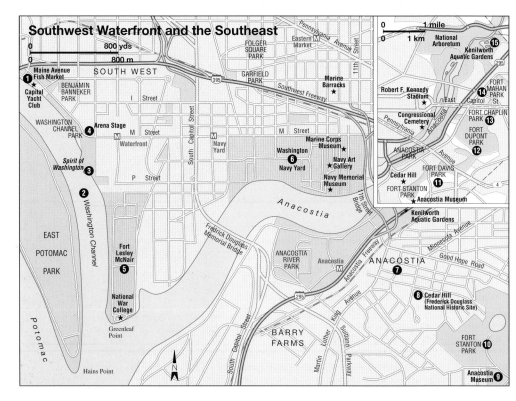

Map on page 234

of the Washington Channel and the Potomac and Anacostia rivers, is **Fort McNair** ❺ (not open to the public), originally an 18th-century fort and arsenal known variously as Turkey Buzzards Point, Greenleaf Point and Fort Humphreys. It is one of the oldest active military posts in the US. In 1865, four of John Wilkes Booth's fellow conspirators in Lincoln's assassination were imprisoned and hanged here. One of them, Mary Surratt, whose crime had been to rent rooms to Booth in the boarding house she operated, was the first American woman executed by federal order. The fort was renamed in 1948 in honor of Lesley J. McNair, commander of the army's ground forces in Normandy in 1944. The fort is now home to the **National War College** and the **National Defense University**.

Naval traditions

Farther up the Anacostia River in Southeast at 9th and M streets is the **Washington Navy Yard** ❻. This historic precinct dates from 1799 and was the navy's first shore facility until the commandant ordered the yard burned to avoid it being captured by the British during the War of 1812. It was rebuilt and served as a naval gun factory during the 19th century and again during both world wars. The yard ceased operating in 1961 and now serves as an administrative center.

The precinct comprises the **Marine Corps Museum** (daily 10am– 4.30pm; closed Jan 1, Thanksgiving, Dec 25; tel: 202-433-3840) and the **Navy Memorial Museum** (Apr–Aug, Mon–Fri, 9am–5pm; Sept–Mar, Mon–Fri, 9am–4pm; weekends and holidays all year, 10am–5pm; closed Jan 1, Thanksgiving, Dec 24–25; guided one-hour tours can be reserved; tel: 202-433-4882; www.historynavy.mil).

The Marine Corps Museum galleries include combat art, uniforms, weapons, and technology. Also on display is memorabilia of John Philip Sousa, master of the Marine Band from 1880 to 1892 and composer of about 140 military marches. The Navy Memorial Museum's 5,000 artifacts trace naval history from the Revolutionary War to the age of space exploration.

Anacostia

On the south side of the Anacostia River is **Anacostia** ❼, one of the city's less affluent neighborhoods. It is best visited by car, only during the day, never alone, and even then with a degree of caution.

Incorporated in 1854 as Uniontown, one of Washington's earliest subdivisions, **Old Anacostia** retains certain historical appeal owing to several restored 19th-century churches and frame houses. But, as the home of Washington's poor and disenfranchised, it is neither the city's most beautiful section nor its safest. This is where much of the city's serious crime occurs.

About the only good reason to go there is **Cedar Hill** ❽ (visit by guided tour only, year round, 9am–4pm; must call ahead, tel: 202-426-5961 www.nps.gov/frdo; or arrange through Tourmobile tel: 202-554-5100, a sightseeing company, which offers tours in summer). Built on a tree-shaded hilltop at 14th and W streets, this is the stately home of Frederick Douglass, a self-educated former slave, presidential advisor, and writer

RIGHT: sailor from the Washington Navy Yard.

(see facing page). The 21-room house speaks volumes of the life and times of Douglass, who lived here with his family from 1877 to 1895. During this period, he served as the US Marshal for DC, the Recorder of Deeds for DC, and Minister to the tiny nation of Santo Domingo, which shared a Caribbean island with Haiti and which, as a possible home for newly freed Southern blacks, had petitioned the US for admittance to the Union after the Civil War.

Many of Douglass's original possessions are on display, including his impressive personal library of 1,200 volumes and gifts from President Lincoln, Harriet Beecher Stowe, author of *Uncle Tom's Cabin*, and abolitionist leader William Lloyd Garrison. Cedar Hill is not a sanitized period home, but a home that keeps Douglass's memory very alive.

At 1901 Fort Place, SE, is the **Anacostia Museum** ❾ (daily 10am–5pm; closed Dec 25; tel: 202-287-3369), one of the Smithsonian Institution's off-the-Mall

museums. This innovative museum documents the history, culture, and contributions of notable African Americans. Created by the Smithsonian in 1967, it serves as combination museum and cultural arts center, focusing on the history of Anacostia from its 16th-century Indian days to the present.

Civil War forts

Around Anacostia are five of the nearly 50 forts built in a ring around the city at the outbreak of the Civil War. **Forts Stanton** ❿, **Davis** ⓫, **Dupont** ⓬, **Chaplin** ⓭, and **Mahan** ⓮ now form a chain of city parks where the forts' original earthworks can be discerned. Fort Dupont, the best preserved, also includes recreational facilities and hiking and biking trails.

With just a little bit of imagination you can visit Monet's gardens at Giverny at **Kenilworth Aquatic Gardens** ⓯ (1900 Anacostia Drive, SE; daily 7am–4pm; closed Jan 1, Thanksgiving, Dec 25; admission free; tel: 202-426-6905). Just off the Anacostia Freeway in DC's northeast corner, this 12-acre (5-hectare) sanctuary, founded in 1882 near the marshlands of the Anacostia River, has pond after pond of exotic water lilies, lotuses, and aquatic plants in a natural outdoor setting.

Summertime promises the most blooms, and early mornings are the best time to visit since you can see the night-blooming flowers before they close and day-bloomers as they open. This garden is a joy just to wander in, filtering color and light as if to please an Impressionist painter.

The gardens are also a treat for amateur naturalists. Behind the visitor center there are three ponds of labeled species where you can familiarize yourself with the tropical lilies and ancient lotuses (the rest of the gardens are label-free). The most extraordinary lilies are the Victoria *amazonica* from South America which have platter-like leaves up to 6 ft (1.8 meters) across.

The quieter you are on your visit, the better your chances for spotting the toads, turtles, muskrats, green herons, red-winged blackbirds, and migrating waterfowl who reside in this tranquil place. ❑

Map on page 234

LEFT: Cedar Hill statuette.

Frederick Douglass

Eloquent voice for the silenced and an unshakable force against the evil that was slavery, Frederick Douglass was not only a great orator but also the embodiment of the American ideal of freedom.

Born Frederick August Washington Bailey in Talbot County, Maryland, in 1818, he acquired an education, forbidden to slaves, by observing his white friends as they were tutored. "With play-mates for my teachers, fences and pavements for my copy books, and chalk for my pen and ink," he later wrote, "I learned the art of writing." From a volume of contemporary speeches, which he bought for 50 cents, the whole of his savings after having been hired out to perform the dirty job of caulking ships on the Baltimore docks, he learned the art of oration and made a life-changing discovery. Enlightened Americans not only opposed slavery, but in the North where blacks were free, they were actively speaking against it.

In 1837, Douglass escaped to New York with his wife Anna Murray, a freed slave whom he met in Baltimore, and then to Massachusetts where he changed his last name to Douglass to avoid capture by his Maryland owner. Educated and articulate, Douglass, whose "diploma was written on his back" as one abolitionist put it, was in demand as a public speaker before anti-slavery gatherings.

In 1845, he published the first volume of his autobiography, in part to document the horrors of slavery for disbelieving Americans, and three years later began writing and publishing *The North Star*, an influential abolitionist periodical that he produced from his home in Rochester, New York.

Douglass threw his support behind Abraham Lincoln and the newly-formed Republican Party, which in its early years championed the cause of freedom, and helped raised a regiment of black soldiers. After the Civil War he cautioned freed slaves, though they were destitute, against accepting the government's post-war handouts of food, clothing and medicine, arguing that it would only enrage the embittered Southern whites and lead to paternalism and an unhealthy dependency. Despite lynchings, oppressive Jim Crow laws, and the rise of the Ku Klux Klan, which combined to drive many Southern blacks to the cities in the North, Douglass urged blacks to stay in the South and claim the land to which they were entitled. "His labor made him a slave," Douglass wrote of his fellow blacks, "and his labor can, if he will, make him free, comfortable, and independent."

Douglass remained committed to the Republican Party, even after it abandoned its promise of restoring civil order to the South, and moved to Washington to accept several political appointments. In 1882, his first wife died and two years later he married Helen Pitts, his white assistant.

Douglass also took on the cause of women's suffrage. In 1895, after completing an address to an audience of Washington women, he died of a heart attack.

Through the sheer force of his character, Frederick Douglass not only transformed himself but also, through the written and spoken word, changed what it means for all Americans, regardless of color or gender, to participate fully in democratic life. ❑

RIGHT: Frederick Douglass.

Map
on page
242

AROUND WASHINGTON

*Within easy reach are the majestic Arlington Cemetery, the quaint
Old Town Alexandria, George Washington's beloved Mount Vernon
estate, and the Potomac's dramatic Great Falls*

The Capital's rapidly expanding Virginia and Maryland suburbs offer almost as many attractions and distractions as the city itself. Surrounding the District on four sides, the suburbs hold dozens of residential neighborhoods, parks, museums, restaurants, parks, and commuter traffic that are urban enough to be within DC's borders.

Southern perspective

This fact often makes the transition between the city and suburb almost indistinguishable, especially inside the Beltway around the city. The most popular sights are Arlington County and the City of Alexandria, and parks in Montgomery County in Maryland and in Fairfax, Virginia. Since Arlington County and a section of Alexandria were part of Washington from the late 18th to the mid-19th centuries, the historic homes and museums here chronicle the history of the nation and of its developing capital city during this vital period, but with a slightly southern perspective. Beyond the Beltway, the suburbs become more rural and the influence of the federal government becomes less pronounced.

You can visit most of the nearby suburban attractions by public transportation, car or bike. Expect traffic congestion on most roads during the morning and evening rush hours, as commuters plunge headlong to or from DC. One way you can usually beat the traffic is on the **W&OD bike trail**, which runs 45 miles (72 km) from Alexandria west into Loudoun County, or on the **Mount Vernon Trail** which parallels the Potomac River from the Lincoln Memorial south to Mount Vernon.

Before leaving DC proper, it's worth stopping in the middle of the Potomac River on the 88-acre (36-hectare) **Theodore Roosevelt Island ❶**. Walking paths criss-cross this wooded wildlife haven and lead to a 23-ft-tall (7-meter) statue of President Theodore Roosevelt.

Just south on the Virginia side of the river is **Lady Bird Johnson Park ❷**, a 121-acre (49-hectare) man-made island planted with some 1,000 flowering dogwood trees and a million daffodils. Within this park is the **Lyndon Baines Johnson Memorial Grove** where white pines, dogwoods, rhododendrons, and azaleas surround a monolith of pink Texas granite.

As you cross **Memorial Bridge** into Arlington, **Arlington House** (Apr–Sept, 9.30am–6pm; Oct–Mar, 9.30am–4.30pm; closed Jan 1, Dec 25; tel: 703-557-0613; www.nps.gov/arho) dominates the hill before you. Arlington House was the home of Robert E. Lee and his wife Mary

Anna Randolph Custis, great-granddaughter of Martha Washington. The couple lived in this gracious Greek Revival mansion with their seven children from 1831 until Lee resigned his commission in the US Army to defend his native Virginia in the Civil War. During the war, hundreds of Union Army soldiers occupied the property and cut most of the 200 acres (80 hectares) of virgin oak forest for fortifications and firewood. In 1864, after Mrs Lee refused to pay the property tax, the Federal government confiscated the estate and established what is now Arlington National Cemetery on the grounds. The house has

been restored to its mid-19th-century appearance to offer a glimpse of pre-war Southern gentility. You can walk the scenic route to the house via 200 easy steps north of the main gate of the cemetery.

Soon after the Civil War broke out, it became apparent that a burial ground would be needed. Charged with the responsibility of establishing one, the Quartermaster-General of the Union Army, Montgomery Meigs, deliberately chose Lee's former plantation as the site where the Union dead would be buried. Meigs considered Lee a traitor and, both in retribution and to discourage Lee's ever

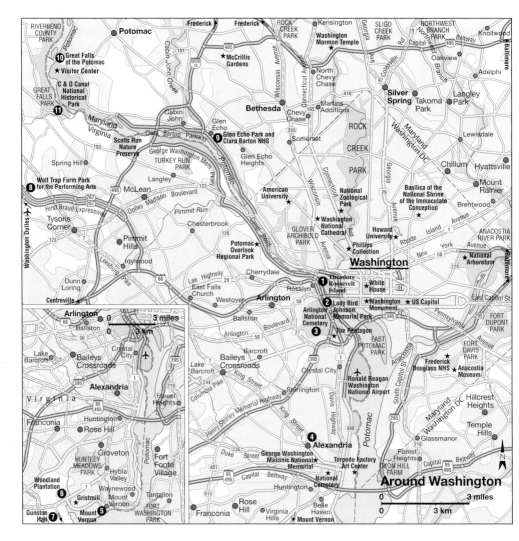

Around Washington

Maps on pages 242, 243

returning, made a point of burying the first war dead within easy sight of the house.

The rolling hills of **Arlington National Cemetery ❸** (daily Apr–Sept 8am–7pm, Oct–Mar, 8am–5pm; tel: 703-697-2131) stretch over 612 acres (248 hectares) and contain the graves of more than 240,000 military personnel and other patriots, including the dead from every war or conflict in which America was a participant beginning with the Revolutionary War. The simple white headstones, set in even rows, stretch as far as the eye can see.

Unless you are visiting the grave of a relative or friend buried here, access to the graves is limited to Tourmobile buses departing from the **Visitors Center Ⓐ** or, even better, your own two feet. Walking is preferable: lines for Tourmobiles are often interminably long and the narrated tour covers only the central cemetery.

RIGHT:
Arlington
National
Cemetery
in winter.

The Tomb of the Unknown Dead of the Civil War ❸ on the former site of Mrs Lee's rose garden marks the grave of the 2,111 unidentified Union soldiers who died on nearby Virginia battlefields. The mast of the **battleship Maine** marks the grave of 229 men who died in the explosion preceding the Spanish-American War. **The Tomb of the Unknowns Ⓒ** (formerly the Tomb of the Unknown Soldier), carved from a 50-ton block of Colorado marble, holds the remains of four US servicemen, one each from World Wars I and II, and the Korean and Vietnam Wars.

The Changing of the Guard, which takes place here every half-hour in summer, every hour in winter and also throughout the night, is on many visitors' list of things to see while in Washington.

Others buried here include Pierre L'Enfant, Oliver Wendell Holmes, Rear Admirals Robert E. Peary and Richard E. Byrd, and presidents Taft and Kennedy. **John F. Kennedy's grave Ⓓ**, with its simple, eternal flame glowing in all weathers, still attracts crowds. The tomb of Robert F. Kennedy is not far away in a grassy plot.

At the cemetery's north end stands the **Iwo Jima Memorial Ⓔ**. The 100-ton

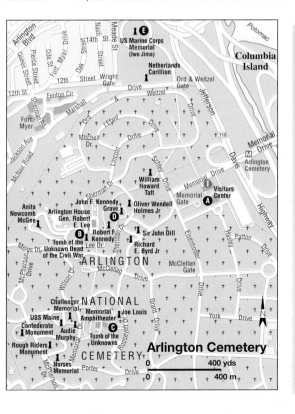

Arlington Cemetery

bronze statue depicts the raising of the US flag on Iwo Jima, the World War II battle site where more than 5,000 marines died. On Tuesday evenings in summer, the Marine Silent Drill Team and Drum and Bugle Corps hold a sunset ceremony here.

Just south of Arlington Cemetery is one of the world's largest office buildings, the **Pentagon**, which is not open to the public. This five-sided headquarters of the Department of Defense is comprised of five concentric circles around a 5-acre (2-hectare) courtyard. It covers 6½ million sq ft (60 hectares) and provides offices, restaurants, and shops for more than 23,000 civilian and military employees.

West of the Pentagon, you'll find Arlington's high-rise office and apartment complexes of **Rosslyn** and **Crystal City**. The immigrant Asian population in Arlington's Clarendon section has created a wealth of Vietnamese, Cambodian, Korean and Thai restaurants, many located in what is called **Little Saigon** on Wilson Boulevard. Worth trying are the **Queen Bee**

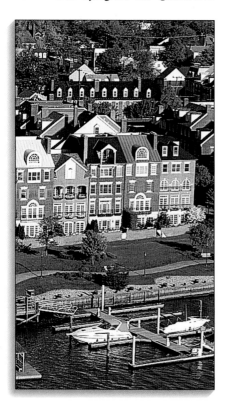

(3181 Wilson Blvd; tel: 703-527-3444), a bit threadbare but an original and one of the best of the Vietnamese restaurants, or the newer and inventive **Tara Thai** (4001 Fairfax Drive; tel: 703-908-4999).

Back along the Potomac River is **Washington Reagan National Airport**, the "Reagan" added in 1999 at the insistence of Congress and in honor of the former president, but to the locals it's still National. Just 3 miles (5 km) from downtown, this airport has been refurbished and expanded to ease the legendary congestion, both air and ground. Conveniently, there's a Metro station in the airport to make it easier to get in and out of the District. For a close-up view of the planes taking off and landing, head south along the GW Parkway, named for George Washington, toward the **Washington Sailing Marina** (1 Marina Drive; tel: 703-548-9027). (Note: Roaches Run is off limits for security reasons.) The marina offers lessons and rentals in sailing and windsurfing as well as renting out bicycles. The marina's **Potowmack Landing** (tel: 703-548-0001) restaurant has outdoor seating with great views of the river and DC.

Old Town Alexandria

The parkway, continuing south to **Alexandria ❹**, becomes Washington Street in **Old Town Alexandria**, a charming eight-square-block district of 18th- and 19th-century buildings with an invigorated and very modern town dock. Pleasure craft of all sorts tie up around a small barge with a red-roofed structure on it. That's the **Alexandria Seaport Foundation** (Mon–Fri, 9am–4pm; for additional hours and programs, tel: 703-549-7078). This is run mostly by volunteers who teach boat-building skills to kids at risk and offer several on-the-water programs to help locals and visitors alike get better acquainted with Alexandria's seaport past and the Potomac's marine and wildlife.

In 1699, the Scottish settler John Alexander bought the land that would eventually become Alexandria from an English ship captain for 6,000 pounds of tobacco. By the 18th century, the area was a thriving center for the export of that profitable

LEFT: Alexandria.

Maps
on pages
242, 245

crop. To facilitate shipping and put it on a par with the likes of New York and Boston, both important seaport centers, Scotsmen William Ramsay and John Carlyle successfully petitioned the Virginia General Assembly to establish a town. By 1749, Alexandria was born.

It's similar in many ways to Georgetown, but this Virginia port city is generally less crowded (though you're likely to be shoulder-to-shoulder if the weekend weather's good), and has managed to preserve more of its colonial heritage. Strict building codes are in force here, so the block upon block of restored Federal-style townhouses you'll see will give you a good indication of what the city looked like in centuries past.

A good place to begin is at the **Ramsay House Visitors Center Ⓐ** on King Street (221 King Street; daily 9am–5pm; tel: 703-838-4200; www.FunSide.com), home of the Scottish merchant. Next door and a bit out of character for Old Town but elegant just the same is the **219 Restaurant** (219 King Street, tel: 703-549-1141) a long-time favorite in Alexandria specializing in Louisiana cuisine. Up King Street, a few blocks from the river, try **La Madeleine** (500 King Street; tel: 703-739-2854), moderately priced country French served cafeteria-style.

To get a background in Alexandria's history, stop by the local history exhibits at the **Lyceum Ⓑ** (201 South Washington Street, tel: 703-838-44994), a Greek Revival structure dating from 1839 where much of the city's history is told in photographs and other artifacts, or the **Carlyle House Ⓒ** (121 North Fairfax Street; Tues–Sat, 10am–4pm, Sun, noon–4pm; tel: 703-549-2997), a grand home in the Scottish-manor style on North Fairfax Street and one of the first homes built in Alexandria. Signs proclaiming "George Washington Slept Here" are common and questionable from New York to Georgia, but Alexandria's claims are bona fide.

In 1765 Washington built a townhouse here at 508 Cameron Street (a replica of

BELOW: in the Lyceum Museum.

the original home now stands in its place) and Washington celebrated his last two birthdays in the ballroom of what is now **Gadsby's Tavern Museum** on North Royal Street (134 North Royal Street; open Mar–Oct, Tues–Sat 11am–4pm, Sun 1pm–4pm; Apr–Sept Tues–Sat 10am–5pm, Sun 1pm–5pm; closed Mon; tel: 703-838-4242). The museum preserves an original tavern (1770), and the City Hotel (1792) that was a center of political, business, and social life in early Alexandria.

Next door, the 200-year-old **Gadsby's Tavern Restaurant** offers hearty 18th-century victuals served by waiters in period dress and character.

Grand mansions and art

Among Alexandria's favorite native sons is Robert E. Lee, who lived in the red-brick house on Oronoco Street, now a private residence, from 1812 until his West Point enrollment in 1825. Across the street, on the corner of Washington Street, is the white-frame **Lee-Fendall House**

(614 Oronoco Street; open Tues–Sat 10am–4pm; Sun noon-4pm; closed Mon; tel: 703-548-1789). It was built in 1785 by one of Robert E. Lee's ancestors and contains an inventory of Lee family furnishings. Cross Washington Street and stop at **Christ Church** (Cameron and North Washington streets; tel: 703-549-1450) which dates from 1767. George Washington's pew is here and, surrounding the church, a pleasant garden.

Alexandria's art scene is as vital as its history, with more than 20 Old Town galleries and the **Torpedo Factory Art Center** (105 North Union Street; open daily 10am–5pm; closed major holidays; tel: 703-838-4565; www.torpedofactory.org). A former munitions factory, the Torpedo Factory houses 150 artists who create, display, and sell their works on the premises.

The Torpedo Factory is also home to the **Alexandria Archaeology Museum** (open Tues–Sat, 10am–3pm, Sat 10am–5pm; closed Mon; tel: 703-838-4399). Here you can watch the urban archaeologists at work, putting together the city's history shard by shard.

Mount Vernon

Farther south, at the end of the GW Parkway, is **Mount Vernon** (open daily including major holidays, Apr–Aug, 8am–5pm; Mar, Sept, Oct 9am–5pm; Nov–Feb 9am–4pm; entrance fee; tel: 703-780-2000; www.mountvernon.org), Martha and George Washington's estate from 1754 to 1799. Situated on a bluff above the Potomac River, the white Georgian mansion was bought by the Mount Vernon Ladies' Association in 1858 and has been meticulously restored to its appearance during the last years of Washington's life. On first glance, the mansion seems to be constructed of stone, but it is actually made of boards that have been grooved and beveled to look like stone.

Washington, an innovative farmer, cultivated just under half of the 8,000-acre (3,200-hectare) estate. He experimented with various crops – tobacco, initially, then a less labor-intensive, more soil-friendly and highly profitable wheat crop. In 1771, he established a gristmill 3 miles (5 km)

LEFT: old-style service at Gadsby's Tavern.

Map on page 242

east of the estate. This has been restored.

An introductory and video is followed by a 20-minute tour, accompanied by docents who explain each of the rooms in turn. The stunningly ostentatious main **parlor** was designed with the deliberate intention of impressing the Washingtons' guests. The walls were finished in vertigis, the furniture is of the finest quality, and Washington's personality projects from six paintings reflecting his love of nature. It was in this room that he planned the Battle of Yorktown, received the news of his presidential election, and lay in state after he died on December 14, 1799.

Washington's personal **study** is furnished with interesting pieces, including his presidential chair and a peculiar but practical chair with pedals that drove an overhead fan. Another highlight is the second-floor **master bedroom**.

You can spend some time exploring the outbuildings (some contain exhibitions) and the two museums, and seeing the impressive coach that belonged to Philadelphia

mayor Samuel Powel. The new tomb, too, is interesting. Washington provided for its creation in his will, and he and Martha were entombed there in 1831. Other family memorials are outside, and nearby is a dignified memorial to the slaves, numbering 316 in 1799 compared to 50 in 1759.

Washington was knowledgeable about livestock. Reportedly, he introduced mules – harder working than horses and donkeys – to America. Animals present today include endangered black and white pigs, 2,000 of which once roamed the estate, and a breed of sheep whose fleece the slaves once spun into yarn for clothing.

About 10,000 tourists a day visit Mount Vernon during peak season – so an early start is recommended.

The **George Washington Pioneer Farmer Site** is also worth seeing. The highlight is a 1996 replica of the innovative 16-sided barn designed by Washington. Around the 12-ft (3.6-meter) wide wooden lane that surrounded a center section, cattle would tread the unthreshed

BELOW:
Mount Vernon.

wheat, separating the grain. The grain would then fall between purpose-designed gaps in the floor boards to the lower floor, there to be gathered for the gristmill. The brown Milking Devon cattle chosen for this chore are still bred here and are used in demonstrations at 11am, 12.30pm, and 3pm daily. The Ladies' Association is planning to open a new museum.

Three miles (5 km) south of the mansion is the refurbished **Gristmill** (tickets available at Mount Vernon). This is part of the original plantation and has a water-powered grinding wheel.

If driving, travel a mile or so to **Woodlawn Plantation** (9000 Richmond Highway; open daily 10–5; closed Jan, Feb, Thanksgiving, Dec 25, admission fee; tel: 703-780-4000). This is the estate George Washington gave his niece as a wedding present and which was designed by William Thornton, architect of the US Capitol. On these grounds is another interesting building: the **Pope-Leighey House**, designed by famous American architect Frank Lloyd Wright, has Wright's contemporary style and was moved here from its original location in a nearby suburb.

Drive another 8 miles (13 km) south along Route One to **Gunston Hall** (10709 Gunston Hall Road, Mason Neck; daily 9.30am–5pm; closed Jan 1, Thanksgiving, Dec 25), home of founding father George Mason, who was responsible for adding the Bill or Rights to the US Constitution. The beautifully restored mansion has several outbuildings, including a schoolhouse, and a boxwood garden dating from the 1760s, all spread over several quiet acres above the Potomac. Although a bit off the beaten track, it's a gem of a house and is seldom crowded.

Fun in the park

An easy drive west of DC toward Vienna, Virginia, is **Wolf Trap Farm Park** (1645 Trap Road; tel: 703-255-1860 or visit www.wolftrap.org or check the *Washington Post* for schedules), America's only

BELOW: Dulles International Airport.

Map
on page
242

national park for the performing arts. The much-loved **Filene Center** stage draws some 6,800 picnic-toting patrons here in summer for concerts (opera, ballet, modern dance, jazz, stand-up comedians) in covered and lawn seating. The intimate **Barns at Wolf Trap** hosts a variety of performances from October through May.

To the west, near the Blue Ridge Mountains, is **Washington Dulles International Airport**, a graceful sweep of a building designed in 1962 by Eero Saarinen.

Between the Potomac River and MacArthur Boulevard in Maryland is **Glen Echo Park ❾** (tel: 301-492-6229; www.nps.gov/glec; free admission into the park, though there are charges for attractions). This is an arts and performance center remembered by generations of Washingtonians as a countryside amusement park at the terminus of DC's old trolley line. It almost closed in the 1960s, but community efforts have restored it and now, as a National Park, Glen Echo mixes applause from theater audiences, squeals from roller-coaster riders, and the plinking tunes of the merry-go-round.

Many of its buildings serve as art studios and classrooms. Glen Echo offers dozens of lectures, craft workshops, dance classes and theatrical performances including puppet shows each year. On weekends you can dance in the ornately redecorated **Spanish Ballroom**, or ride the lovingly restored 1920s-era **Dentzel carousel**.

Nearby is the **Clara Barton National Historic Site** (by guided tour offered on the hour daily 10am–4pm; closed Jan 1, Thanksgiving, Dec 25; free admission; tel: 301-492-6245), the 36-room home of the founder of the American Red Cross.

The Great Falls

No visitor should leave the area without paying homage to the natural feature that determined the location of the capital city. The spectacular and turbulent **Great Falls ❿** of the Potomac can be viewed from parks on both Maryland and Virginia shores. Here, the Potomac River makes its final plunge down rapids and cataracts en route to the Chesapeake Bay.

On the Maryland side, off MacArthur Boulevard, the falls are part of the **C&O Canal Historic Park** that includes the towpath, restored canal locks, and the 1828 **Old Angler's Inn** (10801 MacArthur Blvd.; tel 301-365-2425), known more for its atmosphere than for its food.

On the Virginia side, in Fairfax County off Georgetown Pike, **Great Falls Park ⓫** (open daily 7am–dark; Visitors Center open summer weekdays 10am–5pm, weekends 10am–6pm; winter 10am–4pm; closed Dec 25; www.nps.gov/gwmp) offers more dramatic views of the 76-ft (23-meter) falls from an observation area and from massive boulders above the river.

The best time to visit is during spring floods when the volume of water can exceed that of Niagara Falls. Swift currents and slippery rocks are responsible for several drownings each year; caution is strongly advised when venturing near the water. The park's extensive trails parallel the river, cut through woodland and marshy areas, and offer views of the Potomac River and the chasm of **Mather Gorge**. ❑

RIGHT: fishing near Great Falls.

DAY TRIPS: MARYLAND

*Maryland's reach from mountains to shore offers railroad towns
and 19th-century villages, big city diversions and
the Chesapeake Bay's famous blue crabs*

To the north and east of Washington, DC lie ample opportunities for day trips. You can follow the path of the 17th-century pioneers through the Allegheny Mountains to the Cumberland Gap and Deep Creek Lake. There is a major Civil War site at Antietam, near the old town of Sharpsburg. For summer sun and swimming there are the beaches of the Chesapeake Bay and, if you don't mind the driving, more beaches in the state of Delaware. Along the way are the historic coastal towns of Annapolis and the Eastern Shore, and the port of Baltimore.

Baltimore

The city on the Chesapeake Bay is the site of the nation's oldest monument to **George Washington**, a 178-ft (24-meter) **column** whose spiral staircase you can climb for the best view in Baltimore (Mt. Vernon Place; Wed–Sun, 10am–4pm).

Baltimore's famous sons include **Edgar Allan Poe** (museum home at 203 North Amity Street; call for tour information; tel: 410-396-7932), the waspish man of letters H. L. Mencken, and the baseball hero **Babe Ruth** (museum home at 216 Emory Street; Apr–Sept 10am–5pm, Nov–Mar, 10am– 4pm; closed Sun; admission fee; tel: 410-727-1539). As the city is only an hour's drive from Washington, it can't be too many decades before it becomes a formalized extension of the capital, as the suburbs creep towards one another.

Baltimore ❶ is an elderly, distinguished city that has suffered some drastic revitalizing, particularly around the old run-down **Inner Harbor** area, now a throbbing tourist attraction. The wide pedestrian area that swings round the waterfront hums with action – jugglers, trick cyclists and magicians giving impromptu performances to the crowds that it draws from out of the vast shopping mall behind. You can cover the entire harbor area on foot, but it's more fun to catch one of the water taxis from any one of the 15 landing sites. Tied up in the harbor, and worth seeing, is the ***USS Constellation*** (daily 10am–5pm; admission fee; tel: 410-539-1797), a 22-gun three-masted sloop-of-war and the last surviving ship of the Civil War.

The **National Aquarium** (Pier 3, 501 East Pratt Street; open Sat–Thurs, 10am–5pm; Fri 10am–8pm; tel: 410-576-3800), is expensive ($17.50 adults, $9.50 children at time of writing) but a great introduction to the underwater world. You begin the visit on the top story of the pyramid-shaped building, and gently wind your way down the middle of a glass-fronted, 220,000-gallon **Open Ocean tank** that

LEFT:
Maryland has
many gracious
homes.
RIGHT:
Baltimore's
Inner Harbor,
seen from
Signal Hill.

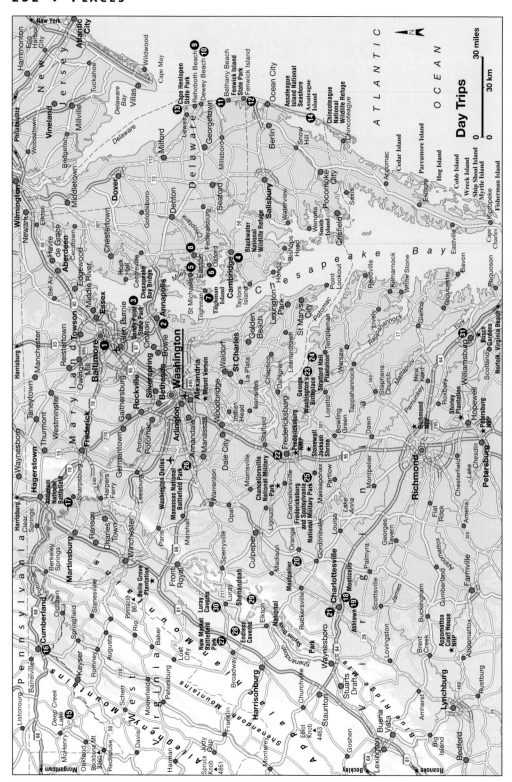

Day Trips

mirrors the marine life you would find at equivalent depths in the sea – brightly colored fish on the top floor, the sharks down in the murky depths on the ground. There is a **Hands-On pool** where children can touch starfish, anemones, shellfish and sea vegetation, as well as other tanks that present specific marine habitats.

Across the other side of the harbor is the Maryland **Science Center** (601 Light Street; open Mon–Fri, 10am–5pm; Sat, 10am–6pm, Sun, noon–5pm; $12 adults, $8 children at time of writing; tel: 410-685-5225) and a planetarium. Although in summer there are lines to get into the aquarium, they are better managed than those at the Science Center, so be prepared for a disorderly wait.

But children generally find it worthwhile, particularly when they can watch – or volunteer for – live demonstrations that set their hair standing on end or show how they can emit electric sparks from their bodies. It is a predominantly hands-on experience, with excellent displays that explain many of modern science's inventions and discoveries in an easy-to-understand fashion. There are also permanent exhibits on the life of the Chesapeake Bay, geology and energy exhibits, plus thought-provoking temporary shows.

Also housed in the building and almost more fun than the scientific section is the awe-inspiring display of stars and planets in the night sky of the **Davis Planetarium**, as well as a cinema screen, five-stories high, showing IMAX theater movies that almost push you into the action.

Take in an **Orioles'** baseball game at **Camden Yards** (333 West Camden Street; tel: 410-685-9800), a new 46,000-seat park made to look old-fashioned. The star-shaped **Fort McHenry** (E. Fort Avenue; open daily in winter 8am–4.45pm; daily in summer 8am–7.45pm; $5 adult; children free; tel: 410-962-4290) is the site of the famous victory of the Americans over the British during the War of 1812 that inspired Francis Scott Key to pen *The Star Spangled Banner*.

Cross the harbor and stop at **Fell's Point**, a funky neighborhood of 18th-century row houses and warehouses

mixed in with early 20th-century storefronts that has held on to its colorful Baltimore character. Yard sales spring up on the square on weekends.

Amid it all is the stately **Admiral Fell Inn** (888 South Broadway; tel: 410-522-7377) with a pleasant dining room.

Nearby is **Little Italy**, chock full of terrific little Italian restaurants. For a taste of authentic Chesapeake Bay fare, stop at **Obrycki's Crab House** (1727 East Pratt Street; tel: 410-732-6399).

Annapolis

A water town only an hour from Washington, **Annapolis ❷** is on the Chesapeake and some consider it the world's sailing capital. Start at the **Historic Annapolis Foundation Museum Store and Welcome Center** at 77 Main Street (tel: 410-268-5576) to pick up maps and brochures useful for a walking tour.

Little has changed in the 18th-century heart of the town, though Annapolis has grown enormously beyond the city cen-

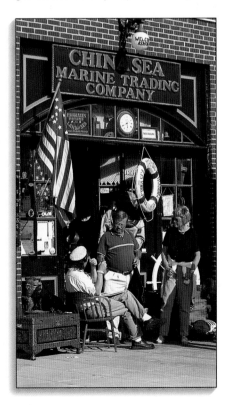

ter. It is a charming and easy place to cover on foot, particularly along the waterfront. This still presents a soothing vista of masts and sails, since the harbor has always been too shallow to accept anything but small vessels. Unless you are a boat enthusiast, it can be a place to avoid on the last weekend in April and in October when the city hosts America's largest sail and power boat shows and all other movement grinds to a halt.

The **Maryland State House** (State Circle; Mon–Fri 8:30am–5pm, Sat–Sun 10-4pm, closed Sun, Jan 1, Thanksgiving, Dec 25; guided tours 11am and 3pm; tel: 410-974-3400), built in 1772, is the oldest state capitol building in continuous use in America. This is where George Washington resigned his commission as commander-in-chief of the Continental Army in 1783; where the Treaty of Paris that officially ended the Revolution was signed in 1784; and where Thomas Jefferson was appointed the first United States ambassador to France. Its cypress-beam dome

is the largest wooden dome in the country. The elegant plasterwork in the central hall cost the life of Thomas Dance, the craftsman who executed it, when he fell 90 ft (27 meters) from the scaffolding.

Other interesting sites include the **Old Senate Chamber**, where a bronze plaque marks the place where George Washington stood to deliver his farewell address, the **House of Delegates** and the **New Senate Chamber**, both with skylights designed by Tiffany.

Many of the city's Federal-style homes are original to the colonial era and there are three you can visit:

● The **house of William Paca** (18 Pinkney Street, mid-Mar–mid-Dec, Mon–Sat, 10am–5pm; Sun noon–5pm; intermittently the rest of the year, admission fee; tel: 410-267-7619) was built in 1765 and is furnished with period furniture and paintings. In the restored terraced garden is a fish-shaped pool. Paca was a signatory of the Declaration of Independence and governor of Maryland.

● **Hammond-Harwood House** (19 Maryland Avenue; by reservation; admission fee; tel: 410-269-1714) was built in 1773–74 and has one of the most beautiful carved doors of the time.

● Opposite is the **Chase-Lloyd House** (22 Maryland Avenue; by reservation; tel: 410-263-2723), was begun in 1769 by Samuel Chase, later a justice on the new nation's Supreme Court. It was continued in 1771 by Edward Lloyd IV, a wealthy planter, when Chase ran out of funds. This house is on three floors with a formidable hall of Ionic pillars and a so-called "floating" staircase.

Main Street has a good choice of eateries. One of the best is **Cafe Normandie** (185 Main Street; tel: 410-263-3382), with lace-curtains, fireplace, and moderately priced French fare. The gourmet **Treaty of Paris Restaurant** (16 Church Circle; tel: 410-269-0990), is in the original 18th-century Maryland Inn; expensive for dinner, but a value at lunch.

If there's time, and security considerations allow, the **Naval Academy** is worth a tour. (Visitors' Center on King George Street open daily 9am–4pm; for tour

LEFT: City Harbor, Annapolis.

Map on page 252

availability, tel: 410-263-6933.) There's lots of naval memorabilia and John Paul Jones, Revolutionary War hero and father of the navy, is buried here.

Chesapeake Bay Area

For some reason, Washingtonians seem to truly believe they have been to the beach only if they have spent three hours – or more frequently, given traffic jams, five hours – getting there. But there are two alternatives to the Atlantic shore haul, where it is just as possible to get sunburnt without the stress of the drive. One is the Eastern Shore, with its small and evocative old colonial towns; the other is the beaches of the Bay.

On the Annapolis side of the **Bay Bridge**, the string of small beaches is less imposing than on the other side, with gentler waves and unthreatening currents. These beaches do attract jellyfish, however, and many have nets to keep them out of the swimming areas. But there is a generous supply of beach amenities, such as paddle-boat rentals, barbecue grills and picnic tables, and swings and playgrounds for children. Right by the Bay Bridge on the last Route 50 exit before the bridge is **Sandy Point State Park ❸** (mid-Sept–Nov, 6am–9pm; Mar–May 6am–9pm; remainder of the year 8am–dusk; tel: 410-974-2149, which is good for picnicking, fishing and swimming.

Cross the bridge and you'll be on Maryland's **Eastern Shore,** one of Washington's best kept secrets. While energetic young people tear across the Bay Bridge in their open jeeps, radios blaring, heading for the Atlantic coast resorts of Rehoboth and Ocean City in Delaware, the relaxed and unassuming towns of the Eastern Shore lie closer to home and are for the most part ignored.

On the far side of the Chesapeake Bay from Baltimore and Annapolis, this is where the gentry of more gentle times came to take the sea air. It's an area of farms and marshlands, wild bird sanctuaries and road-side produce booths,

BELOW:
dress parade
at the US Naval
Academy.

seafood restaurants and small beaches dusted with shells.

Cambridge ❹, a small town of some 11,000 people, is characteristic of the Eastern Shore's watermen's way of life. Founded in the 1680s, this fishing village sits between the **Choptank River** and the **Blackwater National Wildlife Refuge**, a stretch of tidal marshlands and woods favored in winter by tens of thousands of migrating ducks and Canada geese. For touring information, start at the Dorchester County Historical Society (902 McGrange Avenue; Tues–Sat, 10am–1pm; tel: 410-228-7953) in the **Meredith House**. This is a well restored 18th-century manor house beside the **Neild Museum** (Thur–Sat, 10am–4pm) with its smokehouse, antique doll collections, crafts and furniture.

You pass by **Annie Oakley's house** on Bellevue Avenue where the retired cowgirl, Cambridge's most celebrated transplant, kept up her sharpshooting skills by firing out the upstairs windows, much to the dismay of the neighbors.

St Michaels

Less than two hours from Washington is the old but still active village of **St Michaels ❺**, a center of boating activity located where the **Miles River** meets the Chesapeake Bay. Apart from the very attractive **harbor**, the main focus for visitors is the **Chesapeake Bay Maritime Museum** (Mill Street; open daily, hours vary by season, but generally 9am–5pm; admission fee; tel: 410-745-2916). The extensive waterfront campus has a screwpile lighthouse, which first saw service in the Bay in 1879 and which is restored and open for touring. There are also skipjacks, deadrises, dug-out canoes, and other traditional Bay vessels either undergoing restoration, tied up at the dock, or on display in the shed. The staff are well-informed.

Next door is the **Crab Claw**, a laid-back tourist-friendly place where you can watch the watermen unload their crab bushels on the dock, then order a dozen steamers for yourself. If you've never tried the Chesapeake Bay's specialty, blue crab and aren't sure how to go about cracking one open, follow the directions on the paper placemat.

You can take two worthwhile trips from St Michaels – one to **Oxford ❻**, taking the **Oxford-Bellevue ferry** (tel: 410-745-9023), supposedly the oldest privately owned ferry in the US. In the town you can visit the historic **Robert Morris Inn** (tel: 410-226-5111), a restaurant and B&B and the former home of wealthy merchant Robert Morris whose son is often called the "financier of the Revolution."

Tilghman Island

The second trip is to **Tilghman Island ❼**, 10 miles (16 km) farther on the only road out of St Michaels. Once there, you may notice a lighthouse askance and sticking out of the Bay. This is Sharp's Island Lighthouse, abandoned and slowly sinking. The local watermen will be happy to tell you the story as they work on their boats and nets.

St Michaels is not far from **Easton ❽**, an old colonial bay-side town full of antique shops and old bookstores. This is an area of strong religious background. In

LEFT: golf courses are plentiful.

Map on page 252

1777, the Eastern Shore Quakers vowed to "disunite" any member of the congregation who still employed slaves. In Easton, William Penn, the Quaker who founded Pennsylvania, preached at the **Third Haven Meeting House** (405 South Washington Street; tel: 410-822-0293), built in 1682 and possibly the oldest frame house of worship still in use.

There are numerous historical manor houses and dwellings around Easton, but most of them are scattered outside the town itself. Contact the Historical Society of Talbot County at the courthouse on Washington Street (tel: 410-770-8000) for information, guides and maps.

Atlantic Ocean Beaches

Washingtonians head for **Ocean City** in Maryland, and quieter **Rehoboth ❾**, **Dewey Beach ❿**, **Bethany Beach ⓫** and **Fenwick Island ⓬** in Delaware to avoid the heavy humidity of summer. Just 3½ hours from Washington, on the southern Atlantic shore, Rehoboth in summer is a jumping hot bed of noise, music, T-shirt emporia, beach beauties and Adonises.

Along the boardwalk is a run of stores selling saltwater taffy, hot dogs and souvenirs and a small fun fair with airplane rides and dodgem cars. On Saturdays there are band concerts and talent opportunities for summer visitors. As well as the usual fast foods, there are a number of up-market restaurants just behind the main drag, not necessarily charging up-market prices, and striving to serve original and pleasantly presented dishes.

If you prefer your seaside experience a little less brash, the beach at **Cape Henlopen State Park ⓭** (tel: 302-645-8983) is in a protected nature reserve near Lewes, Delaware, and simply offers you hundreds of yards of open beach and empty sand dunes. Those with small children should be aware that the surf along all of this coast is sometimes quite strong and the current can be vigorous. However, all of the beaches are well manned by efficient-looking lifeguards.

RIGHT: a Chesapeake blue crab.

CHESAPEAKE'S BLUE CRABS

A particular Maryland delicacy is *Callinectes sapidus.* The first word means "beautiful swimmer" – a reference to the paddle-shaped pair of hind legs that gracefully propel these blue crabs, members of the crustacean family, up and down the Chesapeake from the salty mouth, which they prefer when young, to the fresh water at the top of the Bay. *Sapidus* means "tasty," which they are. Boiled and seasoned with plenty of Old Bay seasoning, crabs are meant for sharing with friends on a long hot summer afternoon with plenty of beer.

There's an art to "picking" crab. Pull off the claws first and set aside, then open the apron on the underside, scrape away the organs and the "dead man," or the gills, and dig out the meat; now crack open the claws and pull the meat off with your teeth.

Although crabs are found all along the Atlantic coast, they're most plentiful in the Chesapeake, shallow and grassy and a perfect home, and which, in a good year, can yield a harvest of 100 million lbs (45 million kg). In its three-year life span, a crab will molt several times, shedding its old shell for a bigger model. Between shells it's known as "soft-shell" crab, best breaded and fried, and easy to handle with a knife and fork.

For a more peaceful experience, head for Chincoteague Island, just inboard of **Assateague Island** , a national seashore and barrier island, uninhabited except for the birds and the famous ponies that live here (some will remember from their childhood Marguerite Henry's *Misty of Chincoteague*). Chincoteague, actually in Virginia, has the usual tourist joints, but Assateague (National Seashore Visitors Center, daily 9am–5pm; tel: 410-641-1441), which is split between Virginia and Maryland, is pristine, with a wildlife preserve and miles of untouched beaches.

Deep Creek Lake Area

Washington's other water sports center, 4 hours west from the capital and turning itself into a skiing resort in winter, is **Deep Creek Lake** in the heart of the Appalachian Mountains. Built in 1925 as a reservoir, it encompasses 3,900 acres (1,580 hecrares) of fresh water that draws fishermen and watersports enthusiasts. It's a bit of a drive and more than you can rea-

sonably expect to cover in a day, so call ahead (Deep Creek, McHenry, Garrett County, Maryland, tel: 301-387-4386 or visit www.garrettchamber.com) and line up one of the several rental homes or check into an inn (try Lake Pointe Inn; tel: 1-800-523-5253; www.deepcreekinns.com), and rent a boat.

Sailboats, power boats and canoes can be rented at the several marinas around the lake. There are regular weekend sailboat races and water skiing events, including the Barefoot Skiing Championships. From May to October sponsored fishing contests award prizes for the biggest largemouth bass, bluegill, chain pickerel, walleye and more. The woods that reach down to the lake offer cool and quiet trail walks.

East of Deep Creek Lake is the town of **Cumberland** , which began as Fort Cumberland, a crucial frontier outpost during the French and Indian War. George Washington defended the town and you can visit his one-room log cabin headquarters – the only remains of the fort – in **Riverside Park**.

A handsome town of fine buildings, parks and gardens, Cumberland is full of antique and curio shops. **Washington Street** is where the rich coal and rail barons lived, in mansions ornately executed in all styles from Federal to Georgian Revival.

The **History House** (218 Washington Street; Tues–Sat, 11am–4pm; tel: 301-777-8678) is full of household gadgets, technical machinery and toys of the 19th century. Nineteen fully furnished rooms illustrate life in 1867, while another room depicts an early 1900s schoolroom. In the basement is the old servants' quarters, with kitchen and pantry. Elsewhere in the house rooms display 19th-century costumes, veterinary, medical and dental equipment. One of the oldest buildings on the street is the **First Church of Christ Scientist**.

But the railway station, with its displays of memorabilia, is where you should start. The **Western Maryland Scenic Railroad** (13 Canal Street; reservations necessary; call for hours of departure; $19 adults, $13 children; tel: 1-800-872-4650 or visit

LEFT: violent knight life at the Maryland Renaissance Fair.

Map on page 252

www.wmsr.com) offers three-hour excursions. A 1916 Baldwin steam engine hauls you through the Allegheny Mountain's scenic railroad passes, gorgeous in the fall, over an iron truss bridge and through a 900-ft (270-meter) tunnel to Frostburg where you can take a lap around historic Main Street before hoping on board and chugging back to Cumberland.

Antietam National Battlefield

The battle of Antietam, when the South's General Lee tried to follow up his victory at Manassas by invading the northern states, was the scene of the greatest and bloodiest carnage of the entire Civil War. It all took place in a single day, September 17, 1862. More than 23,000 men were killed and "lay in rows precisely as they had stood in their ranks a few moments before," according to Union General Joseph Hooker.

Clara Barton, founder of the American Red Cross, was among those who tended the wounded. After the Manassas battle in Virginia, Lee sent Stonewall Jackson to seize the vital river crossing and arsenal of Harper's Ferry, and lead his soldiers into Maryland in the hope of forcing the evacuation of Washington and swinging the slave-owning state of Maryland to the Southern cause. Unfortunately, his battle plan fell into the hands of Union General George McClellan. When the Confederates marched into the little town of Sharpsburg they walked straight into 87,000 Union soldiers. Although the resulting carnage left both sides wondering who had won, and whether the cost was worth the victory, the battle was a crucial turning point in the Civil War.

After the South's defeat, it lost its very real hope of winning diplomatic recognition and the support of European countries. President Lincoln took the battle as an apposite moment to denounce slavery officially. One week after the battle, on September 23, 1862, a draft of Lincoln's Emancipation Proclamation was published in the press.

The 26-minute film at the **Visitor Center of Antietam National Battlefield** ⓱ (1 mile north of Sharpsburg on Route 65,

open Jun–Aug, 8:30am–6pm, Sept–May, 8.30am–5pm admission fee; tel: 301-432-5124; www.civilwar-va.com/ maryland), set in the rolling Maryland pastureland where the battle occurred, gives the best orientation to start your tour. It takes you from **Dunker Church**, where fighting commenced, to the cornfield which saw most of the military action, and to **Bloody Lane** where, in four hours, 5,000 soldiers died. Unless you are a sturdy hiker (the Visitor Center provides maps if you are), you will need to tour by car as the site is large and enclosed inside a network of minor roads.

Perhaps the most dramatic spot is **Burnside's Bridge**, where a handful of Georgian sharpshooters held off Burnside's infantry division, shooting them down as they tried to cross a narrow stone bridge. Burnside's scouts did not find a ford across the river until it was too late. You can still see the Georgians' fox-holes on the hill above the bridge. A map and brochure detail the events at the 11 stops along the 8½-mile (14-km) drive. ❑

Map on page 252

DAY TRIPS: VIRGINIA

Beyond the Beltway, Virginia's suburban sprawl gives way to a gentle countryside filled with some of the nation's most historic homes and attractions, including Monticello and colonial Williamsburg

Washington was founded as a geographical compromise midway between the Northern and Southern states. And, as a result, it is wonderfully placed for day trips that bring the visitor very different flavors of the surrounding countryside, its turbulent history and cultural progress.

Ninety minutes to the west of the city lie the Blue Ridge Mountains of Virginia. Its Shenandoah Valley is perfect for fall drives to admire autumn leaves the equal of any leaf-color view in New England. To the south and west are the poignantly moving battlefields of the Civil War, each with detailed on-the-spot explanations and diagrams; near Charlottesville, about 120 miles (190 km) southwest of Washington, are monuments to America's more peaceful development: Thomas Jefferson's house and the University of Virginia, two of the most civilized establishments to be found on any continent.

Presidential Trio

Located on the top of a foothill of the Blue Ridge Mountains close to the town of Charlottesville, **Monticello** ⓲ (Route 53, two miles southeast of Charlottesville; open Mar–Oct, 8am–5pm; Nov–Feb, 9am–4.30pm; $13 admission; tel: 434-984-9822; www.monticello.org) is perhaps the only historical house in America that entirely expresses the character of its owner. It's also the only American house on the United Nations' list of World Heritage sites. Jefferson's "essay in architecture," as he thought of his home, is a classical tribute to the architects of ancient Greece and Rome. For this citizen president, it was also a haven after the burdens of life in the White House.

Jefferson built the home and continued to expand between 1769 and 1809. There are 43 rooms altogether, not counting the wings and outbuildings, outfitted with

eight fireplaces and fine timeless furnishing, about 60 percent of them original.

Monticello also incorporates many of Jefferson's own innovative ideas: in his study is a desk-cum-chaise he designed that allowed him to write while lying down; on the long windows are sliding double glass doors to keep in the heat, and nearby are silent "dumb waiters" that travel between floors to carry light parcels upstairs or down. When he died, Jefferson was more than $107,000 in debt and his heirs were forced to sell the house. It passed among a few private owners until, after the Civil War erupted, the Confederate government seized and sold it. It went back into private ownership, fell into serious disrepair, and was then, in 1923,

LEFT: the fall foliage of the Shenandoah Valley.
RIGHT: Monticello.

given to the Thomas Jefferson Foundation and beautifully restored.

Jefferson's garden, restored according to his detailed gardening notes, expresses the civilized nature of the man, laid out in a perfect balance of shape and showing where he practiced his experimental gardening and cultivation techniques. Here the president grew 20 varieties of his favorite English pea, and more than 250 kinds of vegetables. Save time to explore the grounds, especially Mulberry Row, the 1,000-ft (300-meter) long road where about half of Jefferson's 135 slaves lived.

It may not be as well known, but **Ashlawn** ❶ (1000 James Monroe Parkway; 11am–5pm daily; extended hours in summer; closed Jan 1, Thanksgiving, Dec 25; admission fee; tel: 434-293-9539; or visit www.avenue.org/ashlawn), home of the country's fifth president, James Monroe, is cozier. The furnishings are 18th-century elegant, and you can tour the original smokehouse, overseer's cottage, and reconstructed slave quarters.

Next, stop at **Montpelier** ❷ (11407 Constitution Highway near Orange, VA; Apr–Oct, 9.30am–5.30pm; Nov–Mar, 9.30am–4.30pm; admission fee; tel: 540-672-2728; www.montpelier.org), home of president James Madison and a curious mix of old and new. The Dupont family, of Delaware chemical fame, bought the house in 1901 and made extensive renovations, doubling its size. Here, colonial-era trappings nudge modern artifacts.

Charlottesville

For brochures and maps, start at the **Charlottesville Visitors Center** at the intersection of Route 64 and 20 South (daily 9am–5pm; closed Jan 1, Thanksgiving, Dec 25; tel: 434-293-6789), or head straight for the **University of Virginia** (free guided tours of the Rotunda and Lawn are offered year round by the University Guide Service; tel: 434-924-7969; www.virginia.edu).

Charlottesville ❷ is the graceful Virginia town, set against a backdrop of

BELOW: Jefferson's Rotunda at the University of Virginia.

Map on page 252

rolling foothills, where, in 1817, along with James Madison and James Monroe, Jefferson laid the cornerstone for the university. The former president was 75 years old when he drew up the blueprints for these elegant grounds with their classical buildings united by a covered colonnade of lyrical arches and pillars. Here, too, can be seen the influence of ancient Roman architecture – particularly in the **Rotunda**, modeled after the Roman Pantheon – and the designs of the Italian Renaissance architect, Andrea Palladio.

A block from the Rotunda is the **University of Virginia Art Museum** (155 Rugby Road; Tues–Sun, 1pm–5pm; free admission; tel: 434-924-3592, or visit www.virginia.edu/artmuseum), small but impressive with a mix of Old Masters, Chinese porcelains, Native American artifacts, and Roman statuary.

Next, head downtown to the **mall** on Main Street between First and Fifth streets, a blocked-off city thoroughfare for pedestrians only. There are lots of little

cafes, and antique and specialty shops including the **New Dominion Book Shop** (404 East Main Street; tel: 434-295-2552). At 515 East Water Street is the **C&O Restaurant** (tel: 434-971-7044) an old diner transformed into a favorite with the locals whose tastes run toward the sophisticated and international.

Fredericksburg

This peaceful town is famous not only as the place where George Washington grew up, but also for the devastating Civil War battles fought around it. Over 750,000 troops clashed in the fields near here, but the town of Washington's boyhood is a tranquil, civilized place where more than 350 buildings were erected before 1870. It is here, rather than at Mount Vernon, where he was supposed to have thrown a dollar – not across the mile-wide Potomac, but the 300-ft (90-meter) Rappahannock River. What's more, the future president threw not a dollar, but a Spanish doubloon.

Fredericksburg ㉒ is full of history. The

BELOW: Charlottesville.

best place for a pre-tour briefing is the **Visitors Center** on Caroline Street (706 Caroline Street; daily 9am–5pm; closed Jan 1, Thanksgiving, Dec 25; tel: 540-373-1776; www.fredericksburgvirginia.net), which has a film show and maps.

On **Charles Street** is the house of the president's mother, **Mary Washington** (1200 Charles Street; call for tour information; tel: 540-373-1569), in which she lived for the last 17 years of her life. It is furnished with period pieces, and the garden restored to its original 18th-century form. Costumed guides show visitors both. Look out for the ornamental plasterwork in the **Great Room** and Mary Washington's needlepoint. Tea and gingerbread cooked according to Mary Washington's own recipe is served in the kitchen building by folks in period dress.

In the **James Monroe Law Office** at 908 Charles Street (Mar–Nov, daily 9am–5pm; Dec–Feb, 10am–4pm; admission fee; tel: 540-654-1043) can be seen correspondence between Monroe and Jefferson, Washington and Benjamin Franklin; White House china and cutlery from the period, and pieces of furniture President Monroe acquired for the White House, some of them from France and said to have belonged to Marie Antoinette.

The home of another of Washington's relatives, his sister Betty, is **Kenmore** (1201 Washington Avenue; Mon–Sat 9am–5pm; Sun, noon–5pm; admission fee; tel: 540-373-3381) This lovely Georgian mansion, once situated on a 1,200-acre (480-hectare) plantation, is now surrounded by more modern residences. The wealth of Betty's husband, Fielding Lewis, is reflected in the lavishly festooned plaster ceilings and mantelpieces. A brick-pathed and box-wood-lined garden is out back.

A good lunch stop is the **Kenmore Inn** (1200 Princes Anne Street; tel: 540-371-7622), not formally connected with the home but lovely all the same and featuring a classically Southern menu.

The Rising Sun (1304 Caroline Street; Mar–Nov, 9am–5pm; Dec–Feb, 10am–4pm; admission fee; tel: 540-371-1494) was Fredericksburg's first tavern and formerly the home of George Washington's brother Charles. It doesn't offer food or drink but does provide a taste of the 18th century's tavern-centered life.

Slavery remembered

By 2007, the city should have opened the new **National Slavery Museum**, destined to be a major international research center staffed by scholars and historians whose work will spur exhibits and academic courses. In a town full of historic homes and other relics of early American and antebellum life, the museum, which will be designed by Chien Chung Pei, son of the famed architect I.M. Pei, promises to be a more fitting acknowledgement of the divisive and cruel institution of slavery than the small brass plaque in front of an Italian restaurant that marks where the city's slave auction block used to be. Historians and activists have argued over the location of the museum, lobbying for a spot on the Mall in Washington, or at least in Richmond, Virginia, the one-time capital of the Confederacy, but Fredericks-

LEFT: costumed docents lead tours at some Virginia plantations.

Map
on page
252

burg has come out on top in the debate, winning the $200 million project that Pei has said will be something "forward-looking – not something just looking back."

Washington's birthplace

For more history, head about 40 miles (64 km) east of Fredericksburg to **Westmoreland County** in Virginia's so-called **Northern Neck**, a peninsula formed by the Rappahannock and Potomac rivers. This rural jut of land includes the site of **George Washington's birthplace ㉓** at Popes Creek Plantation (daily 9am–5pm; closed Jan 1, Thanksgiving, Dec 25; admission fee; tel: 804-493-0130 or visit www.westmoreland.va.us), a restored brick home and former tobacco farm manor house on 500 peaceful acres (200 hectares). This is the site of the famous "cherry tree incident" – though Washington, born in 1732, lived here only until he was three and at that age was probably incapable of wielding an ax, much less felling a tree.

Stratford Hall Plantation ㉔ (daily

9:30–4pm; closed Jan 1, Thanksgiving, Dec 25, admission fee; tel: 804-493-8038; www.stratfordhall.org) was home to several generations of Lees, including Richard Henry Lee and Francis Lightfoot Lee, both signers of the Declaration of Independence. Francis Lightfoot was a Revolutionary War hero and father of Robert E. Lee, who was born here. The plantation has a restaurant.

Civil War sites

Enthusiasts of the War Between the States will find no shortage of sites to visit within an easy day's trip from Washington. One of the bloodiest battles was at Antietam in Maryland *(see page 259)* but 60 percent of the engagements were fought on Virginia soil. Between December 1862 and May 1864, more than 100,000 soldiers in Confederate and Union forces were found dead or wounded on the fields of what is now the **Fredericksburg and Spotsylvania National Military Park ㉕** (daily, sunrise to sunset, battle-

RIGHT:
Virginia's
green rolling
hills have
obliterated all
trace of many
battles.

field visitors centers 9am–5pm with extended summer hours; admission fee, www.civilwar-va.com).

Four devastating major battles of the Civil War were fought on an area that stretches beyond the 6,000 acres (2,500 hectares) of woods and meadows that make up the military park: the Battle of Fredericksburg, December 11–13, 1862; the Battle of Chancellorsville, April 27–May 6, 1863; the Battle of the Wilderness, May 5–6, 1864; and the Battle of Spotsylvania Court House, May 8–21, 1864, the first confrontation between General Robert E. Lee and General Ulysses S. Grant.

This is an area punctuated with trenches and gunpits. It makes for an immensely moving experience when viewed while listening to the taped narrative available from the Visitors Centers. It is hard to imagine that this gentle countryside was once the scene of the most appalling carnage; it is difficult to tell which is more sobering, the closely-wooded country around **Chancellorsville** where the enemy

got frighteningly close before being seen, or the open killing field of Fredericksburg itself. General Robert E. Lee's troops, sheltered almost invulnerably behind a country wall, mowed down line after line of advancing Union soldiers – almost 10,000 men in a matter of hours.

At Chancellorsville is the small building in which General Stonewall Jackson died after being mortally wounded by his own troops as he reconnoitered Union troop movements. It is now the **Stonewall Jackson Memorial Shrine** (9am–5pm in summer; Fri–Tues in spring and fall; Sat–Mon in winter; closed Jan 1, Dec 25) and commemorates his life and military career.

A mapped driving tour follows a loop that includes most of the major battles, while there are also maps for a 7-mile (11-km) hiking trail, both available at the park Visitors Center.

Manassas (Bull Run)

Twenty-five miles (40 km) west of Washington lies **Manassas (Bull Run) National Battlefield Park** ㉖ (grounds open daily dawn to dusk, Henry Hill Visitors Center daily 8.30am–5pm, closed Jan 1, Thanksgiving, Dec 25; admission fee; tel: 703-361-1339; www.civilwar-va.com), the site of two of the largest and most bloody battles of the war. Both forces sought control over a strategic railroad junction. Known by the Confederates as Manassas after the nearby town and by the Unionists as Bull Run for the local stream, this is where the first major land battle was fought on July 21, 1861.

It is also where General "Stonewall" Jackson earned his nickname for holding firm against advancing Unionists at a critical point in battle – "standing like a stone wall," in General Lee's words. Washingtonians arrived with picnic hampers to witness what they anticipated would be an easy and picturesque victory, certain the Union would never send troops into the South. But they quickly turned tail, and their carriages and coaches joined the desperate flight after what is now thought of as the first example of "modern" warfare.

The Henry Hill Visitors Center is an evocative spot from which to picture the

LEFT: recreating the second battle of Manassas.

Map on page 252

nearly 30,000 soldiers killed or wounded in 1861 and 1862, when the opposing forces on this same spot a second time. Take the driving or walking tour, following the markers which begin at the Visitors Center. There are quite a number of edifices that were part of the battle still to be seen: the **Stone Bridge** over Bull Run stream where the Union soldiers opened fire to launch the first battle; the raised railroad where General Jackson's troops maintained ground during the second battle; and the **Stone House tavern** that was used as a field hospital.

It was to **New Market Battlefield Park** ㉗ (open daily 9am–5pm; closed Jan 1, Thanksgiving, Dec 25, Dec. 31; admission fee; tel: 540-740-3102; or visit www.civilwar-va.com) on May 15, 1864, that 247 young cadets, all under 20 years old, were summoned in desperation from the Virginia Military Institute in Lexington. They were needed to give vital support to Confederate General Breckinridge, whose men were being decimated by the forces

of Union leader General Sigel. The young cadets fought with such courage that they helped the Confederates vanquish the more experienced and larger opposition force. A walking tour takes you round the parkland, following the development of both sides of this important battle.

The battlefield park is smaller than the other two, only 280 acres (113 hectares). Right in the middle of the field is the **Hall of Valor**, in memory of the cadets, a museum that gives both sides of the Civil War story in film, dioramas, and exhibits. Outside in the park are replicas of cannons, plus a farmhouse from the period whose outbuildings have been restored to display a 19-century working farm with garden and blacksmith's. This is where, in May, a re-enactment of the battle is staged in full costume by nearly 1,000 players.

For the battle-weary there are the **Endless Caverns** ㉘ open daily 9am–4pm with extended summer hours; admission fee; tel: 540-896-2283 or 1-800-544-2283), whose extraordinary caves deep inside

BELOW: train trestles guarded by troops at Manassas in 1863.

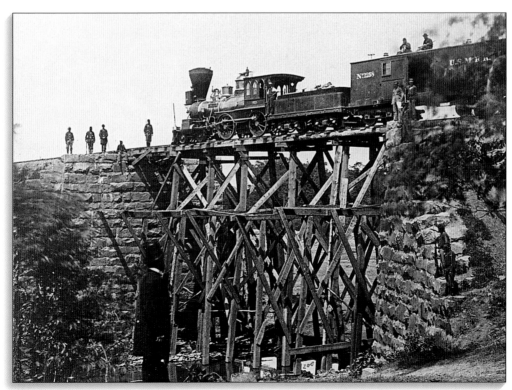

Massanutten Mountain are dramatically lit to display the different caves' intricate and colorful formations.

The Skyline Drive

Between the towns Front Royal and Waynesboro lie 105 twisting miles (170 km) through the tree- and flower-covered summits of the **Blue Ridge Mountains**. There are dozens of overlooks along **Skyline Drive** from which to gaze upon the **Virginia Piedmont** and the **Shenandoah Valley**, to take photographs. There are more than 500 miles (800 km) of trails, all accessible from the Drive. (Enter Skyline Drive at Fort Royal or Luray; open 24 hours year round; enter Shenandoah National Park through the Drive; admission fee; tel: 540-999-3500; www.nps.gov). The speed limit here is a leisurely 35 mph (56 km/h); even that can seem a bit fast.

In summer, these are soft green in a bluish haze, punctuated by flowers. But Skyline Drive really comes into its own in the fall when the colors of the changing leaves are spectacular. This is a relatively recent bonus. Though originally this land had been covered in thick forest, by 1935 when the **Shenandoah National Park** ㉙ was created, the trees had been cleared, the streams fished out and the wild game decimated by the local settlers. In 1976, the efforts of the park's conservationists to reforest and regenerate natural growth were so successful that Congress declared roughly 40 percent of the park an **official wilderness**. Under the protection of the National Park System, the forest has been reclaimed and returned to its natural state.

The Park Services offer several good restaurants, accommodations in overnight log cabins, campgrounds and ledges, as well as many ranger-conducted activities, such as hikes, talks and campfire get-togethers. The **Dickey Ridge Visitors Center** (mid-Apr–late Nov, 9am–5pm; tel: 540-999-3500) can provide information and tips on camping in the Shenandoah Valley. The name Shenandoah, by the way, is an Indian word which means

BELOW: Virginia Beach, at the southeast corner of the state, is the Atlantic coast's longest resort beach.

Map on page 252

"Daughter of the Stars." An ancient belief is that the valley used to be a lake, and that, once every 1,000 years, the stars would sing around it.

Nine miles west (14 km) of the Luray entrance to Skyline Drive are **Luray Caverns** ㉚ (daily 9am; closing times vary with season; admission fee; tel: 540-743-6551), the largest underground caverns in Virginia. The tour covers 1¼ miles (2 km) of some of the weirdest and most impressive icicle-shaped limestone stalagmites (extending up from the cave's floor) and stalactites (hanging from the ceiling), formed over thousands of years by the erosive action of underground streams. As you tour the caves, you'll be treated to a concert by a guide armed with a rubber mallet who'll bang out a tune on a portion of the cave's stalactites that the guides humorously refer to as the Stalac-pipe Organ. This is a perfect outing for one of Virginia's typical scorching summer days, since the temperature underground can be 20 degrees cooler.

RIGHT: yesteryear relived at Williamsburg.

Williamsburg

It's a bit of a drive for one day – 3½ hours south of Washington, if the traffic co-operates – but if you have the time, plan a trip to **Williamsburg** ㉛, the restored colonial-era capital of Virginia. (Grounds open 24 hours; buildings daily 9.30am–4.30pm; $37 adult, $18.50 children admission at time of writing; for information and reservations, tel: 1-800-HISTORY, or visit www.history.org; lodging at the Williamsburg Inn and reservations for dinner at one of the four colonial-style taverns must be made in advance and can all be made by telephoning 1-800-TAVERNS or 757-229-2141; for information on nearby motels, check the website).

From 1699 until just before the end of the Revolutionary War, this quaint town was the essential gathering spot for American notables. Thomas Jefferson, George Washington, George Mason and Patrick Henry (of "Give me liberty or give me death" fame) all either served here in the Virginia legislature, or debated the merits

NAPA EAST

Thomas Jefferson, first in so many things, persuaded an Italian vintner to assess the potential for growing grapes in Virginia. Initial progress was slow, but today, after much experimentation with various grape varieties, the state has become one of the country's notable grape growing regions. In 1985 there were maybe two dozen wineries here; today there are 80.

The Virginia Wine Marketing office (tel: 1-800-828-4637; www.virginiawines.org) provides a list of vineyards and details about touring. Keep in mind that many wineries are small, family-owned operations, so phone ahead.

If you're near Charlottesville, stop at the Barboursville Vineyard (tel: 540-832-3824), credited with reinvigorating the Virginia wine industry when it opened in the late 1970s. Jefferson Vineyards (tel: 434-977-3042) is within easy striking distance of Monticello. Minutes away is the Kluge Estate Winery and Vineyard.

If you're swinging through Williamsburg, stop at the Williamsburg Winery (757-229-0999) whose slightly sweet Governor's White has helped put them on the wine map. And if you're passing through Virginia's Northern Neck, look out for Ingleside Plantation Vineyards (tel: 1-800-747-4645 or 804-244-8687).

of and plotted the course of the war of independence here. By the time the capital moved to Richmond in 1780, the town was headed for serious decline.

As is the case with so many of Virginia's treasured historic sites, the state refused to involve itself in the restoration and upkeep. (George Washington's Mt Vernon is run by a group of dedicated women, Bill of Rights author George Mason's Gunston Hall was saved by volunteers, most Civil War battlefields are the province of the federal government if they aren't in private ownership).

Fortunately, Williamsburg hung on just long enough for Rockefeller money to rescue it early in the 20th century. "That the future may learn from the past," is how John D. Rockefeller put it when he committed millions to the restoration effort that has turned this former rubble into one of the grand examples of American life in the colonial past. Tours start at the **Governor's Palace**, three floors of Georgian grandeur that, during the years that the

succession of royal governors (the king's representatives in Virginia) lived here, took 25 servants and slaves to maintain. The complex includes a stable, carriage house, scullery, and other outbuildings, plus a wine cellar large enough to hold 3,000 bottles.

Duke of Gloucester Street is Williamsburg's main avenue, with red-bricked shops and residences lining both sides. Tradespeople in period costumes work in the shops, spinning and weaving, dipping candles, making furniture, forging, and practicing all the other the crafts of 18th-century life. They'll stop in the midst of their work to give impromptu talks and to answer any questions. One of the nicest things you can do for yourself is to stop in your travels and pause at **Bruton Parish Church**, which dates from 1715. The usual suspects (Jefferson, Washington, Henry, and Mason) all worshiped here, assigned to their respective and enclosed boxed pews, designed for privacy and to keep out winter drafts.

Included in the town are two good art museums, the **Abby Aldrich Rockefeller Folk Art Museum** (daily 11am–5pm) and the **DeWitt Wallace Decorative Arts Museum** (daily 11am–5pm), both offering wonderful collections of American folk art, furniture, paintings, Chinese porcelains, and more.

Period taverns

There are four period taverns, all of them good, but the best are **Christiana Campbell's** and **Chowning's,** both requiring reservations. If these don't suit, try the **Trellis** (tel: 757-229-8610) on Merchant Square, recommended especially the house dessert, Death by Chocolate. Williamsburg can get busy in summer, so it's wise to start early. It's easier to visit during the fall and winter months, when crowds are fewer and the weather this far south is more pleasant than in summer.

Those with children may wish to stop at **Busch Gardens** (tel: 1-800-343-7946 or visit www.4adventure.com/buschgardens), a theme park 3 miles (5 km) east of town. As theme parks go, it's one of the nicer ones and has a European flavor. ❑

Map on page 252

LEFT: demonstrating the techniques of 18th-century wig-making. **RIGHT:** going shopping in Williamsburg.

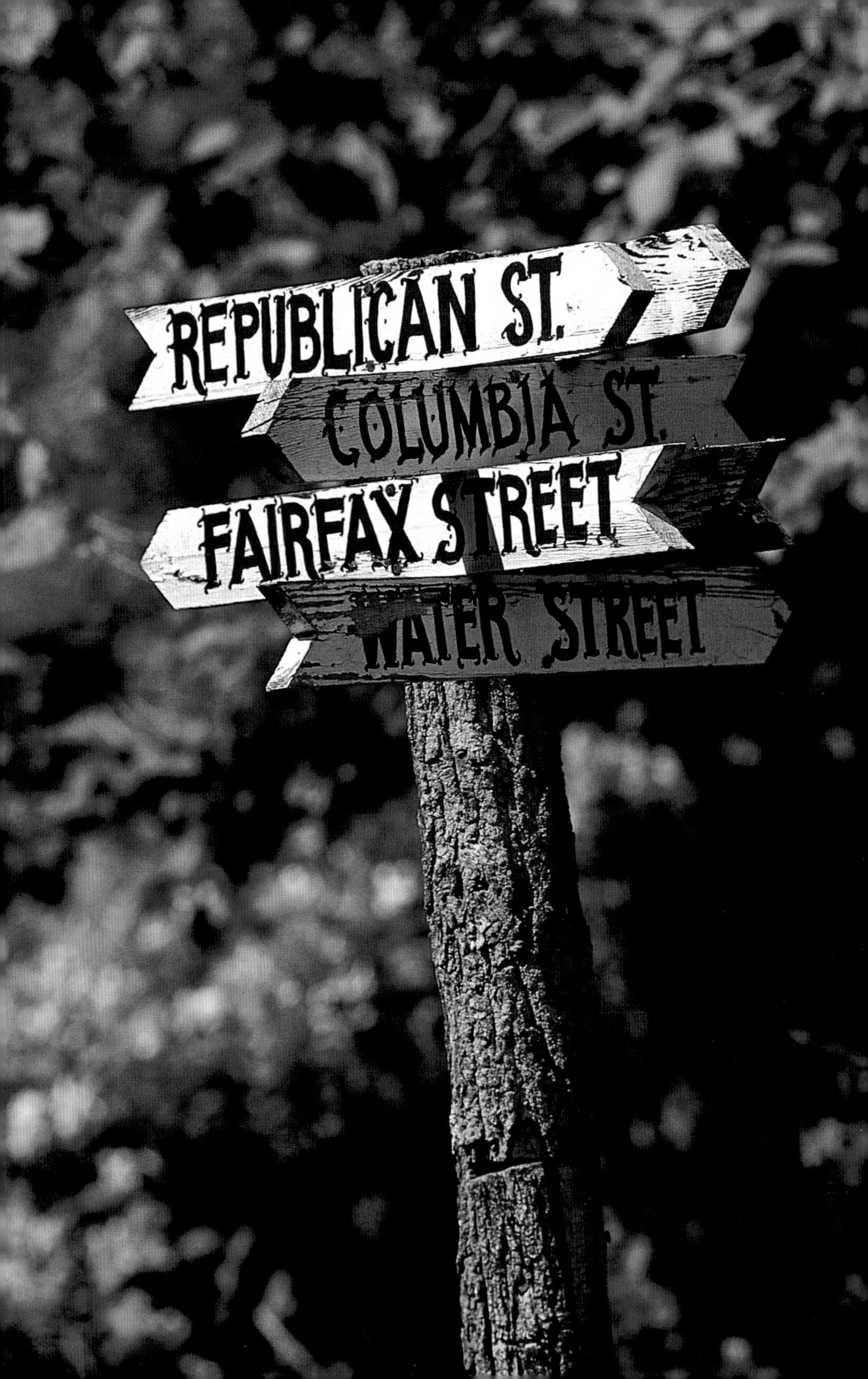
REPUBLICAN ST.
COLUMBIA ST.
FAIRFAX STREET
WATER STREET

INSIGHT GUIDES
Travel Tips

CONTENTS

Getting Acquainted

The Place

Situation: Washington, D. C. lies 30 degrees N, 77 degrees W, along the east coast of North America between the states of Maryland and Virginia and along the Anacostia and Potomac rivers. Washington, DC is the capital of the United States and seat of its federal government. The Potomac River, which separates Washington from Virginia, leads directly to the Chesapeake Bay, the world's largest estuary.

Population: At 600,000 in the city proper, but taking in the suburbs of Maryland and Virginia, the figure rises to 5.5 million. Three-quarters of the city is comprised of African-Americans who live primarily in the Northeast, Southeast, and Southwest quadrants of the city. There is a growing Hispanic population, settled mainly in the Adams Morgan section, comprising roughly 10 per cent of the total population. The remaining white population lives primarily in the Northwest quadrant, which is the city's largest. Immigrants from Asian countries have flocked to the Washington metro area, mostly in Northern Virginia, where they have found jobs in the technology industry that has grown up there.

Area: 61 sq. miles (158 sq. km) divided into four quadrants (Northwest, Northeast, Southwest, Southeast) radiating from the Capitol. The suburban communities surrounding Washington include the close-in Maryland counties of Montgomery and Prince George's, and the cities and counties of Northern Virginia including the cities of Alexandria, Falls Church, and Fairfax, and the counties of Fairfax, Arlington, Prince William, and Loudon.

Time Zone: Eastern Time Zone or GMT minus 5 hours

Currency: US dollars (US$), divided into 100 cents (c)

Weight and measures: imperial

Electricity: Standard is 110-volts; if you plan to use European-made electrical appliances, lower the voltage with a transformer and bring a plug adaptor.

International dialing code: (1)

Local dialing codes: The area code for the District is 202. In Northern Virginia (across the Potomac), the code is 703. The code in suburban Maryland is 301. Whether you are calling within an area or between areas, you must dial the area code. All numbers are 10-digit but are charged as local, not toll, calls.

Climate

Washington's summers are long and notoriously hot. They last generally from May into October and peak in June, July and August. There are often violent thunderstorms in summer. Winters are moderate with occasional snow, usually in January and February and sometimes in March. The best seasons are fall and the brief spring in April, when temperatures are most comfortable.

Monthly Temperature Range

January 34°F–47°F (–1°C–8°C)
February 31°F–47°F (–1°C–8°C)
March 38°F–56°F (3°C–13°C)
April 47°F–67°F (8°C–19°C)
May 58°F–76°F (14°C–24°C)
June 65°F–85°F (18°C–29°C)
July 70°F–88°F (21°C–31°C)
August 68°F–86°F (20°C–30°C)
September 61°F–79°F (16°C–26°C)
October 52°F–70°F (11°C–21°C)
November 41°F–56°F (5°C–13°C)
December 32°F–47°F (0°C–8°C)

Culture and Customs

Washington is not a spontaneous city. Professional life is intense here, and the habit of multi-tasking and filling up days with meetings and obligations starts at the top with "power Washington" and filters down into everyday suburban life.

Washingtonians are not casual people; they work hard and often favor a museum or theater outing and a meal in a good restaurant over a stay-at-home evening. A "native Washingtonian" is a rare thing. Nearly everyone is from somewhere else. Capitol Hill staffers arrive from out-of-town with their states' representatives. Recent college grads with an eye toward civil service come to start their careers in government.

The surrounding suburbs have attracted several major private employers in the hi-tech field and in broadcasting, publishing and other industries which in turn has drawn a burgeoning number of professionals from across the country who now make Washington their home. There are foreign nationals from almost every country to serve in their respective embassies, and also immigrants, escaped from oppressive circumstances in their native countries and here to begin again. The result is a blend of cultures that give Washington no particular character that is truly and locally its own.

English is the official language, though you'll hear everything from Spanish, the most frequently spoken foreign language, to Asian and Middle Eastern tongue.

Taxes

Sales taxes are not included in the price of goods, meals or hotel rooms, but are charged additionally. In the District the sales tax on goods is 5.75 percent, on restaurant meals is 10 percent, and for hotel rooms the tax is 14.5 percent.

In Virginia and Maryland there are similar taxes. In Virginia, the sales tax on goods is 4.5 percent, and on a restaurant meal anywhere from 2 percent to 9 percent, depending on the jurisdiction and the kind of meal. In Maryland the sales tax on goods and on restaurant meals is 5 percent.

Planning the Trip

What to Bring

Washington's long, hot and humid summers mean you should dress in clothing light enough to cope with sightseeing in comfort. Comfortable shoes are also essential for all museum and Mall walking.

But even if the weather is more appropriate to the beach, this is a capital city, and Washington's major hotels and restaurants will expect you to wear appropriate clothing. Women should avoid skimpy outfits, and men may be required to wear jackets and ties in some restaurants. Air conditioning is often set so high in public places, such as restaurants, theaters and cinemas, that a light jacket is advisable. A coat with a zip-in lining, such as a rain or trench coat, is usually sufficient for winter.

Entry Regulations

VISAS AND PASSPORTS

A passport, a visitor's visa and evidence of intent to leave the US after a visit are required for entry into the US by most foreign nationals. Visitors from the United Kingdom and several other countries (including but not limited to Japan, Germany, Italy, France, Switzerland, Sweden, Ireland and The Netherlands) staying less than 90 days may not need a visa if they meet certain requirements. All other foreign nationls must obtain a visa from the US consulate or embassy in their own country. An international vaccination certificate may also be required, depending on your country of origin.

Exceptions to this rule are Canadians entering from the Western Hemisphere, Mexicans with border passes and British residents of Bermuda and Canada. Normally, these people do not need a visa or passport, although it's always best to confirm visa requirements before a visit.

Once admitted to the US, you may visit Canada or Mexico for up to 30 days and re-enter the US without a new visa. If you happen to lose your visa or passport, arrange to get a new one at your country's nearest consulate or embassy.

CUSTOMS

All people entering the US must go through customs. Either by machine or by hand, your bags will be inspected. Follow these guidelines:
● You may bring up to $10,000 in the US. Higher amounts must be declared.
● Adults may enter with a maximum of 200 cigarettes or 50 cigars or 2 kilograms of tobacco and/or 1 liter of alcohol duty-free.
● Gifts valued at less than $400 and objects brought for personal use may enter duty-free.
● Agricultural products, meat and animals are subject to complex restrictions; leave these at home unless absolutely necessary.
● Illicit drugs and drug paraphernalia are strictly prohibited. If you must bring narcotic or habit-forming medicines for health reasons, be sure that all products are properly identified, carry only the quantity you'll need while traveling, plus a prescription or a letter from your doctor.

For additional information or questions, contact the US Customs Service, 1300 Pennsylvania Avenue, NW, Washington, DC, 20229; Tel: 202-354-1000, toll free: 1-877-287-8667, or visit www.ustreas.gov.

EXTENSIONS OF STAY

Visas are usually granted for six months. If you wish to remain in the country longer than six months, you must apply for an extension of stay. Contact the Immigration and Naturalization Service toll free at 1-800-375-5283 for information regarding procedures and for the location of the nearest office, or visit www.ins.usdoj.gov.

Health

In the United States health care is very expensive. If you get sick or need medical attention, the average cost for one night in a hospital in a semi-private room is astronomical. If you live in a foreign country and do not have private health insurance, try to obtain insurance for your stay before coming to Washington. It is not the custom for doctors to refuse care to the sick or injured, but if you don't have health insurance or cannot pay, it is a possibility.

For medical care and an ambulance in a serious **emergency**, tel: **911**. Be aware that response times among DC ambulance drivers are notoriously slow. Don't hesitate to dial 911 more than once, or to go directly to the nearest hospital emergency room.

Money Matters

Cash & Travelers' Checks

US money is based on the decimal system. The basic unit, a dollar ($1), is equal to 100 cents. There are four basic coins, each worth less than a dollar. A penny is worth 1 cent (1¢); a nickel worth 5 cents (5¢); a dime worth 10 cents (10¢); and a quarter 25 cents (25¢). In addition there are several denominations of paper money. They are: $1, $5, $10, $20, $50, $100. Each bill is the same color (green), size and shape, so make sure you check the dollar amount on the face of the bill. It's advisable to arrive with at least $100 in cash (in small bills) to pay for ground transportation and incidentals.

It's always a good idea to carry internationally recognized travelers' checks rather than cash. Travelers' checks are usually accepted by retailers in lieu of cash, or may be exchanged for limited amounts of cash at many banks. Bring your passport with you to the bank.

Money may be sent or received by wire at any Western Union Office (tel: 1-800-325-6000) or American Express MoneyGram office (tel: 1-800-543-4080). The following addresses may also be useful:

Western Union
1762 Columbia Road, NW
Tel: 202-483-6024
American Express
301 4th Street, SW
Tel: 202-488-3375

Credit Cards

Major credit cards are widely accepted at shops, restaurants, hotels and gas stations, and will be necessary if you want to rent a car. Most may be used to withdraw cash from automatic teller machines.

Exchange Bureaux

Currency can be exchanged at some hotels and at Regan National Airport, at Dulles Airport, and at Baltimore Washington Airport. You can also change money at banks and these financial services companies.

Riggs Bank
Throughout Washington, call for the branch nearest you
Tel: 301-887-6000
Thomas Cook Currency Service
1800 K Street, and at Union Station
Tel: 1-800-287-7362
American Express Travel Services
1150 Connecticut Avenue, NW
Tel: 202-457-1300

Getting There

BY AIR

Three airports serve the city: Washington Dulles International Airport (or Dulles), 26 miles (42 km) west of the city; Washington Reagan National Airport (or Reagan National), directly across the Potomac in Virginia; and Baltimore Washington International Airport (or BWI), 20 miles (32 km) north of Washington. Dulles and Reagan National are operated by the

Metropolitan Washington Airports Authority, headquartered at Reagan National and which you can reach by dialing 703-417-8600, or visit www.mwaa.com.The main number for Reagan National is 703-417-8000. The main number for Dulles is 703-572-2700. BWI is operated by the state of Maryland and can be reached by dialing 301-261-1000, or visit www.bwiairport.com. Dulles and BWI offer international and domestic service; Reagan National handles only domestic flights.

Here is a list of the major carriers, all with toll-free numbers, and the airports they serve:
Aeroflot (Dulles) Tel: 1-888-340-6400
Air Canada (Dulles, Reagan National, BWI) Tel: 1-888-247-2262
Air Frances (Dulles) Tel: 1-800-321-4538
Air Tran (Dulles, BWI) Tel: 1-800-247-8726
Alaska Airlines (Dulles, Regan National) Tel: 1-800-252-7522
America West (Dulles, Reagan National, BWI) Tel: 1-800-235-9292
American Airlines (Dulles, Reagan National, BWI) Tel: 1-800-433-7399
Austrian Airlines (Dulles) Tel: 1-800-843-0002
British Airways (Dulles, BWI) Tel: 1-800-247-9297
Continental (Dulles, Reagan National, BWI) Tel: 1-800-525-0280
Delta (Dulles, Reagan National, BWI) Tel: 1-800-221-1212
Ethiopian Airlines (Dulles) Tel: 1-800-445-2733
Frontier (Reagan National, BWI) Tel: 1-800-432-1359
Icelandair (BWI) Tel:1-800-223-5500
JetBlue (Dulles) Tel: 1-800-538-2583
KLM Royal Dutch Air (Dulles) Tel: 1-800-225-2525
Korean Air (Dulles) Tel: 1-800-438-5000
Lufthansa (Dulles) Tel: 1-800-645-3880
Midwest Express (Regan National, BWI) Tel: 1-800-452-2022
Northwest Airlines (Dulles, Reagan National, BWI) Tel: 1-800-225-2525
SAS (Dulles) Tel: 1-800-221-2350
Saudi Arabian Airlines (Dulles) Tel:

1-800-472-8342
Southwest Airlines (BWI) Tel: 1-800-435-9792
Swiss (Dulles) Tel: 1-877-359-7947
United Airlines (Dulles, Reagan National, BWI) Tel: 1-800-241-6522
USAirways (Dulles, Reagan National BWI) Tel: 1-800-428-4322
Virgin (Dulles) Tel: 1-800-862-8621

At Dulles, the International Arrivals area is on the ground floor where you'll pass through customs. A currency exchange office is nearby, as is the baggage claim and access to ground transportation. At Reagan National, the baggage claim is on the ground floor near the exits and close to ground transportation. BWI's customs and currency exchange are on the lower level in International Arrivals, near the baggage claim and ground transportation.

GROUND TRANSPORTATION

For service between Dulles and Reagan National, and between those airports and DC hotels, Washington Flyer operates taxis and buses. Catch them at the airports, or Tel: 1-888-WASH-FLY. There's a Metro subway stop at Reagan National airport that will also take you directly into DC. There's also Amtrak train service between BWI and Washington's Union Station (a free shuttle will take you from the terminal to the train station), Tel: 1-800-USA-RAIL.Taxi or van service between Dulles and DC will run about $40; between Reagan National and DC about $20 for a taxi and about $6 by Metro. From BWI, the taxi or van service will cost about $50; the Amtrak train about $30.

The following shuttle services provide transportation to and from all three airports:
Airport Express
Tel: 301-588-0455
Super Shuttle
Tel: 202-296-6662
toll free: 1-800-258-3826

BY BUS

Greyhound Bus Lines, America's largest bus company, has connections to cities and small towns throughout the country and routinely offers discounts, such as $99 go-anywhere fares (subject to certain restrictions.) For information, Tel: 1-800-231-2222, or visit www.greyhound.com.

Greyhound Bus Lines
1005 1st Street, NW
Tel: 202-289-5154
Greyhound also has three suburban stations.
Silver Spring, Maryland
8100 Fenton Street
Tel: 301-585-8700
Springfield, Virginia
6770 Frontier Drive
Tel: 703-971-7598
Arlington, Virginia
3860 South Four Mile Run Drive
Tel: 703-998-6312

BY CAR

Washington is circled by the Capital Beltway, which is formed by two interstate highway, I-495 on the west, and I-95 on the east. Several major highways intersect the beltway and lead in and out of the city. Interstate 95 also runs north toward Philadelphia and New York, and south toward Richmond, Virginia. The Baltimore-Washington Parkway, also called the BW Parkway, connects Washington to Baltimore. Route 50 runs east and west between DC and Annapolis and the Chesapeake Bay region. Route 66 links the city with the Virginia mountains to the west, and I-270 connects the northern section of the beltway with Frederick, Maryland and beyond. For maps and travel plans, phone your local chapter of the American Automobile Association, or the DC chapter, Tel: 202-331-3000.

BY RAIL

Amtrak offers direct rail service between Washington and New York, and also Philadelphia, plus other cities across the country. Union Station, fully restored with lots of shops and restaurants, is only a few blocks from Capitol Hill. Service between Washington and New York runs at least hourly, and travel time is about three hours. To Philadelphia, travel time is about two hours. For information about fares and schedules, Tel: 1-800-872-7245 or visit www.amtrak.com.

For Travelers with Special Needs

For people with disabilities or special needs, Washington is one of the easiest cities to get around in. All of the curbs have cuts in them to accommodate wheelchairs, every museum and attraction has an entrance to handle visitors who need assistance, and the parking spaces nearest many sites are reserved for the disabled. Contact the Washington DC Convention and Visitors Association, 1212 New York Avenue, NW, Suite 600, for details about accessibility, or tel: 202-789-7064. There are also travel organizations that can help those with special needs plan and enjoy their trips.

Accessible Journeys
toll free 800-TINGLES
The Guided Tour
toll free 800-783-5841

Travelers' Aid

If you find yourself stranded in Washington, ask for help from:
Travelers' Aid International
1612 K Street, NW
Tel: 202-546-1127

There are Travelers' Aid booths at Reagan National Airport (open Monday–Friday, 7am–pm; Saturday; 7am–7pm; Sunday, 8am–8pm), at Dulles Airport (open Monday–Friday, 10am–9pm; Saturday and Sunday, 10am–6pm) and at Union Station (open daily 9:30am–5pm).

Practical Tips

Medical Services

For a medical emergency, call 911 or go to the nearest hospital emergency room. There are several excellent hospitals in the District:
Children's National Medical Center
111 Michigan Avenue, NW
Tel: 202-884-5000
The George Washington University Hospital
901 23rd Street, NW
Tel: 202-715-4000
Georgetown University Hospital
3800 Reservoir Road, NW
Tel: 202-784-2000
Howard University Hospital
2041 Georgia Avenue, NW
Tel: 202-865-6100

PHARMACIES

Pharmacies sell over-the-counter drugs without a doctor's prescription, For controlled drugs, including narcotics and antibiotics, you must obtain a written doctor's prescription or you can ask the doctor to phone in the prescription directly to the pharmacy. There are several CVS pharmacies, or drug stores, across the city, some open 24-hours. They are:

CVS Pharmacy
Tel: 1-800-SHOP-CVS
1121 Vermont Avenue, NW
Store: 202-737-3962
Pharmacy: 202-628-0720
6-7 Dupont Circle, NW
Store: 202-833-5704
Pharmacy: 202-785-1466
4555 Wisconsin Avenue, NW
Store: 202-537-1459
Pharmacy: 202-537-1587

Tourist offices

Contact the following organizations for maps, travel brochures, information about sites and attractions, and special seasonal events.

Washington DC Convention and Tourism Corporation
1212 New York Avenue, NW
Suite 600
Washington, DC 2005
Tel: 202-789-7000
www.washington.org

Washington DC Visitors Information Center
Ronald Reagan International Trade Center Building
1300 Pennsylvania Avenue, NW
Washington, DC 20004
Tel: 202-328-4748
www.dcvisit.com

Alexandria Convention and Visitors Association
421 King Street
Suite 300
Alexandria, Virginia 22314
Tel: 703-838-4200
www.funside.com

Arlington Visitors Center
735 South 18th Street
Arlington, Virginia 22202
Tel: 800-677-6267
www.stayarlington.com

Public Holidays

All government offices, banks and post offices are closed on public holidays, many of which are observed on the closest Monday, creating several three-day weekends. Public transportation does not run as often on these days.

January: New Year's Day
January: (third Monday) Martin Luther King Day
February: President's Day
March or April: Easter Sunday
May: (last Monday) Memorial Day
July: Independence Day
September: (first Monday) Labor Day
October: Columbus Day
November: Thanksgiving Day and the day after
December: Christmas Day

Embassies

Australia
1601 Massachusetts Avenue, NW
Tel: 202-797-3000
Canada
501 Pennsylvania Avenue, NW
Tel: 202-682-1740
Ireland
2234 Massachusetts Avenue, NW
Tel: 202-462-3939
Italy
1601 Fuller Street, NW
Tel: 202-328-5500
Singapore
3501 International Place
Tel: 202-537-3100
UK
3100 Massachusetts Avenue
Tel: 202-588-6500

Business hours

Museums are generally open 10am to 5.30pm. Smaller museums usually close one day a week, often Monday. Most businesses and offices are open 9am to 5pm, though some government offices open and close earlier. Major stores and shopping malls generally open at 9.30am and close at 9pm. Most stores are open on Sundays. Many large supermarkets and some drug stores stay open 24 hours. Banks are usually open between 9am and 2pm, then re-open at 3pm or 4pm for a few hours. Many have Saturday hours until noon. All are closed on Sunday.

Security and Crime

Washington has one of the highest murder rates of any city in the United States, but homicides are for the most part confined to the depressed areas of the city. For most travelers, the city is reasonably safe, but it's best to follow some simple precautions. Avoid carrying large sums of cash and wearing expensive jewelry. Hang on to your purse, and keep your wallet in your front pocket. No one should go into a strange or deserted area alone at night, especially women. Your best bet is to know where you are, know where you're going, and how you're going to get there. Serious crime on the Metro is rare, but if you're alone at night, stick with other people on the platform and ride in the car with the conductor. If you're on a Metrobus, sit close to the driver. And don't argue with a mugger. He's likely to be armed, so give him your cash, get away as quickly as you can, and dial 911 and report the crime to the police.

LOSS OF BELONGINGS

If any of your possessions are lost or stolen while you're visiting Washington, report it to the police at once. It's unlikely your property will be returned, but at the very least the police report will be necessary for an insurance claim. Car theft and the much more dangerous car-jackings are a problem here. Keep your car doors locked even when you're in the car, and lock all your belongings in the trunk where they'll be out of sight. If you have anything particularly valuable, ask the concierge at your hotel to lock it in the safe.

If you lose or have your credit cards or travelers' checks stolen, report the loss to your credit card company.
American Express
Tel: 1-800-528-4800
Visa
Tel: 1-800-227-6800
Mastercard
Tel: 1-800-825-2181

Media

NEWSPAPERS AND MAGAZINES

The *Washington Post* began publishing in 1877 and is the capital's oldest daily and Sunday newspaper, with a slightly liberal editorial bent. The *Washington Times* was established in 1982 and has a decidedly conservative slant. It also publishes on Sundays. Almost every suburban Washington neighborhood has its own free newspaper, delivered weekly.

The *City Paper* is a free weekly

news, arts and information sheet, available from street corner machines and some outlets, such as book stores, restaurants and cafés. The *Washingtonian* is a glossy monthly magazine, covering local news, politics and gossip, with reviews of new art and culture presentations, restaurants and specialty shops. Foreign newspapers are available at many bookstores, including Barnes & Noble and some Olsson's stores. The newsstand at the National Press Building, 14th and F streets, NW, carries a wide range of US regional dailies.

TELEVISION & RADIO

Washington, DC has at least a dozen television stations, and with expanded cable service the number can exceed 100. The three original commercial network stations with their call letters are NBC (WRC-TV), ABC (WJLA-TV), and CBS (WUSA-TV). There are also three public broadcasting stations, WHUT, which is affiliated with Howard University, WETA an independent public system located in Virginia but serving the entire region, and MPT, or Maryland Public Television. Of the three, WETA is most prominent. PBS is viewer-supported and commercial-free. There are at least two dozen AM and FM radio stations service the DC and suburban area. Below are some of the most popular:

88.5FM WAMU: news and public affairs; National Public Radio
89.3FM WPFW: jazz and public affairs
90.9FM WETA: classical music and news; National Public Radio
93.9FM WKYS: urban contemporary
97.1FM WASH: soft rock
98.7FM WMZQ: county and western
103.5FM WGMS: classical
105.9FM WJAZ: contemporary jazz
107.7FM WTOP: all news, weather and traffic; also 1500 AM

POSTAL SERVICES

Stamps can be purchased at any post office from 9am to 5pm,

Monday through Friday, and 9am to noon on Saturday. Stamps are also available in vending machines located in airports, hotels, stores and bus and train stations. They are also available at the check out of most supermarkets. The headquarters for the entire US Postal System is at 475 L'Enfant Plaza, SW, Tel: 202-268-2000. Other post offices in the city include:
Ben Franklin
1200 Pennsylvania Avenue, NW
Tel: 202-523-2386
Farragut
1800 M Street, NW
Tel: 202-523-2506
Georgetown
3050 K Street, NW
Tel: 202-523-2405
Pavilion Postique
1100 Pennsylvania Avenue, NW
Tel: 202-523-2571
Union Station
50 Massachusetts Avenue, NE
Tel: 202-523-2057

TELECOMMUNICATIONS

Payphones can be found throughout the city: at restaurants, bars, hotels, gas stations, public buildings and many street corners. To operate, drop in at least 35¢ and dial your local 10-digit number. For long distance calls made from a payphone, you may use your dialing card. Follow the instructions on your card and on the payphone. If you need to deposit more change, a recorded voice you tell you the amount. The prefix, or area code, in Washington is 202. The prefix for Northern Virginia is 703. The prefix for Maryland is 301. You must dial the prefix for all calls even though these are local. To make a long-distance call, you must dial a 1 before the prefix for all areas outside of the area code from which you are dialing.
If you need assistance, dial 0 and wait for the operator. For directory information, dial the area code and 555-1212 or 411. Take advantage of toll-free 1-800 numbers whenever possible. Some toll-free numbers have the prefix

877 or 888. Long-distance rates tend to be lower on weekends and after 5pm.
To dial other countries, first dial the international access code 011, then the country code: Australia (61); France (33); Germany (49); Italy (39); Japan (81); Mexico (52); Spain (34); United Kingdom (44). If using a US phone credit card such as Sprint or AT&T, dial the company's access number, then 01, then the country code and the number you want to read.

TELEGRAMS AND FAXES

Most hotels can arrange telegram, telex, and facsimile transmission, as can Western Union (tel: 1-800-325-6000). Office supply shops all have fax machines, as do some convenience stores.

Tipping

As elsewhere, service personnel in Washington depend on tips for a large part of their income. With few exceptions, tipping is left to your discretion; gratuities are not automatically added to the bill.
In most cases 20 per cent is the going rate for tipping waiters, taxi drivers, bartenders, barbers and hairdressers. Porters and bellmen usually get $1 per bag, but never less than $1 total.

Religious Services

Just about every religion is represented here, with places of worship spread across the city. For a detailed list, check the Yellow Pages of the phone directory or see the concierge in your hotel. The following are some of the largest and oldest congregations.

Christian
The Church of the Epiphany
1317 G Street, NW
Tel: 202-347-2635
www.epiphanydc.org
St. John's Episcopal Church
Lafayette Square
16th and H Streets, NW

Tel: 202-347-8766
Washington National Cathedral
3001 Wisconsin Avenue, NW
Tel: 202-664-6616
Basilica of the National Shrine of the Immaculate Conception
Michigan Avenue and 4th Street, NE
Tel: 202-526-8300
www.nationalshrine.com
St. Matthew's Cathedral
1725 Rhode Island Avenue, NW
Tel: 202-347-3215
www.stmatthewscathedral.org
Universalist National Memorial Church
1810 16th Street, NW
Tel: 202-387-3411
First Congregational Church
10th and G streets, NW
Tel: 202-628-4317

Jewish
Adas Israel Congregation
Connecticut and Porter streets, NW
Tel: 202-362-4433
Tifereth Israel Congregation
7701 16th Street, NW
Tel: 202-882-1605
Temple Sinai
3100 Military Road, NW
Tel: 202-363-6394

Islamic
Islamic Center
2551 Massachusetts Avenue, NW
Tel: 202-332-8343

Emergency Numbers

For police, medical or fire emergencies, dial **911** from any phone anywhere in the city.

Getting Around

Washington is really the nation's first planned city and is laid out, theoretically at least, in a logical fashion. It is divided into four basic sections: Northwest, Northeast, Southwest and Southeast. Numbered streets run north–south, beginning on either side of the Capitol, with East 1st Street to the map's right of the Capitol Building and West 1st Street to the map's left. Lettered streets run east–west, becoming name streets in alphabetic order as you go farther from the center. Avenues, named after states, dissect the city at angles across the grid, often meeting at traffic circles, such as Dupont Circle. It all makes perfect sense until you try driving in DC, which can be a nightmare. The best way to get around is by Metro, the city's clean, efficient, and reasonably priced subway system with stations near all the major attractions. Or you can simply walk.

On Arrival

There is shuttle and taxi service between the airports and the hotels, and some hotels offer complimentary service. There's also a Metro station at Reagan National Airport where you can catch the subway directly into the city. (See page 276; Ground Transportation).

Driving

In a word, don't. DC has the second worst traffic congestion in the nation behind Los Angeles. While the layout of the city may appear to be sensible, the reality is that DC is a bewildering place to navigate, especially for the uninitiated. The circles where major avenues converge can be a challenge to enter and exit. Streets carrying two-way traffic will suddenly switch to one-way during rush hour to accommodate either in-bound or out-bound traffic.

Street parking is at a premium, and garage parking expensive, $15 and more for the day. There are restricted lanes for carpools only, and streets and highway exits that are not adequately marked (just try reading the signs in Rock Creek Park at the same time you're trying to steer the car).

As if that wasn't enough, DC has one additional complication that no other city in America has: diplomats. Exempt from having to obey traffic laws, including speed limits and parking regulations, reckless diplomats behind the wheel routinely add to DC's traffic woes. If you're involved in a traffic accident with a diplomat, you may receive some minimal compensation in lieu of insurance payment, but don't count on it. Even if you're seriously injured or worse, diplomats enjoy full immunity and you'll be left bearing the burden. The State Department has issued red-white-and-blue license plates to the diplomatic corp, so be on the lookout for them and keep your distance.

With those cautions in mind, you'll need a car if you plan to maky any excursions much beyond the city. Here's a list of some of the top rental car agencies.
Avis
Tel: US 1-800-331-1212
Tel: International +1-918-664-4600
Budget
Tel: US 1-800-527-0700
Tel: International +1-214-404-7600
Dollar
Tel: US 1-800-800-4000
Tel: International +1-813-877-5507
Enterprise
Tel: US 1-800-325-8007
Tel: International +1-314-781-8232
Hertz
Tel: US 1-800-654-3131
Tel: International +1-405-749-4424

National
Tel: US 1-800-227-7368
Tel: International: +1-612-830-2345
Thrifty
Tel: US 1-800-331-4200
Tel: International: +1-918-669-2499

Public Transportation

To it's credit, Washington has one of the most efficient, clean and reasonably priced subway systems anywhere in the world. For a map and guidance on buying tickets, see the last page of this book.

Metro also operates a companion bus system around the city which links up with several suburban bus lines. Unlike Metrorail, the bus routes are complicated and subject to schedule changes. If you want to take the bus, ask for a current schedule at your hotel, or phone Metro for more information.

Metro
Washington Metropolitcan Area Transit Authority
600 5th Street, NW
Tel: 202-637-7000
www.wmata.com

TAXIS

Long-distance taxi fares are steep, but traveling inside Washington is cheaper than in other American cities and much cheaper than in some European capitals. This is because Washington cabs operate on fixed sums for each zone of the city they have to go through, and the zones were designed by the US Congress for maximum convenience and minimum expense to themselves. So a 2½-mile ride from Dupont Circle to Capitol Hill counts as one zone, and is a very reasonable fare.

Although you save money, you gain confusion. Drivers often speak little English, and once away from "Tourist Washington" rarely seem to know where they are going. Most do not carry maps.

The boundaries of each fare zone are not clearly displayed (there is occasionally, but not always, a zone map dangling on a chain from the back seat) so passengers have no way of knowing how many zones have been traveled or how much the fare should be. It must be said, however, that fares into central Washington from the airports are clearly marked, and most drivers are relatively honest. Besides, the city is not large enough for much "clockable" extra driving.

Taxis tend to be decrepit and usually without air-conditioning. Washington is one of the few Western cities in the world in which it is perfectly acceptable for drivers to pick up several fares during the course of one journey – an alarming practice if you are unaware it might happen or are in a hurry. This occurs mainly during rush hours, so leave plenty of time to arrive at your destination.

For planned trips, reserve a taxi at least an hour in advance of departure. Some of the major taxi cab companies are:

Yellow Cab Tel: 202-544-1212
Diamond Cab Tel: 202-387-6200
Capitol Cab Tel: 202-546-2400

PRIVATE TRANSPORTATION

Limousine Services
For a more comfortable, luxurious form of transportation, try limousines. The price of limousine service does not vary much from company to company, with a two-to-three-hour minimum and a 15 percent mandatory tip for the driver.
Admiral Limousine Tel: 202-554-1000
Capital Limousine Tel: 1-800-900-4469
Congressional Limousine Tel: 301-656-0133

Tour Operators

The following private operators provide regular, scheduled sightseeing tours to the general public, with some schedule variations during the high season.

All About Town
Tel: 301-856-5556
www.allabouttowntours.webatonce.com
Arranges group and individual tours
DC Ducks
Tel: 202-966-3825
Amphibious carrier tours the Mall then splashes into the Potomac
Gray Line Tours
Tel: 202-289-1995
Bus and trolley tours starting at Union Station
Guide Servcie of Washington
Tel: 202-628-2842
Licensed guides conduct tours for groups or individuals in all languages
Tourmobile Sightseeing
Tel: 202-554-5100
www.tourmobile.com
Narrated sightseeing shuttle tours with stops across the city that allow you to get on and off at each attraction.

Water Cruises

Spirit Cruises
Pier 4, 6th and Water streets SW
Tel: 202-554-8000.
www.spiritofwashington.com
Board for a narrated round-trip cruise between DC and George Washington's Mt. Vernon. Lunch and dinner cruises available.

Dandy Restaurant Cruise Ship
City dock, Old Town Alexandria
Tel: 703-683-6076
www.dandydinnerboat.com
Dinner and dancing aboard a climate-controlled cruise ship to see all the monuments from the river.

Where to Stay

DC has lots of hotel rooms, concentrated in the sections of the city of most interest to tourists and business travelers. Spring and summer are peak seasons for tourists, and weekdays are heaviest for those here on business. During autumn and winter the weather is moderate and the tourist crowds have diminished, so it's a good time to plan a trip.

Many hotels often offer reduced rates on weekends and off-peak seasons. Also, prices can drop in late August and early September, and again between mid-December and mid-January when Congress is in recess.

Occasionally you can negotiate a lower room rate by calling the individual hotel directly, as opposed to making arrangements through the chain's main reservation line. In either case, reservations are essential, and you'll need a credit card to secure your room. Hotels are listed in the same order as the Places section and from most to least expensive.

THE MALL

Expensive
Loew's L'Enfant Plaza Hotel
480 L'Enfant Plaza, SW
Tel: 202-484-1000
toll free 1-800-23LOEWS
www.loewshotels.com
You're within easy walking distance of the mall from this top-notch hotel with 392 spacious rooms. Expect all the usual amenities, plus a good dining room, health club and rooftop swimming pool.

CAPITOL HILL

Expensive
Best Western Capitol Skyline
10 I Street, NW
Tel: 202-488-7500
toll-free 1-800-458-7500
This hotel within sight of the Capitol has 200 rooms, an outdoor pool, and free underground parking. A good choice for families and, since it's centrally located, you can walk to all the attractions on the Mall.

Capitol Hill Suites
200 C Street, SE
Tel: 202-543-6000
toll-free 1-800-424-9165
www.capitolhillsuites.com
Converted apartment building near the Library of Congress on the House side of the Capitol with 152 suites decorated in 18th-century style and outfitted with kitchens. Convenient to all attractions, pleasant decor, and good for families.

DOWNTOWN

Deluxe
Hay Adams
One Lafayette Square
Tel: 202-638-6600
toll-free: 1-800-424-5054
www.hayadams.com
A block from the White House across Lafayette Square, this is where diplomats and other foreign dignitaries check in. Fine art and silk fabrics are the hallmarks of these 136 luxurious English country-style rooms.

Hotel Monaco
700 F Street, NW
Tel: 202-628-7177
toll-free 1-877-202-5411
www.monaco-dc.com
The city's old post office converted to a hotel in all its marble glory complete with luxury extras, modern art in the lobby and first-class service.

The Jefferson
1200 16th Street NW
Tel 202-347-2200
toll-free 1-800-368-5966
www.loewshotel.com

A small top-notch hotel with fine service, antiques everywhere, and canopied beds in the 100 rooms. The hotel has an good restaurant and there's access to the nearby University Club fitness center.

Renaissance Mayflower
1127 Connecticut Avenue NW
Tel: 202-347-3000
toll-free 1-800 HOTELS-1
www. renaissancehotels.com/WASSH
This 10-story hotel is on the National Register of Historic Places and is a Washington landmark. Presidential inaugural balls are held here, and politicians and lobbyist make the bar and restaurant a regular gathering spot. The 565 rooms are spacious and outfitted in mahogany with tasteful fabrics and art.

St. Regis Washington
923 16th Street, NW
Tel: 202-638-2626
toll-free 1-800-325-3535
www.starwood.com
Located three blocks from the White House. Elegantly decorated hotel with 193 tastefully appointed rooms, office amenities, and free shuttle service within a 5-mile radius.

Willard Intercontinental Hotel
1401 Pennsylvania Avenue NW
Tel: 202-628-9100
toll-free 1-800-327-0200
www.hotels.washington.interconti.com
Steeped in Washington history, this is the grandest of the capital's grand hotels, with sumptuous rooms and scrupulous service. Lincoln stayed here, as have other legendary figures. The rooms are generously sized and decorated in 19th-century style, but with all the modern comforts. Amenities include two restaurants, bar, and health club.

Expensive
Capital Hilton
16th Street, between K and L streets
Tel: 202-393-1000
toll-free: 1-800 HILTONS
www.hilton.com
An easy walk to the White House and many of the monuments. The 543 rooms are spacious, convenient and modern. There's a

fitness center, and the shuttle bus service next door will deliver you to and from Reagan National and Dulles airports.

Governor's House
1615 Rhode Island Ave
Tel: 202-296-2100
toll-free 1-800-821-4367
www.governorshousehotewdc.com
This small but elegant hotel has 149 rooms, 24 with kitchens, and features a health club and indoor valet parking. You can walk to the White House and other downtown attractions.

Grand Hyatt
1000 H St
Tel: 202-582-1234
toll-free 1-800-233-1234
www.grandwashington.hyatt.com
A busy central meeting place, this hotel has 888 rooms and caters to business travelers. Rooms are nicely appointed, what you'd expect from an upscale chain. There's a sports bar, and the hotel is near the Convention Center and the MCI Center, a sports and concert venue.

Henley Park
926 Massachusetts Avenue, NW
Tel: 202-638-5200
toll-free 1-800-222-8474
www.henleypark.com
Tudor-style apartment building transformed into a comfortable, clubby hotel with lots of period antiques. Weekday limo service offered around town, afternoon tea, and a pleasant restaurant. Health club facility nearby.

Hotel Washington
515 15th Street, NW
Tel: 202-638-5900
toll-free 1-800-424-9540
www.hotelwashington.com
This is the oldest continuously operating hotel in Washington. Caters to families and has an old-line hotel feel to it with two-poster beds. There's a health club, and the top-floor terrace cafe has a wonderful view of Washington, especially at night, where you can have dessert. 344 rooms.

JW Marriott Hotel
1331 Pennsylvania Avenue, NW
Tel: 202-393-2000
toll-free 1-800-228-9290
www.marriott.com

The Marriott chain's flagship hotel, in a good central location. Caters to families and to business travelers. Rooms are modern with office amenities, and there's a pool, fitness center, and four restaurants.

Morrison-Clark Inn
1015 L Street, NW
Tel: 202-898-1200
toll-free 1-800-332-7898
www.morrisonclark.com
Lovely Victorian inn with lace curtains and carved armoires and a 19th-century flavor. Listed in the National Register of Historic Places. Some rooms have porches, others overlook a pleasant courtyard. Amenities include an exercise room and continental breakfast.

Hotel Sofitel Lafayette Square
806 15th Street, NW
Tel: 202-730-8800
toll-free 1-800-SOFITEL
www.sofitel.com
Close to the White House, this new French-style hotel features 237 cozy rooms, a fitness center, and a restaurant that's drawing interest from the food critics.

Price Guide

The price guide below gives approximate rates for a standard double room.
$$$$ Deluxe: over $300
$$$ Expensive: $175–300
$$ Moderate: $125–175
$ Budget: under $125

Moderate
Hamilton Crowne Plaza
14th and K Streets, NW
Tel: 202-682-0111
toll-free: 1-800-2CROWNE
www.sixcontinentshotels.com
Right in the heart of downtown and overlooking Franklin Square, this hotel features has 318 rooms nicely appointed in earth-tone florals. Some of the rooms have a view of the Washington Monument.

Budget
Red Roof Inn
500 H Street, NW
Tel: 202-289-5959
toll-free 1-800-THE-ROOF

Right in the heart of Chinatown and within an easy walk or Metro ride of all the major attractions including the Convention Center and the MCI Center. Rooms are recently renovated and comfortable. There's an exercise room, coin-operated laundry facility, and a small cafe.

FOGGY BOTTOM AND GEORGETOWN

Deluxe
Four Seasons
2800 Pennsylvania Avenue, NW
Tel: 202-342-0444
toll-free 1-800-332-3442
www.fourseasons.com/washington
Extravagance with the first-class service the chain is known for. There are 196 rooms, all very tasteful, with down comforters, lots of overstuffed furniture, and all the extras such as bathrobes. As it's on the very edge of Georgetown, you can walk and enjoy the sights.

Park Hyatt
1201 24th Street NW
Tel: 202-789-1234
toll-free 1-800-922- PARK
www.parkwashington.hyatt.com
One block from Rock Creek Park, a quiet spot between downtown attractions and all the shopping and restaurants of Georgetown. The rooms are subtly decorated, and have sofa beds.

Swissotel Watergate
2650 Virginia Avenue, NW
Tel: 202-965-2300
toll-free 1-888-737-9477
www.swissotel.com
Not the site of the infamous break in – that was the apartment building next door. Rooms are large and outfitted with French Provincial furnishings. There's a lap pool and a health club, and you can walk to the Kennedy Center, which is next door.

Washington Fairmont
2401 M Street, NW
Tel: 202-429-2400
toll-free 1-800-441-1414
www.fairmont.com
This hotel, with 415 rooms, has an impressive fitness center on the lower level. The rooms are small,

but elegant and overlook a nicely landscaped courtyard.

Expensive
The Latham
3000 M Street, NW
Tel: 202-726-5000
toll-free 1-800-295-2003
www.thelatham.com
You're right in the heart of Georgetown in this small hotel with 97 rooms. There's an excellent view of the river, and an outdoor pool. Home of the acclaimed Citronelle restaurant.

Savoy Suites
2505 Wisconsin Avenue
Tel: 202-337-9700
toll-free 1-800-944-5377
Quiet upper Georgetown neighborhood. The 152 rooms are large, some with jacuzzis or kitchenettes. There is complimentary parking, and shuttle service to the Metro.

St. Gregory
2033 M Street
Tel: 202-223-0200
toll-free 1-800-829-5034
www.stgregoryhotelwdc.com
New hotel with 154 rooms, some of them suites. The suites include oven, dishwasher and stocked pantry. Stop at Donna's Coffee Bar in the hotel. Plenty of business amenities, and indoor valet parking.

St. James Suites
950 24th St, NW
Tel: 202-457-0500
toll-free 1-800-852-8512
www.stjamessuiteswdc.com
Rooms in this hotel are generously sized, tastefully decorated, and include a fully equipped kitchen. There's a fitness center, and also an outdoor pool.

Washington Marriott
1221 22nd Street, NW
Tel: 202-872-1500
toll-free 1-800-228-9290
www.marriott.com
Four hundred-plus comfortable, modern rooms with all the amenities including a pool and health club and a restaurant known best for its steaks.

Price Guide

The price guide below gives approximate rates for a standard double room.
$$$$ Deluxe: over $300
$$$ Expensive: $175–300
$$ Moderate: $125–175
$ Budget: under $125

Moderate
Doubletree Hotel Guest Suites
801 New Hampshire
Tel: 202-785-2000
toll-free 1-800-222-8733
www.doubletree.com
A good choice if you plan to be in town for a while. There are 105 one-bedroom suites with full kitchen. Located on a quiet residential street within an easy walk of the Kennedy Center.

Embassy Suites
1250 22nd Street, NW
Tel: 202-857-3388
toll-free 1-800-EMBASSY
www.embassysuites.com
Caters to business travelers and offers 318 suites around a 9-story atrium. Bedrooms are comfortably furnished and the living rooms have sofa beds. Rates include breakfast.

Holiday Inn Georgetown
2101 Wisconsin Avenue
Tel: 202-338-4600
toll-free 1-800-HOLIDAY
www.hightown.com
You'll be a short downhill walk to Georgetown from this well-kept hotel with 296 rooms. Top floors have good views of the Potomac River and Washington Monument.

Hotel Lombardy
2019 Pennsylvania Avenue NW
Tel: 202-828-2600
toll-free 1-800-424-5486
www.hotellombardy.com
The rooms are spacious in this hotel that features Italian-style decor. There are 50 suites and 80 guest rooms, a beautiful garden, and you're within an easy walk of all the attractions of Georgetown.

Hotel Monticello
1075 Thomas Jefferson Street, NW
Tel: 202-337-0900
toll-free 1-800-388-2410
www.monticellohotel.com

Offers 47 suites of one- or two-bedroom and caters to foreign diplomats and celebrities. Rate includes continental breakfast, and use of a nearby health club.

Melrose Hotel
2430 Pennsylvania Avenue, NW
Tel: 202-955-6400
toll-free 1-800-955-6400
www.melrosehotel.com
The Melrose has 240 comfortable rooms, each with a sitting area. Attracts an international clientele and is a block from Georgetown.

Topaz Hotel
1733 N Street, NW
Tel: 202-393-3000
toll-free 1-800-424-2950
www.topazhotel.com
A small hotel with 99 rooms on a quiet side street and in a prime location. Rooms have queen or king beds and kitchenettes.

Washington Suites Georgetown
2500 Pennsylvania Avenue, NW
Tel: 202-333-8060
toll-free 1-877-736-2500
There are 124 comfortable suites complete with kitchens, and the hotel also features a health club. Walk to the Kennedy Center.

Budget
Brickskeller Inn
1523 22nd St
202-293-1885
Modest, no-nonsense, but right on the edge of Georgetown, and good if budget is a consideration. This inn bills itself as a b&b, but without the breakfast. There are 44 rooms, but only 2 have a private bath.

DUPONT CIRCLE AND ADAMS MORGAN

Deluxe
Churchill Hotel
1914 Connecticut Avenue, NW
Tel: 202-797-2999
toll-free 1-800-424-2464
www.thechurchillhotel.com
This hotel has 144 rooms and is close to DC's embassies and within an easy walk to all the restaurants and attractions of Dupont Circle. The rooms are comfortable, and the lobby is classy.

Washington Hilton and Towers
1919 Connecticut Avenue
Tel: 202-483-3000
toll-free 1-800-445-8667
www.washington-hilton.com
A huge curving structure situated on a hillside, this is the site of the Reagan assassination attempt. There are more than 1,100 rooms, all of them conveniently appointed as you'd expect from this chain. Lots of business amenities and hotel services, including a large fitness center.

Expensive
Omni Shoreham
2500 Calvert Street, NW
Tel: 202-234-0700
toll-free 1-800-228-2121
www.omnihotels.com
Built in the 1930s, this has been one of the settings for the inaugural balls. This 800-room hotel is near Rock Creek Park and has been upgraded with traditional furnishings, office amenities and a new fitness center. If you believe in ghosts, asks about Room 800G.
Westin Embassy Row
2100 Massachusetts Avenue
Tel: 202-293-2100
toll-free 1-800-WESTIN1
www.westin.com
This stately hotel along Embassy Row was once the boyhood home of Al Gore. Rooms are sumptuously outfitted with rich fabrics and marble baths. All the usual office amenities, and excellent service from the staff. Rooms on the higher floors overlook Georgetown.

Moderate
Carlyle Suites
1731 New Hampshire Avenue
Tel: 202-234-3200
toll-free 1-800-964-5377
www.carlylesuites.com
A prime Dupont Circle location decorated in an Art Deco style. The 176 suites with sofa beds are good for families. The restaurant serves a good Sunday brunch.
Embassy Inn/Windsor Inn
1627 and 1842 16th Street, NW
Tel: Embassy 202-234-7800
Tel: Windsor 202-667-0300
toll-free 1-800-423-9111

www.virtualcities.com
These twins are on the edge of Dupont Circle. Combined, there are 85 rooms, small but nicely appointed. Rate includes continental breakfast.
Radisson Barcelo Hotel Washington
2121 P Street, NW
Tel: 202-293-3100
toll-free 1-800-333-3333
www.radisson.com
Large rooms recently renovated, each with a fridge. There's a courtyard pool, and a restaurant with good Sunday brunch.

SOUTHWEST WATERFRONT

Moderate
Channel Inn
650 Water St
Tel: 202-554-2400
toll-free 1-800-368-5668
www.channelinn.com
On the waterfront with the marina restaurants, including its own popular Pier 7. Rooms are simple but comfortable, there's an outdoor pool, and you're near Arena Stage.

B&Bs

Kalorama Guest House
1854 Mintwood Place, NW
Tel: 202-667-6369
Also, 2700 Cathedral Avenue
Tel: 202-328-0860
www.inntravels.com
Four houses furnished with eclectically form this quirky and inexpensive B & B popular with young travelers. Altogether there are 30 rooms, half of which have private baths. Rate includes continental breakfast.

To book a room in one of the many smaller guest houses and bed-and-breakfasts across the city and in Virginia, call either of the following organizations. They can help you make a reservation for an inn close to where you want to be.
Bed and Breakfast Accommodations, Ltd.
Tel: 202-328-3510

toll-free 1-877-893-3233
www.bedandbreakfast.com
Alexandria and Arlington Bed & Breakfast Network
Tel: 703-549-3415
www.aabbn.com

IN VIRGINIA

Expensive
Hilton Alexandria Old Town
1767 King Street
Alexandria, Virginia
Tel: 703-837-0440
toll-free 1-800-445-8667
www.hilton.com
One of Alexandria's newest hotels, next door to the King Street Metro station and an easy trip into DC.
Holiday Inn of Old Town
480 King Street
Alexandria, Virginia
Tel: 703-549-6080
toll-free 1-800-HOLIDAY
www.oldtownhis.com
Right on King Street, Alexandria's main avenue, with 227 nicely appointed rooms and a good restaurant. Easy walk to Alexandria's attractions and restaurants.
Morrison House
116 Alfred Street
Alexandria, Virginia
Tel: 703-838-8000
toll-free 1-800-367-0800
www.morrisonhouse.com
Nicely done inn with a good restaurant on a side street, but still near the heart of Old Town Alexandria.
Ritz-Carlton Pentagon City
1250 South Hayes Street
Alexandria, Virginia
Tel: 703-415-5000
toll-free: 1-800-241-3333
www.ritzcarlton.com
Charming 18-story hotel in an upscale shopping mall and on the Pentagon City Metro stop for convenience into DC. All the usual amenities, first-class service, health clup and lap pool, and tea in the elegant dining room.

Moderate
Key Bridge Marriott
1401 Lee Highway

Arlington, Virginia
Tel: 703-524-6400
toll-free 1-800-228-9290
www.marriott.com
At the foot of Key Bridge in Arlington, directly across from Georgetown, this 588-room hotel has all the amentities, including pools and exercise rooms, and, from the rooftop, a wonderful view of the city.
Sheraton Suites Alexandria
801 North St. Asaph Street
Tel: 703-836-4700
toll-free 1-800-325-3535
www.sheraton.com
Located on the north end of Old Town, not far from Reagan National Airport. Modern two-room suites with complimentary breakfast and lots of office amenities.

IN CHARLOTTESVILLE, VIRGINIA

Expensive
Boar's Head Inn
200 Ednam Drive
Charlottesville, Virginia
Tel: 434-296-2181
toll-free 1-800-476-1988
www.boarsheadinn.com
Romantic 172-room hotel with a historic gristmill that serves as the restaurant. Near the wine country.
Keswick Hall
701 Club Drive
Keswick, Virginia
Tel: 434-979-3440
toll-free 1-800-174-5391
www.keswick.com
Romantic country inn outfitted in the English country style with Laura Ashly prints and a golf course designed by Arnold Palmer.

IN WILLIAMSBURG, VIRGINIA

Expensive
Williamsburg Inn
136 Francis Street
Williamsburg, Virginia
Tel: 757-229-1000
toll-free 1-800-HISTORY
www.colonialwilliamsburg.com
A romantic inn with the feel of a country estate and excellent dining. There are 180 rooms, either in the main inn or in the adjacent colonial houses.

Moderate
Four Points by Sheraton
351 York Street
Williamsburg, Virginia
Tel: 757-229-4100
toll-free 1-800-962-4743
www.fourpoints.com
Adjacent to Colonial Williamsburg, and close to Busch Gardens. There are 199 rooms and suites with kitchens and an indoor pool.

IN CHINCOTEAGUE, VIRGINIA (NEAR ASSATEAGUE ISLAND)

Moderate
Island Motor Resort
Main Street
Chincoteague, Virginia
Tel: 757-336-3141
Small, immaculate family-run hotel with comfortable rooms, all with small balconies facing Chincoteague Bay and the mainland. Health club, pool, small restaurant.

IN BALTIMORE, MARYLAND

Expensive
Admiral Fell Inn
888 South Broadway
Tel: 410-522-7377
toll-free 1-800-292-4667
This 80-room inn is a restored sailors' lodging and is right on Baltimore's harbor at Fell's Point.
Harbor Court Hotel
550 Light Street
Tel: 410-234-0550
toll-free 1-800-824-0076
www.harborcourt.com
At the revitalized inner harbor with 204 rooms decorated with reproduction furniture.

Price Guide

The price guide below gives approximate rates for a standard double room.
$$$$ Deluxe: over $300
$$$ Expensive: $175–300
$$ Moderate: $125–175
$ Budget: under $125

IN DEEP CREEK LAKE, MARYLAND AREA

Lake Pointe Inn
174 Lake Pointe Drive
Deep Creek Lake, Maryland
Tel: 1-800-523-5253
www.deepcreekinn.com
Fifteen rooms in a refurbished 19th-century country lodge near skiing and golf and decorated in the Arts and Crafts style. Rate includes breakfast.

Youth Hostels

Washington International Youth Hostel
1009 11th Street, NW
Tel: 202-737-2333
www.hiwashingtondc.org
There are 125 beds here. Accommodations are spartan, but it's clean and comfortable.

Campsites

Capitol KAO Campground
768 Cecil Avenue
Millersville, Maryland 21108
Tel: 410-923-2771
toll-free 1-800-KOA-0248
www.koakampgrounds.com

Where to Eat

What to eat

Washington is seldom thought of as a gourmet capital, though the truth is that it's full of excellent restaurants. These are often the meeting ground of choice among politicos who prefer the neutral territory of a congenial restaurant setting to the less friendly environment of the office. An excellent meal is essential, then, along with perhaps some fine wine, in making matters go smoothly.

Of course, not everyone who eats out is a deal maker. Most Washingtonians simply enjoy good food and are willing to pay for it, which brings up price. Although you can find bargain meals, eating out in the city can be expensive. Budget accordingly, check for *prix fixe* meals, make a less expensive lunch your biggest meal, then enjoy a splurge or two in one of the fancier restaurants. As for cuisine, go by what you like. Because Washingtonians hail from everywhere, there are no regional dishes, no distinctly local flavor.

Restaurants are arranged by neighborhood following the order of this book and then alphabetically.

CAPITOL HILL

America
Union Station
50 Massachusetts Avenue, NE
Tel: 202-682-9555
Celebration of the ethnically inspired cooking of the US, each dish on the huge menu taking its name from its American hometown. Lunch and dinner daily. $–$$
Barolo Ristorante
223 Pennsylvania Avenue, SE

Tel: 202-547-5011
Elegant Northern Italian cuisine with exceptional seafood dishes. The menu changes often and features hand-made pastas. Lunch and dinner, Mon–Fri; dinner only Sat. $$$–$$$$
B. Smith's
Union Station
50 Massachusetts Avenue, NW
Tel: 202-289-6188
Upscale Southern Creole with lots of gumbos, andouille sausages, ribs, jambalaya, crawfish and sweet potato-pecan pie. Lunch and dinner daily. $$–$$$
Bullfeathers
410 First Street, SE
Tel: 202-543-5005
Come for the burgers and inventive sandwiches and the consistently well done seafood and chicken dinner specials. The restaurant owes its name to Teddy Roosevelt, who'd cry "Bullfeathers!" when he was hungry. The bar is a favorite hang-out with Hill staffers. Lunch and dinner, daily; brunch on Sunday. $$
Dubliner
520 North Capitol Street, NW
Tel: 202-737-3773
A smoky Irish pub with the usual favorites: fish and chips, shepherd's pie and beef stew. A hang-out for Capitol Hill staffers. Wide range of lagers and ales. Lunch and dinner daily. $$
Hawk and Dove
329 Pennsylvania Avenue, SE
Tel: 202-543-3300
Long-time favorite of congressmen and their staffers, within sight of the Capitol. Excellent burgers and a good choice of brews. Lunch and dinner daily. $$
Jimmy T's
501 East Capitol Street, SE
Tel: 202-546-3646
Eccentric diner with mismatched dishware and a colorful staff. Waffles, omelets, eggs any style, grilled sandwiches and authentic milkshakes. Expect a crowd, especially on weekend mornings. Breakfast and lunch, Tues–Sun. Cash only. $
La Colline
400 North Capitol Street, NW

Tel: 202-737-0400
Schmoozers and celebs passing through town gather here and enjoy French fare that's consistently good and reasonably priced. Try the bouillabaisse. Breakfast, lunch and dinner, Mon–Fri; dinner only Sat. $$–$$$
Market Lunch
225 7th Street, SE
Tel: 202-547-8444
Try this walk-up counter in Eastern Market, which serves the fresh meats and fish for sale in the market. Tues–Fri, 7:30am–3pm; Sat and Sun, 11am–3:30pm. $
The Monocle
107 D Street, NE
Tel: 202-546-4488
A long-time favorite with politicos located near the Senate side of the Capitol and known for its good food. Try the crab cakes. Lunch and dinner, Mon–Fri. $$
Taverna the Greek Islands
305 Pennsylvania Avenue, SE
Tel: 202-547-8360
Greek fare done more elegantly than usual and delivered in an unpretentious storefront setting. Moussaka and pastichio are, of course, house specialties, but the salads are exceptional. Lunch and dinner, Mon–Sat. $–$$
Two Quail
320 Massachusetts Avenue, NE
Tel: 202-543-8030
Romantic Victorian setting with comforting American dishes such as stuffed pork chops and grilled chicken. Lunch and dinner, Mon–Fri; dinner only Sat–Sun. $$
White Tiger
301 Massachusetts Avenue, NE
Tel: 202-546-5900
Outstanding Indian fare in an elegant setting, known especially for its breads and curries. Lunch and dinner daily. $$

DOWNTOWN

Bombay Palace
2020 K Street, NW
Tel: 202-331-4200
All the Indian standards served in a luxurious setting and at reasonable prices. Lunch and dinner daily. $$

Butterfield 9
600 14th Street, NW
Tel: 202-BU9-8810
The name is taken from the telephone exchange of the *Thin Man* movies. Located in the old Garfinkel's building, a defunct department store, this restaurant is a mix of retro-cool and way cool modern. Wins consistent raves and the crab cakes are exceptional. Lunch and dinner, Mon–Fri; dinner only Sat–Sun. **$$–$$$**

Café Atlantico
405 8th Street, NW
Tel: 202-393-0812
Inventive and upscale Latin cuisine with frequently changing menu. Lunch and dinner, Mon–Sat; dinner only Sun. **$$$**

Café 15
806 15th Street, NW
in the Sofitel Hotel near Lafayette Square
Tel: 202-730-8800
New, daring (frog legs and ravioli), delicious, though a bit "uneven," according to the *Washington Post* food critic. Breakfast, lunch, dinner daily. **$$$$**

Café Mozart
1331 H Street, NW
Tel: 202-347-5732
Down home European with kielbasa and knockwurst, good soups and a good selection of beer. Lunch and dinner daily. **$**

Capital Grille
601 Pennsylvania Avenue, NW
Tel: 202-737-6200
For those who love their beef bloody and their political bar talk conservative. Steaks are huge and

Liquor Laws

The legal drinking age in the District, as well as in Virginia and Maryland is 21. You may be "carded," that is, asked to show a picture identification, such as a driver's license, to verify your age. Wine, beer and mixed drinks are generally available by the glass or drink at most restaurants. Larger quantities of beer, wine and spirits are available at liquor stores.

aged in a see-through meat locker; a cigar smoker's paradise. Lunch and dinner, Mon–Fri; dinner only, Sat–Sun. **$$$$**

The Caucus Room
401 9th Street, NW
Tel: 202-393-1300
Who says Reps and Dems don't get along? A Democratic politico and the former head of the RNC teamed up to open this swank new hotspot where power brokers come to rest their elbows, when they're not rubbing them, on the white linen tablecloths. Lunch and dinner, Mon–Fri; dinner only Sat. $$$$

CF Folks
1225 19th Street, NW
Tel: 202-293-0182
Nutricious, delicious, Cajun, fish, sandwiches and more. Lunch only, Mon–Fri. Cash only. **$**

Fado
808 7th Street, NW
Tel: 202-789-0066
Hearty Irish fare near the MCI Center, the sports arena, and adjacent to Chinatown. Good stews, and every beer you can imagine. Especially busy when games are on. Lunch and dinner daily. **$$**

15 Ria
in the Washington Terrace Hotel
1515 Rhode Island Avenue, NW
Tel: 202-742-0015
New restaurant with a new twist on Mom's old comfort foods such as ribs, tomato soup, burgers stuffed with blue cheese, plus grilled seafood and a dessert that's a little like Jell-O. The locals have taken to it and are making this a regular. Breakfast, lunch, dinner daily. **$–$$**

Galileo
1110 21st Street, NW
Tel: 202-293-7191
Refined Italian featuring several "tasting menus" that combine special entrées such as wild boar with blackcurrant sausages, octopus, sweetbreads and eel. A long-time Washington favorite. Lunch and dinner, daily. **$$$–$$$$**

Georgia Brown's
950 15th Street, NW
Tel: 202-393-4499
What your grandmother would serve if she were Southern: fried chicken and biscuits and smothered pork

chops and gravy, gravy, gravy. Very homey, and very popular. Lunch and dinner, Mon–Fri; dinner only Saturday; brunch and dinner, Sunday. **$$–$$$**

Jaleo
480 7th Street, NW
Tel: 202-628-7949
Jaleo is Spanish for "commotion," which is what you'll find in this loud spot, popular for its tapas. Takes no reservations, and a wait for a table can exceed 90 minutes. Near the Shakespeare Theatre. Lunch and dinner daily. **$$–$$$**

The Jefferson
in the Jefferson Hotel
1200 16th Street, NW
Tel: 202-833-6206
Cozy and elegant hotel dining room with a fireplace, top-notch service and an excellent menu. With notice and for special events will serve some favorites of Thomas Jefferson's. Afternoon tea is elegant. Breakfast, lunch and dinner daily. **$$–$$$**

Lafayette
in the Hay-Adams Hotel on Lafayette Square
800 16th Street, NW
Tel: 202-638-2570
Elegant, as is everything in this hotel. Let the waiter tell you the daily special in this dress-up place. Breakfast, lunch and dinner daily. **$$$$**

Les Halles
1201 Pennsylvania Avenue, NW
Tel: 202-347-6848
Well done and totally French with classics such as cassoulet, seafood and steak. There's also a cigar bar. Daily noon to midnight. **$$$–$$$$**

McCormick & Schmick's Seafood
1652 K Street, NW
Tel: 202-861-2233
A Pacific Northwest chain known for its excellent fish and crustaceans flown in fresh from everywhere and served by a knowledgeable waitstaff. Salmon and rockfish are exceptional. Lunch and dinner daily. **$$$**

Occidental Grill
1475 Pennsylvania Avenue, NW
Tel: 202-783-1475
A clubby Washington watering hole

Capitol Hill Cafeterias

Yes, you can have lunch with senators and congressmen and even supreme court justices. Usually overlooked by visitors, these government-operated cafeterias offer food that is better than standard cafeteria fare, and at prices that are reasonable. Hours are generally 8am to 2pm on weekdays, depending on whether or not Congress is in session. It's a good idea to call ahead for details. In some cases you may need to contact your state's senator or congressman first. Security concerns and congressional schedules may restrict availability to the public.

Dirksen Senate Office Building Cafeterias
First Street and Constitution Avenue, NE
Tel: 202-224-4249
There are two cafeterias here, including the Senators' Dining Room, buffet-style but with tablecloths and friendly service.
Library of Congress
James Madison Building

First Street and Independence Avenue, SE
Tel: 202-707-8300 or 202-707-5000
The Capital Market Members Dining Room
US Capitol Building, Senate Side
Tel: 202-224-4870
House of Representatives Restaurant
US Capitol Building, Room H118
Tel: 202-225-6300
Canon House Office Building Cafeteria
First Street and Independence Avenue, SE
Tel: 202-225-1406
Longworth House Office Building Cafeteria
South Capitol and C streets, SE
Tel: 202-225-4410
Rayburn House Office Building Cafeteria
Indepedence Avenue and South Capitol Street, SW
Tel: 202-225-7109
Supreme Court Cafeteria
Supreme Court Building
First Street, NE
Tel: 202-479-3246

701
701 Pennsylvania Avenue, NW
Tel: 202-393-0701
One of the newer upscale establishments in revitalized downtown, known for its excellent way with fish. Near the theaters and drawing the supper crowd. Lunch and dinner, Mon–Fri; dinner Sat–Sun. **$$$**
Smith and Wollensky
1112 19th Street
Tel: 202-466-1100
Where the politicos and company execs gather for all steak all the time. Lunch and dinner daily. **$$$$**
Teaism
400 8th Street, NW
Tel: 202-638-6010
Japanese-inspired menu features bento boxes and a good selection of teas. Order at the counter and enjoy your meal in the Zen-like dinining room. Breakfast, lunch and dinner, daily. There's another branch at Dupont Circle, 2009 R Street, NW. **$**
Ten Penh
1001 Pennsylvania Avenue, NW
Tel: 202-3939-4500
Asian-Pacific inspired menu including some inventive desserts served in a classy setting by a very attentive waitstaff. Lunch and dinner, Mon–Fri; dinner Sat. **$$$**
Tony Cheng's Seafood Restaurant and Mongolian Barbecue
621 H Street, NW
Tel: 202-371-8669
This is Washington's real Chinatown, and an institution. Downstairs is a circus, but for the serious Chinese food fan, head upstairs where the seafood is exceptional. Lunch and dinner daily. **$$**

since 1906 with sandwiches and meat-and-potatoes dishes and everything well done. Lunch and dinner daily. **$$$**
Old Ebbitt Grill
675 15th Street, NW
Tel: 202-347-4800
Established in 1856 and a long-time favorite with a good selection of seafood and excellent burgers. You can't get any more Washington. Lunch and dinner daily. **$$**
Oodles Noodles
1120 19th Street, NW
Tel: 202-293-3138
Eat and get out. Not much on atmosphere, but the portions are big and everything's delicious. Lunch and dinner, Mon–Sat. Cash only. **$**
Oval Room
800 Connecticut Avenue, NW
Tel: 202-463-8700
Dress up for the power lunches here where the menu changes often

and everything's done to perfection. Lunch and dinner daily. **$$$$**
The Palm
1225 19th Street, NW
Tel: 202-293-9091
Legendary gathering spot for politicos and their followers who come for steaks, lobsters, martinis, cigars and gossip. Lunch and dinner daily. **$$$**
Red Sage
60514th Street, NW
Tel: 202-638-4444
Distinctly Southwestern flavor and an adjacent market with grab-and-go sandwiches, coffee and snacks. Lunch and dinner daily. **$$**
Reeves Restaurant and Bakery
1306 G Street, NW
Tel: 202-628-6359
Famous since 1886 for its pies, cakes, comfort foods and characters behind the counter. Cash only. Breakfast and lunch, Mon–Sat. **$**

Price Guide

The price guide indicates approximate cost of dinner for one, excluding beverages, tax and tip.

$	under $20
$$	$20–$30
$$$	$30–$45
$$$$	over $45

Vidalia
1990 M Street, NW
Tel: 202-659-1990
Totally Southern with chicken and
dumpling, pork chops, and their
signature sweet roasted vidalia
onions. Lunch and dinner, Mon–Fri;
dinner only Sat and Sun. **$$**

The Willard Room
1401 Pennsylvania Avenue, NW
Tel: 202-637-7440
Very elegant and in the grand old
Willard Hotel. This is where 19th-
century lobbyists got their start,
and they're still coming for the
steaks and chops. For something a
little less pricey, try the coffee shop
across the hall. Breakfast, lunch
and dinner daily. **$$$$**

Zaytinya
701 9th Street, NW
Tel: 202-638-0800
The name means "olive oil" in
Turkish, perfect for this restaurant's
focus on Middle Eastern and Greek
appetizers, or, as they're called,
"mezze," which you combine to
make a meal. Lunch and dinner
daily. **$$**

Zola
800 F Street, NW
Tel: 202-654-0999
In the new International Spy
Museum and billed by the chef as
"straightforward American,"
(meatloaf, steak) though with a
decided international twist. Lunch
and dinner, Mon–Fri; dinner only Sat
and Sun. **$$$$**

FOGGY BOTTOM, WEST END AND GEORGETOWN

Austin Grill
2404 Wisconsin Avenue, NW
Tel: 202-337-8080
Tex-Mex with all the usual including
enchiladas, fajitas, and an
exception crabmeat quesadilla.
Lunch and dinner daily. **$**

Bistro Francais
3128 M Street, NW
Tel: 202-338-3830
Good French fare, especially the
lamb, onion soup and omelets. It's
open until 4am, making it perfect
for a late-night supper. Lunch and
dinner daily. **$**

Price Guide

The price guide indicates
approximate cost of dinner for
one, excluding beverages, tax
and tip.

$	under $20
$$	$20–$30
$$$	$30–$45
$$$$	over $45

Blackie's
1217 22nd Street, NW
in the Washington Marriott
Tel: 202-333-1100
A classic long-time Washington
steakhouse that's gone upscale
after a facelift. Steaks are as good
as ever, and the big desserts even
bigger. Dinner daily. **$$$**

Booeymonger's
3265 Prospect Street, NW
Tel: 202-333-4810
A popular hang-out for students who
come for the inventive sandwiches
and the big breakfasts. Breakfast,
lunch and dinner daily. **$**

Bread and Chocolate
2301 M Street, NW
Tel: 202-833-8360
A French bakery that, besides
excellent pastries, serves good
sandwiches, soups, light fare.
Lunch and dinner daily. **$**

Café Milano
3251 Prospect Street, NW
Tel: 202-333-6183
Upscale Italian with everything from
veal to pasta to grilled fish. Open
until 1am Thur–Sat, making it a
good choice for a late supper.
Lunch and dinner daily. **$$$–$$$$**

Ching Ching Cha
1063 Wisconsin Avenue, NW
Tel: 202-333-8288
Peace and quiet in the fray of
Georgetown with ceremonious tea
service and light Asian meals.
Lunch and tea daily. **$$**

Citronelle
in The Latham Hotel
3000 M Street, NW
Tel: 202-625-2150
Considered the best in the city, the
experience is part theatrical, part
gourmet heaven. Caviar, squab and
more. Leave room for dessert.
Breakfast, lunch, dinner daily. **$$$$**

Clyde's
3236 M Street, NW
Tel: 202-333-9180
Long-time pub favorite, with
excellent burgers and steaks, and
other menu items that change
frequently. A Georgetown landmark.
Lunch and dinner daily. **$–$$**

Fettoosh
3277 M Street, NW
Tel: 202-342-1199
What your mother would make if
she were Lebanese. Great kebabs,
hummus, and 32 appetizers, plus
salads, fish, and terrific desserts.
Lunch and dinner daily. **$**

The Guards
2915 M Street, NW
Tel: 202-965-2350
A Frank Sinatra hang-out in the
Kennedy years, and not much has
changed since. Very good-value
Steaks, seafood, veal, and a
fireplace. Dinner daily. **$–$$**

Jeffrey's at the Watergate
in the Swissotel Watergate
2560 Virginia Avenue
Tel: 202-298-4455
A favorite among White House
Republicans who have several
Texas-inspired menu items named
for them. Breakfast, lunch and
dinner daily. **$$$–$$$$**

Kinkead's
2000 Pennsylvania Avenue, NW
Tel: 202-296-7700
Quality seafood simply prepared
with some inventive and frequently
changing dishes on the menu. You
can have a light meal downstairs in
the bar. Lunch and dinner daily. **$$**

Mendocino Grille
2917 M Street, NW
Tel: 202-333-2912
California east, with salads, pizza,
sandwiches, an Asian-influenced
dinner menu, and a good wine bar.
Lunch and dinner, Mon–Sat; dinner
only Sun. **$$–$$$**

Miss Saigon
3057 M Street
Tel: 202-333-5545
Fine Vietnamese with exceptional
seafood, including the shrimp rolls
and the mixed seafood entrée.
Lunch and dinner daily. **$–$$**

Nathan's
3150 M Street, NW
Tel: 202-338-2000

A clubby Georgetown landmark with white tablecloth service, good burgers and excellent fish dishes. Sunday brunch is best. Lunch and dinner daily. **$$–$$$**

Old Glory All American Barbecue
3139 M Street, NW
Tel: 202-337-3406
Ribs, brisket, pulled pork with plenty of biscuits and cornbread and homemade root beer. Lunch and dinner daily. **$–$$**

Roof Terrace Restaurant
The Kennedy Center
2700 F Street, NW
Tel: 202-416-8555
There are three restaurants here, but this is the nicest. Perfect for before or after the show, with a gorgeous view of the city, and some well done standard entrées, such as chicken, tuna, duck, lamb. Lunch on matinee days; dinner daily; Sunday brunch. **$$–$$$**

Sequoia
3000 K Street, NW
Tel: 202-944-4200
Tony restaurant on the Potomac with a great view from the terrace that wins raves. Go for Sunday brunch. Lunch and dinner daily. **$$$**

1789
1226 36th Street, NW
Tel: 202-965-1789
New-American food in an elegant Federal-style townhouse. A Washington classic, one of the best in the city. Dinner daily. **$$$–$$$$**

The Tombs
1226 36th Street, NW
Tel: 202-337-6668
Casual eats for the college crowd (it's near Georgetown Univeristy) with a good Sunday brunch menu. Lunch and dinner daily. **$–$$**

West End Café
One Washington Circle, NW
Tel: 202-293-5390
Good choice after a Kennedy Center performance with a jazz pianist in the bar. Menu is internationally influenced, and the desserts delicious. Lunch and dinner, daily. **$$–$$$**

Zed's
1201 28th Street, NW
Tel: 202-333-4710
Excellent Ethiopian fare including doro watt (spicy chicken) and gomen (collard greens) in an upscale setting. Lunch and dinner daily. **$–$$**

DUPONT CIRCLE AND ADAMS MORGAN

The Diner
2453 18th Street, NW
Tel: 202-232-8800
Upscale menu with salmon, rosemary roasted chicken, but it's still a diner and great for burgers, hot dogs, mac and cheese, and breakfast. Open 24 hours. **$**

Florida Avenue Grill
1100 Florida Avenue, NW
Tel: 202-265-1586
The neighorhood can be a bit iffy, but this diner has been here since 1944, unchanged and serving authentic Southern food (smothered pork chops, chicken livers), and everything's made from scratch. Breakfast is especially popular. Breakfast, lunch and dinner, Tues–Sat. **$**

Food for Thought at the Black Cat Club
1831 14th Street, NW
Tel: 202-797-1095
Organic and primarily vegetarian menu with a casual waitstaff and a schedule of live music. Dinner, Mon–Sat. **$**

Franklin's
2000 18th Street, NW
Tel: 202-319-1800
A trendy coffee spot with good light fare, interesting art and some colorful characters. Open daily. **$**

Grill from Ipanema
1858 Columbia Road
Tel: 202-986-0757
Very hip and lively spot with Brazilian fare featuring their version of a comforting stew and a daring alligator dish. Dinner daily; brunch Saturday and Sunday. **$$**

I Ricchi
1220 19th Street, NW
Tel: 202-835-0459
Tuscan style grill famous for its fish steaks, lamb, game and shrimp with some excellent pasta dishes. One of the city's favorites. Lunch and dinner, Mon–Sat **$$$–$$$$**

Iron Gate Inn
1734 N Street
Tel: 202-737-1370
The atmosphere is definitely European, but the menu is American with a decidedly Mediterranean flair. Try the lamb or the chicken. **$$**

Johnny's Half Shell
2002 P Street, NW
Tel: 202-296-2021
Their motto: "Seafood Specialities, Strong Drink." That said, stay after dinner for the lively bar scene. Lunch and dinner, Mon–Sat. **$$–$$$**

Nora
2132 Florida Avenue, NW
Tel: 202-462-5143
Organic and free-range ingredients carefully prepared by renowned chef Nora Pouillon and served in an elegant townhouse. Dinner, Mon–Sat. **$$$$**

Obelisk
2029 P Street, NW
Tel: 202-872-1180
Only a dozen tables and Italian inspired. The menu depends on what the chef finds fresh that day, but whatever it is, it'll be delicious. Dinner, Tues–Sat. **$$$$**

Pasta Mia
1790 Columbia Road, NW
Tel: 202-328-9114
Thirty-plus varieties of pasta and sauces to enjoy, all at reasonable prices and in a pleasant atmosphere. Dinner daily. **$**

Sam & Harry's
1200 19th Street, NW
Tel: 202-296-4333
All dark wood and deep booths with dry-aged meats, buttery mashed potatoes, and legendary desserts. Lunch and dinner, Mon–Fri; dinner only Sat–Sun. **$$$$**

Tabard Inn
1739 N Street, NW
Tel: 202-331-8528
A long-time favorite with an eclectic menu in a peaceful garden setting and mismatched furniture. Lunch and dinner daily; Sun brunch. **$$**

Teaism
2009 R Street, NW
Tel: 202-667-3827
With at least three dozen teas, this is a good alternative to coffee bars.

Japanese, Indian and Thai food, plus tasty scones. There's another branch downtown at 400 8th Street, NW. Breakfast, lunch and dinner daily. **$**

Topaz Restaurant
in the Topaz Hotel
1733 N Street, NW
Tel: 202-393-3000
One of the city's newer offerings with an Asian-inspired menu and a nicely appointed bar with an eye toward drawing a local following, which so far hasn't been disappointed. Lunch and dinner daily. **$–$$**

Tryst
2459 18th Street, NW
Tel: 202-232-5500
A good place to stop for coffee and relax with a sandwich. Enjoy the art on the walls and people-watch. **$**

LOGAN CIRCLE

Colonel Brooks Tavern
901 Monroe Street, NE
Tel: 202-529-4002
All Southern and all fried, except for the pulled pork sandwiches. On a Tuesday, the Federal Jazz Commission, on stage there for the past two decades, is still swinging. **$**

Ben's Chili Bowl
1213 U Street, NW
Tel: 202-667-0909
A legendary American diner since 1958 serving real fries, real shakes, real burgers. Home of the "Famous Chili Dog." Enormous breakfasts, and good no-fancy-stuff coffee. Attracts all kinds. Breakfast, lunch and dinner, Mon–Sat; dinner only Sun. Cash only. **$**

SOUTHWEST WATERFRONT

Hogate's
800 Water Street, SW
Tel: 202-484-6300
Huge restaurant on the waterfront catering mostly to tourists, but locals like it too, especially for its excellent seafood and Sunday brunch. The rum buns are legendary and you can buy them to

take home. Lunch and dinner daily; brunch on Sunday. **$**

Le Rivage
1000 Water Street, SW
Tel: 202-488-8111
Nicely done French fare and a favorite with the theatergoers at nearby Arena Stage. Pleasant water view and friendly staff. Lunch and dinner daily. **$$**

Phillip's Flagship
900 Water Street, SW
Tel: 202-488-8515
Serious about its seafood, this is an old-line family-run and family-oriented operation. Good for Sunday brunch; try for a table on the terrace. **$$–$$$**

Pier 7
in the Channel Inn Hotel
650 Water Street, SW
Tel: 202-554-2500
Caters mostly to hotel guests, but locals come for a table near the water and seafood simply prepared. There's also a bar. **$$**

Zanzibar
700 Water Street, SW
Tel: 202-554-9100
Part restaurant, part nightclub, the menu is a mix of unlikely tropically-influenced flavors that seem somehow to work. Busy bar scene after dinner. **$$**

AROUND WASHINGTON

Arlington, Virginia
Mediterranee
3520 Lee Highway
Tel: 703-527-7276
French by way of Algeria, this tiny spot has a dedicated local following. The menu is always inventive. (Catfish with bananas? It works.) Let the staff choose for you, sit back and enjoy. Dinner, Mon–Sat. **$$–$$$**

Tara Thai
4001 Fairfax Drive
Tel: 703-908-4999
Upscale Thai in an ocean motif atmosphere complete with aquarium and excellent seafood. Lunch and dinner daily. **$$**

Queen Bee
3181 Wilson Blvd.
Tel: 703-527-3444

Price Guide

The price guide indicates approximate cost of dinner for one, excluding beverages, tax and tip.

$	under $20
$$	$20–$30
$$$	$30–$45
$$$$	over $45

One of the first Vietnamese restaurants to open in Arlington with the influx of immigrants. Authentic, delicious. Go early to avoid the line. Lunch and dinner daily. **$**

Alexandria, Virginia
Café Mariana
1201 North Royal Street
Tel: 703-519-3776
Excellent Cajun-inspired dishes, especially the crawfish etoufee. A homey atmosphere with mismatched furniture and dishes. For dessert, make a trip to the pastry case. Lunch and dinner, Tues–Sun; Sunday brunch. **$$–$$$**

Gadsby's
134 North Royal Street
Tel: 703-838-4242
Open since GW's days and where he and Martha enjoyed the social season. Waitstaff in period costumes serve 18th-century American fare such as turkey, venizon, chicken. Lunch and dinner daily. **$–$$**

Potowmack Landing
Washington Sailing Marina on the George Washington Parkway
Tel: 703-548-0001
Right on the water and near the airport, so the view is terrific, though the food (American) is a bit predictable. There's a bar. Better for lunch. Lunch and dinner, daily; brunch on Sunday. **$–$$**

La Bergerie
218 North Lee Street
Tel: 703-683-1007
Elegant and quiet and long-time Old Town favorite with all the usual French choices, but some interesting Basque specialties and a frequently changing menu. Lunch and dinner, daily. **$$–$$$**

La Madeleine
500 King Street
Tel: 703-739-2854
Cafeteria-style French with excellent roasted chicken, delicious soups, quiches, pastries, and the best coffee. **$**

Mount Vernon Inn
at the Mount Vernon Estate
Tel: 703-780-0011
Early American fare plus seafood and beef, elegantly prepared and served in a peaceful setting. Tour the mansion, then plan to stay for a meal. Lunch and dinner, daily. **$$–$$$**

219
219 King Street
Tel: 703-549-1141
New Orleans-style menu with linen tablecloths, real silver, and a friendly staff. Sunday brunch is a treat (try any of the egg dishes), and there's a small bar downstairs with good lighter fare. Lunch and dinner, daily; brunch on Sunday. **$$**

Union Street Public House
121 South Union Street
Tel: 703-548-1785
Good seafood and an oyster bar with a Cajun inspired menu that includes Po' Boy sandwiches and good seafood stews. Busy bar scene for young professionals. **$–$$**

MARYLAND

Old Angler's Inn
10801 MacArthur Blvd.
Potomac
Tel: 301-365-2425
Across the road from the C&O Canal and best for its view from the deck in summer. Entrées are simply prepared with fresh ingredients. Lunch and dinner, Tues–Sun. **$$$–$$$$**

Inn at Glen Echo
6119 Tulane Avenue
Glen Echo
Tel: 301-229-2280
A feel of the country without the long drive with a good American-style menu including grilled fish and pork roast, and jazz on Sundays. Lunch and dinner, daily; brunch on Sunday. **$$**

Day Trips

BALTIMORE, MARYLAND

Obrycki's Crab House
1727 East Pratt Street
Tel: 410-732-6399
Order Maryland's famous Chesapeake Bay crabs here. No need for anything else. Lunch and dinner, daily; closed during the winter. **$$–$$$**

Paolo's
310 Light Street
Harborplace
Tel: 410-539-7060
Creative pastas and salads in this Italian trattoria with a view of the harbor. Pizzas comes from the wood-burning oven with some unusual but tasty toppings. **$$**

ANNAPOLIS

Café Normandie
185 Main Street
Te: 410-263-3382
Cozy French bistro with excellent stews and soups, and good crêpes. Fireplace makes it perfect on a cold day, but can get crowded. Lunch and dinner daily. **$**

Treaty of Paris
in the Maryland Inn
16 Church Circle
The inn is as old as the country. Excellent seafood with white-tablecloth service. Near the dock and the statehouse. **$$**

EASTERN SHORE

The Crab Claw
156 Mill Street
St. Michaels, Maryland
Tel: 410-745-2900
Watermen unload their catch on the dock so the steamed crab, dumped on a brown paper-covered picnic table for you to tackle, are really fresh. There's an indoor dining room upstairs, but the dock is more fun. **$–$$**

Robert Morris Inn
314 North Morris Street
Oxford, Maryland
Tel: 410-226-5111
Original 18th-century inn still serving Eastern Shore crab cakes to perfection. Near the ferry dock. **$$**

VIRGINIA

Fredericksburg

Kenmore Inn
1200 Princess Anne Street
Tel: 540-371-7622
Colonial fare in an elegant period setting. Have lunch on the porch. Menu includes Southern classics such as grits, and meat and cheese pies. Lunch and dinner, daily. **$$**

Charlottesville

C&O Restaurant
515 East Water Street
Tel: 434-971-7044
Unpretentious but inspired French, Cajun and Thai entrees. The upstairs dining room in this otherwise rustic spot is a bit more formal. Lunch and dinner, Mon–Fri; dinner only Sat–Sun. **$$**

Williamsburg

Christiana Campbell's Tavern
Waller Street
Tel: 757-229-2141
Restored colonial-era tavern within the Williamsburg complex features seafood, early American fare and strolling musicians. Must reserve in advance. Lunch and dinner, Tues–Sat. **$$**

Trellis
403 Duke of Gloucester Street,
Merchants' Square
Tel: 757-229-8610
Regional American dishes with a French influence served in several dining rooms, each with a different feel. Order Death by Chocolate for dessert. Reserve well in advance. **$$$**

Culture

Art Galleries

Many of the galleries in the Dupont Circle area coordinate openings and showings on "First Fridays," between 6pm and 8pm on the first Friday of every month, except during August and September. For more information, visit their website, www.artgalleriesdc.com. The following list gives details for the most popular galleries and the city's smaller museums.

Aaron
1700 Connecticut Avenue, NW, Suite 300
Tel: 202-483-9644
Mon–Fri, 10am–5pm
Contemporary American folk art.
Affrica
2020-1/2 R Street, NW
Tel: 202-745-7272
Tues, 2–6pm; Wed–Sat, noon–6pm
Burdick Gallery
2114 R Street, NW
Tel: 202-986-5682
Tues–Fri, noon–6pm; Sat
11am–5pm
Inuit (Eskimo) sculpture and works on paper.
Conner Contemporary
1730 Connecticut Avenue, NW
2nd floor
Tel: 202-588-8750
Fine arts in all media including photography, painting, sculpture, and digital.
Espacio Cultural Salvadoreno
1724 20th Street, NW
Tel: 202-256-6542
Mon–Fri, 9am–3pm
Emerging Latin American artists.
Foundry Gallery
9 Hillyer Court
Tel: 202-387-0202
Wed–Sat, 11am–5pm; Sun,

1pm–5pm
21st-century artists.
Gallery 10, Ltd.
1519 Connecticut Avenue, NW
Tel: 202-232-3326
Wed–Sat, 11am–5pm
Gallery K
2010 R Streeet, NW
Tel: 202-234-0339
Tues–Sat, 11am–6pm
Contemporary fine art in all media.
Geoffrey Diner
1730 21st Street, NW
Tel: 202-488-5005
Sat, 1pm–6pm
Decorative arts from 1860-1950; Arts and Crafts; Tiffany studio lighting.
Gary Edwards
9 Hillyer Court
Tel: 202-232-5926
Tues–Sat, 11am–5pm
19th and 20th-century photography.
Kathleen Ewing
1609 Connecticut Avenue
Tel: 202-328-0955
Wed–Sat, noon–5pm
Vintage and contemporary photography.
Marsha Mateyka Gallery
2012 R Street, NW
Tel: 202-328-0088
Contemporary painting, sculpture, and works on paper.
By appointment.
Robert Brown Gallery
2030 R Street, NW
Tel: 202-483-4383
Tues–Sat, noon–6pm
Paintings, works on paper, and Chinese antiquities.
Studio Gallery
2108 R Street, NW
Tel: 202-232-8734
Wed–Sat, 11am–5pm
Sun, 1–5pm
TARTT/Washington
1710 Connecticut Avenue, NW
Tel: 202-332-5652
By appointment.
Troyer Gallery
1710 Connecticut Avenue, NW
Tel: 202-328-7189
Tues–Sat, 11am–5pm
Contemporary photography, painting, sculpture, works on paper.
Washington Printmakers Gallery
1732 Connecticut Avenue, NW, 2nd floor

Tel: 202-332-7757
Tues–Thur, noon–6pm; Sat and Sun, noon–5pm
Artist-pulled prints.

Small museums

Washington has several smaller museums worth exploring. Visit those listed below and check the listings in the "Weekend" section of the *Washington Post* on Fridays for details about traveling and temporary shows in the public spaces of various office and trade association buildings across town.
B'nai B'rith Klutznick Museum
2020 K Street, NW
Tel: 202-857-6583
Mon–Thur, noon–3pm by advance reservation only.
Features Jewish folk and ceremonial art.
Dolls' House and Toy Museum
5236 44th Street, NW
Tel: 202-244-0024
Tues–Sat, 10am–5pm; Sun, noon–5pm; $4 admission
Wonderful for miniature enthusiasts.
Kreeger Museum
2401 Foxhall Road, NW
Tel: 202-338-3552
Mon–Fri, 10am–5pm; $5 admission
19th and 20th-century European paintings and sculptures of the private collection of insurance tycoon David Kreeger.
National Museum of Health and Medicine
16th Street and Georgia Avenue, NW on the grounds of Walter Reed Army Medical Center
Tel: 202-782-2200
Daily, 10am–5.30pm, free admission
Fascinating exhibits on the human body, including malformations, organs, skeletons and various tissues. Exhibits on medicine in the Civil War.
The Octagon
1799 New York Avenue
Tel: 202-638-3105
Tues–Fri, 10am-4pm; Sat and Sun, noon–4pm
Elegant Federal-style home designed by William Thornton, architect of the Capitol, and now headquarters for the American Association of Architects.

Cinema

For details about what's showing, see the daily movie listings in the *Washington Post*, or phone the theaters directly.

American Film Institute (AFI) Theater
The Kennedy Center
Tel: 202-833-2348
Film classics at the Kennedy Center.

AMC Mazza Gallerie 7
5300 Wisconsin Avenue, NW
Tel: 202-537-9553
First-run movies; inside shopping mall.

AMC Union Station 9
Union Station
Tel: 703-998-4AMC
First-run movies; at Union Station with shops and restaurants.

American City Movie Diner
5532 Connecticut Avenue, NW
Tel: 202-244-1949
Dinner and old classic movies.

Cineplex Odeon Inner Circle 3
2301 M Street, NW
Tel: 1-800-555-TELL
First-run movies.

Cineplex Odeon Wisconsin Avenue Cinemas
4000 Wisconsin Avenue, NW
Tel: 1-800-555-TELL
First-run movies.

Cineplex Odeon Cinema
5100 Wisconsin Avenue
Tel: 1-800-555-TELL
First-run movies.

Cineplex Odeon Dupont Circle 5
1350 19th Street, NW
Tel: 1-800-555-TELL
First-run movies.

Cineplex Odeon Outer Circle
4849 Wisconsin Avenue, NW
Tel: 1-800-555-TELL
First-run movies.

Cineplex Odeon Uptown
3426 Connecticut Avenue, NW
Tel: 1-800-555-TELL
First-run movies

Loews Theatres Georgtown
3111 K Street, NW
Tel: 1-800-555-TELL
First-run movies.

Smithsonian Theaters
Tel: 202-633-4629
Johnson Imax Theater in the National Museum of Natural History and Lockheed Martin Imax Theater in the National Air and Space Museum and Einstein Planetarium in the National Air and Space Museum. Feature films and documentaries about space, science and exploration.

Visions
1927 Florida Avenue
Tel: 202-667-0090
A cinema bistro/lounge with first-run features.

Dance

Besides the permanent dance companies listed below, several national and internationally acclaimed dance troupes make Washington a regular stop on their tours. They perform at the Kennedy Center, but also at venues across the city. Check the *Washington Post* for listings and ticket information.

The Dance Place
3225 8th Street, NE
Tel: 202-269-1600
Modern, African and other ethnic dance performances.

The Washington Ballet
Tel: 202-432-SEAT
www.washingtonballet.org
Washington's permanent ballet company performs at the Kennedy Center and elsewhere in the city.

Music

Washington is home to several exceptional musical groups, listings for which you can find in the *Washington Post*. The more prominent groups perform at the Kennedy Center and elsewhere in the District, but there are also several groups affiliated with universities and community organization in the suburbs that are on par with the professionals. Here are some of the most popular musical groups:

Air Force, Army, Navy, and Marine Bands
Tel: 202-433-4011
The military bands perform concerts on the Capitol steps, at Sylvan Theater near the Washington Monument, and at DAR Constitution Hall throughout the year. Concerts are free, but tickets are required. Call for an up-to-date schedule and for ticket information.

Folger Concert
201 East Capitol Street, SE
Tel: 202-544-7077
Early music ensemble in residence at the Folger Shakespeare Library.

National Gallery Orchestra
National Gallery of Art
Tel: 202-842-6941
www.nga.gov
Sunday evening concert series
A favorite in Washington for over 60 years; free tickets on a first-come-first-served basic beginning at 6pm; concerts begin at 7pm.

National Symphony Orchestra
The Kennedy Center Concert Hall
Tel: 202-467-4600
The world-class National Symphony under the direction of Leonard Slatkin.

Washington Opera
The Kennedy Center Opera House
Tel: 202-462-4600
www.kennedy-center.org
Classical and contemporary opera.

Wolf Trap for the Performing Arts
1624 Trap Road
Vienna, Virginia
Tel: 703-255-1800
www.wolftrap.org
Musical acts including orchestras, rock and jazz bands, folk singers, plus ballet, dance, and comedy acts in an outdoor setting; perfect for summer evening picnics. Also includes The Barns, an on-site indoor venue.

Theater

Washington has several resident theater companies that stage everything from Broadway musicals to Shakespeare to new works by established and new playwrights. DC is also a regular stop for theatrical touring companies, comedy acts, and solo performers. Check the daily listings in the *Washington Post*.

Arena Stage
6th Street and Maine Avenue, SW
Tel: 202-488-3300

www.arenastage.org
American classics, Broadway musicals, new works.

Church Street Theater
1742 Church Street, NW
Tel: 202-265-3748
Experimental.

Discovery Theater
Smithsonian Arts & Industry Building
900 Jefferson Drive SW
Tel: 202-357-1500
Geared especially for children.

Eisenhower Theater
The Kennedy Center
Tel: 202-467-4600
www.kennedy-center.org
Classics, musicals, touring companies.

Folger Theatre
201 East Capitol Street, SE
Tel: 202-544-7077
www.folger.edu
Elizabethan-era plays staged in a period theater.

Ford's Theatre
511 10th Street, NW
Tel: 202-347-4833
www.fordstheatre.org
Comedies, musicals in the theater that Lincoln made famous.

Metro Stage
1201 North Royal Street
Alexandria, Virginia
Tel: 703-218-6500
www.metrostage.org
New works by an inventive and dedicated troupe; Café Mariana, a good restaurant, next door.

The National Theater
1321 Pennsylvania Avenue, NW
Tel: 202-628-6161
Touring companies, musicals, comedies.

Shakespeare Theatre
450 7th Street, NW
Tel: 202-547-1122
www.shakespearetheatre.org
World-class resident Shakespeare company.

Signature Theatre
3806 South Four Mile Run Drive
Arlington, Virginia
Tel: 703-218-6500
New plays, musicals, comedies presented by an award-winning company.

Studio Theatre
14th and P streets, NW

Tel: 202-332-3300
www.studiotheatre.org
Experimental and always worth the ticket price.

Warner Theatre
13th and E Streets, NW
Tel: 202-783-4000
Touring companies, comedy, solo performers in a downtown theater.

Nightclubs, Comedy Clubs, Music Venues

Birchmere
3701 Mount Vernon Avenue
Alexandria, Virginia
Tel: 703-549-7500
Folk, blue grass and faded rock stars in a restaurant setting with full menu.

Black Cat
1811 14th Street, NW
Tel: 202-667-7960
Rock and alternative musical groups with bar service and a full menu.

Blues Alley
1073 Wisconsin Avenue, NW, rear
Tel: 202-337-4141
Classic blues in the heart of Georgetown with full dinner service.

Bohemian Caverns
2001 11th Street, NW
Tel: 202-299-0800
Jazz and blues, for the dedicated follower.

Crush
2323 18th Street, NW
Tel: 202-3191111
All that the name implies. Loud with non-stop dancing and for the younger set.

Felix
2406 18th Street, NW
Tel: 202-483-3549
For those who love to dance; French/Asian menu.

The Improv
1140 Connecticut Avenue
Tel: 202-296-7008
The city's comedy club draws nationally-ranked performers.

Madam's Organ
2461 18th Street, NW
Tel: 202-667-5370
A lively club in the heart of Adams Morgan. Look for the striking wall mural of the redhead and you'll know you're in the right place.

9:30 Club
815 V Street, NW
Tel: 202-393-0930
A favorite among Washington's trendiest with lost of loud rock from some of the big name performers.

State Theater
220 North Washington Street
Falls Church, Virginia
Tel: 703-237-0300
Rock, blues and jazz performers, big names and lesser-knowns.

Zanzibar
700 Water Street
Tel: 202-554-9100
Jazz club in a hotel on the waterfront.

GAY AND LESBIAN VENUES

Dupont Circle is ground zero for the gay and lesbian scene. For details, stop in at Lambda Rising, a bookstore at 1625 Connecticut Avenue, NW (tel: 202-462-6969) and pick up a copy of the *Washington Blade* for men, or *Woman's Monthly*, where you can find the latest on clubs and more. Here are two of the more popular night spots:

JR's
1519 17th Street, NW
Tel: 202-328-0090

Club Chaos
1603 17th Street, NW
Tel: 202-232-4141

Buying Tickets

To buy tickets by phone, call Ticketmaster tel: 202-432-7328, or visit www.ticketmaster.com. You can also purchase half-price tickets on the day of the show from Ticket Place located in the Old Post Office Pavilion, 1100 Pennsylvania Avenue, NW, open Tues–Sat, 11am-6pm; Tel: 202-842-5387.

Things To Do

Washington for Kids

When George Washington and the original committee of planners designed the city, they didn't have children in mind. That, however, doesn't stop the busloads of school kids on their class annual trip who arrive to see the sights each spring. It certainly doesn't halt the flow of families in mini-vans who make the summer pilgrimage to Washington. And it shouldn't stop you. While most sights weren't intended for children, many are of interest to them and some of the newer attractions and those recently renovated have been devised with children especially in mind.

Here are some answers to kids' most frequently asked questions:

"Are we there yet?"

The Mall is a great place to start with kids, who'll love the National Air and Space Museum *(see page 136)*, The National Museum of Natural History *(see pages 136–8)* and the National Museum of American History *(see page 142)*, probably in that order. Just outside the Smithsonian Castle *(see page 132)*, there's an old-fashioned carousel that makes a perfect rest stop between jaunts through the museums.

The tour of the Bureau of Engraving and Printing *(see page 148)* is especially popular with families, and the gift shop sells bags of shredded money.

Among the monuments, the Washington Monument *(see page 147)* is a must, though if you miss out on the required timed tickets, there is a Plan B. Head for the Old

Post Office *(see page 181)* and your kids will be just as impressed with the view from the top of the tower where there's seldom a line. And don't forget the Lincoln Memorial *(see page 152)*, or Ford's Theatre *(see page 178)*, where Lincoln was shot.

Most kids and grown-ups will love the National Zoo *(see page 222)* and National Geographic's Explorers Hall *(see page 176)*, and the younger kids will especially like the National Aquarium *(see page 251)*. The Rock Creek Park Nature Center *(see page 218)* offers special programs just for kids.

The Capital Children's Museum *(see page 169)* is all hands-on and loads of fun. So is the National Postal Museum *(see page 168)*, where kids can make their own stationery. The Navy Memorial Museum *(see page 235)* gives kids a real sense of life aboard ship. And if you tour nearby Alexandria, make a point of stopping at George Washington's Mount Vernon *(see page 246)*, beautiful and chock full of history interesting to kids.

"I'm tired and hot and thirsty!"

Lines for Washington's most popular attractions can be very long during the peak summer season when it's hot and humid. Start out as early in the day as you can to avoid as much of the heat and the crowds as possible. You'll be doing a lot of walking, so outfit yourself and your kids with comfortable shoes. And carry bottled water. While there are street vendors, they're usually expensive, and naturally there won't be a single one in sight when you need one.

As for strollers, leave them at home. Many sites don't permit them and they can be dangerous in crowds, in the Metro cars and on the stations' platforms and escalators, especially during rush hour. If you must bring a stroller, make it collapsible, or opt for a backpack.

"I'm lost!"

Even for grown-ups, Washington, with its confusing criss-cross

streets and traffic circles and its hulking marble buildings and their maze of corridors inside can be challenging. It's very easy to get lost. A good idea is to decide beforehand on a central meeting place in case you're separated. The information booth at any of the museums makes a good choice. You may want to equip your child with some form of identification and a few dollars, just in case. And if you're separated in the Metro, instruct your child not to panic but to get off at the next station and go directly to the stationmaster's booth near the exit where Metro authorities can page you.

"I'm hungry!"

Most Washington restaurants do not offer special menus for children, but there are a couple of good alternatives to the usual round of fast-food joints. Try the food courts at Union Station and the Old Post Office Building, both of which offer fun and inexpensive selections. The Ice Cream Parlor in the National Museum of American History *(see page 142)* is always a hit with kids. If you'd like something a little more substantial, try Oodles Noodles (1120 19th Street, NW, tel: 202-293-3138), noodles for them, good Asian food for you, or the Old Ebbitt Grill (675 15th Street, NW; tel: 202-347-4800), good hamburgers in a nice restaurant setting.

Scan the *Washington Post*, especially the Friday "Weekend" section, for a list of celebrations and special events. The *City Paper* and *Washingtonian* magazine will also have up-to-date listings. Below are some of the annual traditions you'll find in Washington and in the close-in suburbs.

Festivals and Events

January

Presidential Inauguration Day. Every four years the city turns out for the swearing in of the president followed by the parade along Pennsylvania Avenue and the inaugural balls.

Robert E. Lee's Birthday Celebration. Held at the Confederate General's former home, Arlington House, in Arlington National Cemetery. Open house features 19th-century music, samples of period food and exhibitions of restoration work.

Martin Luther King Jr's Birthday Observance. Wreath-laying ceremony at the Lincoln Memorial, accompanied by the presentation of King's "I Have a Dream" speech, local choirs and guest speakers.

February

Chinese New Year Parade. Firecrackers, lions, drums and dragon dancers make their way through Chinatown.

Abraham Lincoln's Birthday. Wreath-laying ceremony and reading of the Gettysburg Address at the Lincoln Memorial.

George Washington's Birthday Parade. Parade through Old Town Alexandria, plus special activities at George Washington's home, Mount Vernon.

March

St Patrick's Day Parades. Parades through Old Town Alexandria, and downtown along Constitution Avenue: dancers, bands, bagpipes and floats.

Smithsonian Kite Festival. Kite makers and flyers gather at the Washington Monument to compete for prizes and trophies.

April

National Cherry Blossom Festival. The Cherry Blossom Festival Parade celebrates the blooming of 6,000 Japanese cherry trees, with princesses, floats and VIPs. Events include fireworks, free concerts in downtown parks, the Japanese Lantern Lighting Ceremony, the Cherry Blossom Ball and an annual Marathon Race.

Thomas Jefferson's Birthday. A wreath laying ceremony and military drills at the Jefferson Memorial.

Duke Ellington Birthday Celebration. Musical celebration of this native Washingtonian's contribution to American music.

Historic Garden Week. Tour privately owned homes and gardens during Virginia's Historic Garden Week.

William Shakespeare's Birthday. A day of music, theater, children's events, food and exhibits at the Folger Shakespeare Library.

Wash Festival. Dozens of American and foreign films are screened in theaters across the city during this annual event.

Georgetown House Tour. Tour six Georgetown houses during this week-long event.

Easter. Children, each accompanied by an adult, gather on the White House South Lawn for a traditional Easter Egg Roll. For tickets and details, tel: 202-208-1631.

May

Cathedral Flower Mart. Washington National Cathedral holds its annual flower mart and crafts show.

Chesapeake Bay Bridge Walk. Stroll the 4.3 miles (6.9 km) across the bay.

National Zoo Zoofari. Annual fundraiser features area restaurant tastings, live entertainment and animal demonstrations at the National Zoo.

Mount Vernon Wine Festival. Taste wines from Virginia vineyards and learn about George Washington's attempts to make wine.

Memorial Day Weekend. Ceremonies at Arlington Cemetery include a wreath laying at the Tomb of the Unknowns, the Kennedy gravesite, and services at the Memorial Amphitheater with military bands and a presidential keynote address. Also, a wreath laying at the Vietnam Veterans' Memorial. The National Symphony Orchestra performs on the West Lawn of the US Capitol.

June

Kemper Open. The world's top pro golfers compete.

Alexandria Red Cross Waterfront Festival. Tour ships tied up at the dock in Alexandria. Food, crafts, exhibits, games, and fun.

Capital Pride. A parade and street fair with good food, crafts and concerts to celebrate the area's gay and lesbian community.

Smithsonian Festival of American Folklife. This annual celebration of folk culture takes place on the Mall and features concerts, craft demonstrations, food and more.

July

National Independence Day Celebration. Parade along Constitution Avenue, plus day-long concert event on the Mall followed by the evening fireworks display over the Washington Monument.

Virginia Scottish Games. Highland music, dancing, bagpipe band and athletic competitions along with sheep herding demonstrations, crafts, and food held on the grounds of Alexandria's Episcopal High School.

August

Maryland Renaissance Festival. 16th-century England comes to life with jousting tournaments, food, crafts, and more in Crownsville, Maryland.

National Frisbee Festival. On the National Mall near the Smithsonian's National Air and Space Museum.

September

Labor Day Weekend Concert. National Symphony Orchestra in concert on the West Lawn of the US Capitol.

Annual International Children's Festival. Three-day outdoor arts celebration at Wolf Trap Farm Park for the Performing Arts.

Kennedy Center Open House. Celebrate the arts with free concerts and other performances.

Adams Morgan Day. Lively street festival with music, crafts and cooking, along 18th Street, Columbia Road and Florida Avenue NW.

October

Columbus Day Ceremonies. The Knights of Columbus and other groups in a tribute to Christopher Columbus at the Columbus Memorial Plaza in front of Union Station, with wreath laying, speeches and music.

Theodore Roosevelt's Birthday Celebration. Festivities at Theodore Roosevelt Island and Reserve.

November

Armistice Day Celebration. Tour of home of President Woodrow Wilson, with music from the World War I era.
Veterans' Day Ceremonies. Service in the Memorial Amphitheater at Arlington National Cemetery and wreath laying at the Tomb of the Unknown Soldier.

December

Festival of Music and Lights. More than 80,000 tiny lights sparkle on the trees and shrubs of the Washington Mormon Temple in Kensington, Maryland, with live nativity scene and concerts.
Holidays at Mount Vernon. The recreation of an authentic 18th-century holiday season, with a tour of the the mansion.
National Christmas Tree Lighting and Pageant of Peace Ceremony. Holiday celebration marked with seasonal music performed by military bands and the lighting by the president of the National Christmas Tree on the Ellipse.
US Botanic Gardens' Christmas Poinsettia Show. Massive display of red, white and pink plants amid Christmas wreaths and trees.
Washington National Cathedral Christmas Celebration and Services. Christmas carols and seasonal choral performances.
Basilica of the National Shrine of the Immaculate Conception. Christmas Masses and carols.
New Year's Eve Celebration at the Old Post Office Pavilion. Live entertainment.

Shopping

Where to Shop

There's no shortage of shopping malls in Washington where you can find the usual array of familiar chain stores mixed in with specialty shops and souvenir stands. Serious shoppers should begin at Georgetown Park (tel: 202-298-5577), a Victorian parlor-looking mall at the intersection of Wisconsin Avenue and M Street filled with over 100 upscale shops.

As you head up Wisconsin Avenue toward the Maryland border, you'll come to two shopping malls, Chevy Chase Pavilion (tel: 202-686-5335) and Mazza Gallerie (tel: 202-966-6114). Chevy Chase Pavilion has several exclusive women's clothing stores, and Mazza Gallerie has the upscale Neiman Marcus (5300 Wisconsin Avenue, NW, tel: 202-966-9700) department store. Other department stores in this neighborhood, known as Friendship Heights, include Lord and Taylor (5255 Western Avenue, NW tel: 202-362-9600), and Saks Fifth Avenue (5555 Wisconsin Avenue, NW; tel: 301-657-9000).

Head back toward Capitol Hill to Union Station (Massachusetts Avenue at North Capitol Street; tel: 202-371-9441), a three-level mall complex, all marble and vaulted ceiling inside the beautifully restored train station. Besides the chain stores, there's a good food court, and in the east hall a collection of vendors who sell unique ethnic clothing, jewelry and decorative art. The malls downtown include the Old Post Office Pavilion (1100 Pennsylvania Avenue, NW; tel: 202-289-4224) where you'll find

a food court, lots of souvenirs and 17 specialty shops, and the Shops at National Place (13th and F Streets, NW; tel: 202-662-1250) with several clothing stores.

Downtown's only department store is Hecht's (12th and G streets, NW; tel: 202-628-6661), an old favorite of Washingtonians with a good selection of men and women's clothing and housewares.

Museum Shops

All the museums lining the Mall have gift shops where you'll find unusual jewelry and stationery, silk scarves, ceramics, unique toys for children, and a good selection of books. The National Museum of American History (tel: 202-357-1527) gift shop is the largest of the Smithsonian's shops and carries a good selection of American crafts, art reproductions, books and recordings. Try the National Gallery of Art (tel: 202-737-4215) for art reproductions. The Indian Craft Shop at the Department of the Interior (1849 C Street, NW; tel: 202-208-4056) has a wide range of work from several tribes including jewelry and collector-quality paintings and sculptures,

Specialty Shops

Every city has a shopping mall, but there are some shops that are unique to Washington and worth including on a shopping excursion.

Al's Magic Shop
1012 Vermont Avenue, NW
Tel: 202-789-2800
In business for over 50 years, Al's has been outfitting magicians, and probably some politicians, too, with card tricks, top hats, magic wands and more. Go for the fun of it.
Beadazzled
1507 Connecticut Avenue, NW;
Tel: 202-265-2323
Beautiful beads of every shape and size ready to string into jewelry.
Commander Salamanders
1420 Wisconsin Avenue, NW
Tel: 202-337-2265
Trendy fashion emporium; a long-time favorite with Washington's hip.

Eastern Market
7th and C streets, SE
The city's oldest market featuring fresh produce and deli and meat counters, plus the Market Five art gallery. On weekends, there's a combination farmers'/flea market outside.

Fahrney's
1317 F Street, NW
Tel: 202-628-9525
Exquisite fountain pens and other writing instruments.

Penn Camera
401 9 Street, NW
Tel: 202-347-7777
Washington's photojournalists have depended on this store for decades.

Political Americana
1331 Pennsylvania Avenue, NW
Tel: 202-731-7730
Campaign buttons and more political paraphernalia that says you've been to Washington.

Book Stores

Washington is a city of readers. In addition to the national chains, there are several thriving independent bookstores filled with books of interest to readers whose tastes run from literature to history to science to the latest fiction thriller.

Barnes and Noble
3040 M Street, NW
Tel: 202-965-9880

Borders
1801 K Street, NW
Tel: 202-466-4999

Chapters
1512 K Street, NW
Tel: 202-347-5495
A literary book store with a regular schedule of events.

Idle Time Books
2410 18th Street, NW
Tel: 202-232-4774
Good stock of used books.

Kramerbooks and Afterwords
1517 Connecticut Avenue, NW
Tel: 202-387-1400
Small but important inventory and a good café.

Lambda Rising
1625 Connecticut Avenue, NW
Tel: 202-462-6969

Specializes in books for gays and lesbians.

Olsson's Books and Music
1200 F Street, NW
Tel: 202-347-3686
Also:
1307 19th Street, NW
Tel: 202-785-1133
Also:
418 7th Street, NW
Tel: 202-638-7610
Washington's most popular local chain, with an excellent selection of music and literary titles and a knowledgeable staff.

Politics and Prose
5015 Connecticut Avenue, NW
Tel: 202-364-1919
Excellent for literature and current events; has a regular schedule of readings and other events.

Second Story Books
2000 P Street, NW
Tel: 202-659-8884
Good selection of used books in the Dupont Circle area.

Trover Books
221 Pennsylvania Avenue, SE
Tel: 202-547-2665
Quality paperbacks and a good selection of newspapers; on Capitol Hill.

Beyond DC

Besides the shops and malls in the District, the surrounding suburbs offer lots of shopping. Take Metrorail to King Street, the Old Town Alexandria station, and browse the shops up and down King, the main shopping avenue. You'll find everything from antiques to trendy clothing.

At the foot of King, right on the Potomac, be sure to stop at the Torpedo Factory (King and Union streets, Tel: 703-838-4565), previously a factory where war armaments were made and now home to 150 painters, sculptors and artisans who craft everything from textiles to pottery, jewelry to bookbinding.

In Arlington you'll find the upscale four-story Fashion Centre mall at Pentagon City (1100 South Hayes Street; tel: 703-415-2400) with Macy's and Nordstroms and a good food court. Stop for tea at the adjoining Ritz-Carlton hotel, which is over Metrorail's Pentagon City stop.

If you have a car and are a real bargain hunter, then head 45 minutes south along I-95 to Potomac Mills (tel: 703-490-5948) in Woodbridge, Virginia, a mile-long discount shopper's paradise with 200-plus outlet stores including Nordstrom's Rack and, in a new building nearby, IKEA, the Swedish furniture retailer.

Other malls in Virginia include Tysons Corner Center (1961 Chain Bridge Road, tel: 703-893-9400) and Galleria at Tysons II (2001 International Drive; tel: 703-827-7730) with Bloomingdales, Macy's and all the familiar retail shops. In Maryland, go to White Flint Mall (11301 Rockville Pike, Bethesda, Maryland; tel: 301-231-7467) which you can reach by Metrorail on the White Flint station and which features all the upscale regulars, including Coach, Eddie Bauer, Bloomingdale's and Lord and Taylor.

Sport

Spectator Sports

Baseball

The Baltimore Orioles
Camden Yard
Baltimore, Maryland
Tel: 410-685-9800 for information,
or 202-432-7328 for tickets
Washington doesn't have its own
professional baseball team, but the
Baltimore Orioles are the next best
thing. The ballpark resembles an
old-time stadium that's within an
easy walk of the harbor.

Bowie Baysox
Prince George's County Stadium
Bowie, Maryland
Tel: 301-805-6000
The Orioles' Class AA farm team
plays in suburban Maryland, 30
minutes from Washington.

Frederick Keys
Harry Grove Stadium
Frederick, Maryland
Tel: 301-662-0013
The Orioles' Class A team plays in
Frederick, Maryland, an hour's drive
north of Washington.

Potomac Cannons
Richard Pfitzner Stadium
Manassas, Virginia
Tel: 703-590-2311
St. Louis Cardinals' Class A affiliate
plays in Virginia, an hour west of
Washington.

Basketball

Washington's two professional
basketball teams and the
Georgetown University team play in
the MCI Center (601 F Street, NW;
tel: 202-628-3200). Phone for
schedules and ticket information, or
see the *Washington Post*.
Washington's universities also have
teams. Call for ticket information.

The Washington Wizards
Men's professional NBA team.

The Washington Mystics
Women's professional WNBA team.

Georgetown Hoyas
NCAA champion team of
Georgetown University.

American University
Tel: 202-885-1000

George Washington University
Tel: 202-994-6650

Howard University
Tel: 202-806-7198

University of Maryland
in College Park, Maryland
Tel: 310-314-7029

Football

Tickets for the Washington Redskins,
the city's popular football team, are
nearly impossible to get. There's a
waiting list for season tickets that's
now 10 years long. Games are
played at the ultra-modern FedEx
Field in Landover, Maryland (tel: 301-
276-6050). Occasionally, individual
tickets are advertised in the
classified section of the *Washington
Post*, though at exorbitant prices.
There may also be a slim chance of
getting a ticket to a preseason game
in August. College football is a good
alternative. Call for ticket and
schedule information.

Howard University
Tel: 202-806-7198

University of Maryland
in College Park, Maryland
301-314-7070

US Naval Academy
in Annapolis, Maryland
410-293-1000

Hockey

Washington Capitals
MCI Center
601 F Street, NW
Tel: 202-628-3200

Horse Racing

Laurel Park
Laurel, Maryland
Tel: 301-725-0400

Pimlico Race Course
Hayward and Winner avenues
Baltimore, Maryland
Tel: 410-542-9400

Rosecroft
Fort Washington, Maryland
Tel: 301-567-4000

Charles Town Races
Route 340
Charles Town, West Virginia
Tel: 1-800-795-7001

Soccer

DC United
RFK Stadium
22nd and East Capitol streets
Tel: 703-478-6600

PARTICIPANT SPORTS

Bicycling

For information about touring by
bicycle, contact the Washington
Area Bicyclist Association (tel: 202-
628-2500). Bikes may be rented at
these locations:

Big Wheel Bikes
1034 33rd Street, NW
Tel: 202-337-0254

City Bikes
2501-Champlain Street, NW
Tel: 202-265-1564

Fletcher's Boathouse
4940 Canal Road, NW
Tel: 202-244-0461

Thompson's Boathouse
2900 Virginia Avenue, NW
Tel: 202-333-9543

Boating and Sailing

Atlantic Canoe and Kayak
1201 North Royal Street
Alexandria, Virginia
Tel: 703-838-9072
Canoe and kayak rentals with river
guides.

Fletcher's Boathouse
4940 Canal Road, NW
Tel: 202-244-0461
Canoe, rowboat, kayak rentals.

Swain's Lock
River Road
Potomac, Maryland
Tel: 301-299-9006
On the C&O Canal. Canoe or kayak
rentals.

Thompson's Boathouse
2900 Virginia Avenue, NW
Tel: 202-333-4861
Canoes, rowboats, kayaks, rowing
sculls rentals.

Tidal Basin
near the Jefferson Memorial
Tel: 202-479-2426
Paddleboat rentals.

Golf courses

East Potomac Park
Hains Point, East Potomac Park
Tel: 202-554-7660
Two 9-hole courses and driving
range, and a miniature golf course.
Langston Golf Course
26th Street and Benning Road, NE
Tel: 202-397-8638. Has 18 holes.
Rock Creek Park Golf Course
16th and Rittenhouse streets, NW
Tel: 202-882-7332. Has 18 holes.

Health clubs

Many hotels either have a health
club, or offer privileges to their
guests at nearby facilities. In
addition, Washington, DC hotel
guests may use the Fitness
Company West End in the Monarch
Hotel (2401 M Street, NW; tel: 202-
457-5070) by showing a room key
and paying a fee.

Horseback riding

Rock Creek Horse Center
Military and Glover roads, NW
Tel: 202-362-0117

Ice Skating

National Gallery Sculpture Garden
Constitution Avenue
Tel: 202-737-4215
Pershing Park
Pennsylvania Avenue, near 14th
Street, NW
Tel: 202-737-6938

Running

Popular routes for runners include
the C&O Canal towpath, the Mall
between the US Capitol and the
Washington Monument, Rock Creek
Park, and in Virginia the Mount
Vernon Trail, which stretches
between Roosevelt Island and the
Mount Vernon Estate. For more
details and information about group
runs, contact Road Runners Club of
America (tel: 703-836-0558).

Tennis

East Potomac Park
Hains Point, East Potomac Park
Tel: 202-554-5962
Rock Creek Tennis Center
16th and Kennedy streets, NW
Tel: 202-722-5949

Further Reading

**AIA Guide to the Architecture of
Washington DC**, by Christopher
Weeks. Johns Hopkins University
Press (1994). Details history and
architecture of notable buildings.
America's Smithsonian: Celebrating
150 Years, Kay Fleming, ed.
Smithsonian Institution Press
(1996). An anthology featuring the
museum's treasures and highlights.
American Sphinx: The Character of
Thomas Jefferson, by Joseph J.
Ellis. Alfred A. Knopf (1997).
Award-winner Ellis makes the minor
god more accessible.
Beautiful Swimmers, by William W.
Warner. Penguin (1976)
The Pulitzer prize-winner about the
blue crab and its home waters of
the Chesapeake Bay.
Democracy, by Henry Adams. New
American Library (1983)
Written in 1880 by the descendant
of two presidents, this novel
examines the scandal-ridden
presidency of Ulysses Grant.
Founding Brothers: The
Revolutionary Generation, by Joseph
J. Ellis. Alfred A. Knopf (2000).
Absorbing account by historian and
respected college professor of the
men who made America.
The Great Experiment: George
Washington and the American
Republic by John Rhodehamel. Yale
University Press (1998)
How Washington's character shaped
the new republic.
John Adams, by David McCullough.
Simon and Schuster (1991).
Pulitzer Prize-winning author on the
often overlooked second president.
Katharine Graham's Washington,
Katharine Graham, ed. Alfred A.
Knopf (2002).
Anthology of 100-plus articles,
essays, and humor pieces chosen
by the late *Post* publisher.
On This Spot: Pinpointing the Past
in Washington, DC, by Douglas E.
Evelyn and Paul Dickson. National
Geographic Press (2002)

Reveals the city's history address-
by-address with short and engaging
entries and photos.
Passionate Sage: The Character
and Legacy of John Adams, by
Joseph J. Ellis. W. W. Norton (1993)
Builds the case for the often
overlooked visionary and second
president John Adams.
Personal History, by Katharine
Graham. Vintage (1998)
Pulitzer Prize-winning account of the
rise of the *Washington Post* and the
woman behind it.
Theodore Rex, by Edmund Morris.
Random House (2001).
Roosevelt's bully years in the White
House from 1901 to 1909.
Truman by David McCullough.
Simon and Schuster (1992).
Pulitzer-prize winning author
examines the post-war presidency of
Truman as the Cold War begins.
Washington, by Meg Greenfield and
Katharine Graham and Michael R.
Beschloss. Public Affairs Press
(2002). Sharp observations about
the city and its people by the
Pulitzer Prize-winning *Post* editorial
writer completed just before her
death with the help of her friends.
Washington: The Indispensable
Man, by James Thomas Flexner.
Little Brown (reissued edition 1994).
Classic biography of first president.
Washington's Monuments: A Guide
to the Monuments and Memorials of
the Nation's Capital, by Alex Padro.
Monument Books (2000).
This exhaustive and fascinating
inventory covers all of the city's
750-plus monuments.

Other Insight Guides

*Insight Pocket Guide: Washington,
D.C.* contains personal
recommendations from a local
author on how to make the most of
a short visit. Contains fold-out map.
*Insight Compact Guide: Washington,
D.C.* is the perfect book to carry
with you as you explore. Text,
pictures and maps are all cross-
referenced for ease of use.
Insight FlexiMap: Washington, D.C.
is laminated to make the map
durable, waterproof and easy to
fold. Contains useful travel details.

ART & PHOTO CREDITS

Agence France-Presse 58, 61, 69, 127
AKG-images 36, 116, 161
Art Archive 118
Bettmann/Corbis 143
Clive Barda/PAL 76
Corbis Sygma 94
Getty Images 80, 82, 92/93
Hulton Getty 43, 45
C.M. Glover 266
Catherine Karnow 1, 6/7, 8/9, 10/11, 12, 18, 25, 44, 46, 47,48, 52/53, 54/55, 56, 57, 62, 63, 64, 65, 68, 70, 71, 72, 73, 74, 77, 79, 83, 84/85, 86, 100, 101, 102, 103, 104, 105, 106/107, 108/109, 110/111, 117, 128/129, 130, 131, 132, 135, 137, 140, 146, 147, 156/157, 159, 163, 168, 170/171, 172, 173, 177, 184/185, 186, 190L, 193R, 194/195, 196, 197, 199, 202/203, 204, 205, 209, 211, 213, 214/215, 216, 220, 222, 223, 224/225, 226, 227, 229, 232, 233, 235, 236, 238/239, 248, 249, 250, 256, 260, 265, 271, 272
Lyle Lawson 262
Robert Llewellyn 263

Peter Newark's American Pictures 26
Richard T. Nowitz all cover pictures, 2, 3, 4T, 4L, 5, 50, 66, 67, 78, 112, 134, 136, 138L, 138R, 139, 141, 142, 144/145, 148, 149, 150, 155, 158, 164, 165166L, 166R, 167, 169, 175, 176, 178, 179L, 179R, 180, 181, 182, 183, 187, 189, 190R, 191, 192, 193L, 200, 201,207, 208, 210, 212, 217, 221, 230, 231, 240, 241, 243, 244, 245, 246, 247, 251, 253, 254, 255, 258, 264, 268, 269, 270
Office of the Vice President 219
Vivian Ronay 90
Topham/Associated Press 41, 49, 89, 95, 96, 97, 99
Topham/ImageWorks 39, 51, 59, 60, 81, 91, 122, 123, 151, 152, 153, 154, 162, 237, 257
Topham/PA 75
Topham/Photri 125, 126
Topham Picturepoint 16, 17, 24, 42, 124, 261

Map Production Laura Morris

Index

Numbers in italics refer to photographs

A
B
C

E
F
G
H
I
J
a
b
c
d

f
g
h
i
j
k
l

☀ INSIGHT GUIDES

The world's largest collection of visual travel guides

A range of guides and maps to meet every travel need

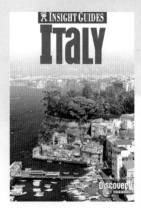

Insight Guides

This classic series gives you the complete picture of a destination through expert, well written and informative text and stunning photography. Each book is an ideal background information and travel planner, serves as an on-the-spot companion – and is a superb visual souvenir of a trip. Nearly 200 titles.

Insight Pocket Guides

focus on the best choices for places to see and things to do, picked by our local correspondents. They are ideal for visitors new to a destination. To help readers follow the routes easily, the books contain full-size pull-out maps. 120 titles.

Insight Maps

are designed to complement the guides. They provide full mapping of major cities, regions and countries, and their laminated finish makes them easy to fold and gives them durability. 60 titles.

Insight Compact Guides

are convenient, comprehensive reference books, modestly priced. The text, photographs and maps are all carefully cross-referenced, making the books ideal for on-the-spot use when in a destination. 120 titles.

Different travellers have different needs. Since 1970, Insight Guides has been meeting these needs with a range of practical and stimulating guidebooks and maps

Insight FlexiMaps

Maps in Insight Guides are tailored to complement the text. But when you're on the road you sometimes need the big picture that only a large-scale map can provide. This new range of durable Insight Fleximaps has been designed to meet just that need.

Detailed, clear cartography
makes the comprehensive route and city maps easy to follow, highlights all the major tourist sites and provides valuable motoring information plus a full index.

Informative and easy to use
with additional text and photographs covering a destination's top 10 essential sites, plus useful addresses, facts about the destination and handy tips on getting around.

Laminated finish
allows you to mark your route on the map using a non-permanent marker pen, and wipe it off. It makes the maps more durable and easier to fold than traditional maps.

The world's most popular destinations
are covered by the 125 titles in the series – and new destinations are being added all the time. They include Alaska, Amsterdam, Bangkok, Barbados, Beijing, Brussels, Dallas/Fort Worth, Florence, Hong Kong, Ireland, Madrid, New York, Orlando, Peru, Prague, Rio, Rome, San Francisco, Sydney, Thailand, Turkey, Venice, and Vienna.

INSIGHT GUIDES
The world's largest collection of visual travel guides

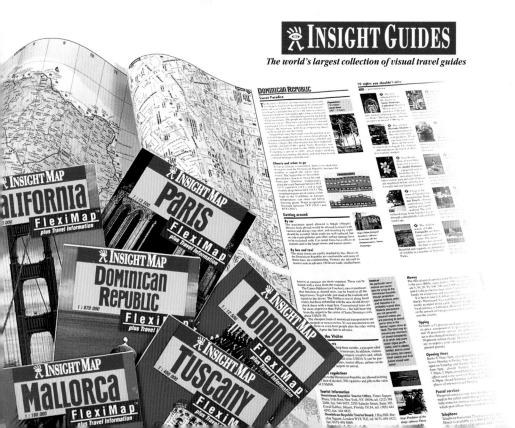

INSIGHT GUIDES

The classic series that puts you in the picture

Alaska
Amazon Wildlife
American Southwest
Amsterdam
Argentina
Arizona & Grand Canyon
Asia, East
Asia, Southeast
Australia
Austria
Bahamas
Bali
Baltic States
Bangkok
Barbados
Barcelona
Beijing
Belgium
Belize
Berlin
Bermuda
Boston
Brazil
Brittany
Brussels
Buenos Aires
Burgundy
Burma (Myanmar)
Cairo
California
California, Southern
Canada
Caribbean
Caribbean Cruises
Channel Islands
Chicago
Chile
China
Continental Europe
Corsica
Costa Rica
Crete
Cuba
Cyprus
Czech & Slovak Republic
Delhi, Jaipur & Agra
Denmark

Dominican Rep. & Haiti
Dublin
East African Wildlife
Eastern Europe
Ecuador
Edinburgh
Egypt
England
Finland
Florence
Florida
France
France, Southwest
French Riviera
Gambia & Senegal
Germany
Glasgow
Gran Canaria
Great Britain
Great Railway Journeys
  of Europe
Greece
Greek Islands
Guatemala, Belize
  & Yucatán
Hawaii
Hong Kong
Hungary
Iceland
India
India, South
Indonesia
Ireland
Israel
Istanbul
Italy
Italy, Northern
Italy, Southern
Jamaica
Japan
Jerusalem
Jordan
Kenya
Korea
Laos & Cambodia
Las Vegas
Lisbon

London
Los Angeles
Madeira
Madrid
Malaysia
Mallorca & Ibiza
Malta
Mauritius Réunion
  & Seychelles
Melbourne
Mexico
Miami
Montreal
Morocco
Moscow
Namibia
Nepal
Netherlands
New England
New Orleans
New York City
New York State
New Zealand
Nile
Normandy
Norway
Oman & The UAE
Oxford
Pacific Northwest
Pakistan
Paris
Peru
Philadelphia
Philippines
Poland
Portugal
Prague
Provence
Puerto Rico
Rajasthan

Rio de Janeiro
Rome
Russia
St Petersburg
San Francisco
Sardinia
Scandinavia
Scotland
Seattle
Sicily
Singapore
South Africa
South America
Spain
Spain, Northern
Spain, Southern
Sri Lanka
Sweden
Switzerland
Sydney
Syria & Lebanon
Taiwan
Tenerife
Texas
Thailand
Tokyo
Trinidad & Tobago
Tunisia
Turkey
Tuscany
Umbria
USA: On The Road
USA: Western States
US National Parks: West
Venezuela
Venice
Vienna
Vietnam
Wales
Walt Disney World/Orlando

𝌗 INSIGHT GUIDES

The world's largest collection of visual travel guides & maps

USING METRORAIL

The Metrorail system is made up of five color-coded lines that wind through the city and extend into the suburbs. There's a Metrorail station at or near every DC attraction, and the system is easy to master. Transit fees are collected by magnetic card, which you buy from the farecard machines in each station. Run your card through the faregate when you enter the station, then again when you exit at your destination, and your fare will be automatically subtracted. The only trick is to make sure you don't lose your card. If you do, you'll have to pay the maximum fare. Fares differ between stations, and at various times of day, when higher rush-hour fees apply. You can buy several trips' worth of fares with one stop at the farecard machine, which will save time. Farecard machines will take paper money, including $20 bills, but be forewarned that they make change only in coins.

Washington Metro